HOLT

LIFE
SCIENCE

HOLT
LIFE
SCIENCE

Patricia A. Watkins
Science Curriculum Specialist
Dean of College Preparatory Programs
Incarnate Word College
Principal, Incarnate Word High School
San Antonio, Texas

Glenn K. Leto
Biology Teacher
Barrington High School
Barrington, Illinois

SENIOR EDITORIAL ADVISOR
Curriculum and Multicultural Education

John E. Evans, Jr.
Science Education Specialist
Philadelphia, Pennsylvania

HOLT, RINEHART AND WINSTON

Austin • *New York • Orlando • Chicago • Atlanta*
San Francisco • Boston • Dallas • Toronto • London

ACKNOWLEDGMENTS

Content Advisors

Jerry Brand, Ph.D.
Professor, Department of Botany
University of Texas
Austin, Texas

John R. Bristol, Ph.D.
Professor of Biology
Department of Biological
    Sciences
University of Texas at El Paso
El Paso, Texas

Robert Fronk, Ph.D.
Head, Science Education Department
Florida Institute of Technology
Melbourne, Florida

Georgia E. Lesh-Laurie, Ph.D.
Professor of Biology
Vice Chancellor for Academic
    Affairs
University of Colorado
Denver, Colorado

Jan M. Ozias, Ph.D., R.N.
Coordinator of Health Service
Austin Independent School
    District
Austin, Texas

Barbara Rothstein, Ph.D.
Environmental Education
    Consultant
Hollywood, Florida

Leon J. Zalewski, Ph.D.
Professor of Science Education
Dean, College of Education
Governors State University
University Park, Illinois

Curriculum Advisors

Charles Kish
Science Teacher
Saratoga Springs
    Junior High School
Saratoga Springs, New York

Lynn Mederos
Life Science Teacher
Glenridge Middle School
Winter Park, Florida

Jim Pulley
Science Teacher
Oak Park High School
North Kansas City, Missouri

Dana Ste. Claire
Curator of Science and History
The Museum of Arts and Sciences
Daytona Beach, Florida

Thomasena H. Woods, Ed.D.
Science Supervisor
Newport News City Schools
Newport News, Virginia

Reading/Literature Advisors

Edward C. Turner, Ph.D.
Associate Professor
College of Education
University of Florida
Gainesville, Florida

Philip E. Bishop, Ph.D.
Professor, Department of
    Humanities
Valencia Community College
Orlando, Florida

Cover Design: Didona Design Associates

Cover: Bluestriped Snappers. Photo by Fred McConnaughey/Photo Researchers, Inc.

For permission to reprint copyrighted material, grateful acknowledgment is made to the following sources:

Ballantine Books, a division of Random House, Inc.: From *Dances with Wolves* by Michael Blake. Copyright © 1988 by Michael Blake.

Bantam Books, a division of Bantam Doubleday Dell Publishing Group, Inc: From *Fantastic Voyage* by Isaac Asimov. Copyright © 1966 by Bantam Books. A novel by Isaac Asimov based on the screenplay by Harry Kleiner from original story by Otto Klement and Jay Lewis Bixby.

Bonnier Fakta Bokförlag AB: From *Close to Nature: An Exploration of Nature's Microcosm* by Lennart Nilsson. Copyright © 1984 by Lennart Nilsson and Bonnier Fakta Bokförlag AB, Stockholm.

Children's Better Health Institute, Benjamin Franklin Literary & Medical Society, Inc., Indianapolis, IN: Adaptation of "An Eagle to the Wind" by Nancy Ferrell from *Young World*, June/July 1979. Copyright © 1979 by The Saturday Evening Post Company.

Current Science ®: Adapted from "I Slept for Science: A Study of Sleep Problems" by Andy T. McPhee from *Current Science ®*, October 7, 1988. Copyright © 1988 by Weekly Reader Corporation. Published by Weekly Reader Corporation.

Danbury Press, a division of Grolier Inc.: From *The Ocean World of Jacques Cousteau: The Quest for Food*. Volume 3. Copyright © 1973 by Jacques-Yves Cousteau.

Farrar, Straus & Giroux, Inc.: "earthworms" from *More Small Poems* by Valerie Worth. Copyright © 1976 by Valerie Worth.

Greenwillow Books, a division of William Morrow & Company, Inc.: Illustrations and adaptation of text from *How the Forest Grew* by William Jaspersohn, illustrated by Chuck Eckart. Text copyright © 1980 by William Jaspersohn; illustration copyright © 1980 by Chuck Eckart.

Harcourt Brace & Company: Illustration and text from *The Jungle* by Helen Borton. Copyright © 1968 by Helen Borton. Illustration and text from *Lumberjack* by Stephen W. Meader, illustrated by Henry C. Pitz. Copyright 1934 by Harcourt Brace & Company; copyright renewed © 1961 by Stephen Meader.

Holt, Rinehart and Winston, Inc.: Adapted text from "Controlling Algal Blooms," adapted text from "Growing Plants in Space," and adapted text from "Saving Endangered Wildlife" from *Biology Today*, Annotated Teacher's Edition, by Thomas C. Emmel, Harvey Goodman, Linda E. Graham, and Yaakov Shechter. Copyright © 1991 by Holt, Rinehart and Winston, Inc.

Little, Brown and Company: From *Life on Earth: A Natural History* by David Attenborough. Copyright © 1979 by David Attenborough Productions Ltd. "The Octopus," "The Purist," and "The Termite" from *Verses from 1929 On* by Ogden Nash. Copyright 1935, 1942 by Ogden Nash.

Lodestar Books, an affiliate of Dutton Children's Books, a division of Penguin Books USA Inc.: From *If You Lived On Mars* by Melvin Berger. Copyright © 1989 by Melvin Berger.

Gina Maccoby Literary Agency: "Spider" from *Bugs: Poems* by Mary Ann Hoberman. Copyright © 1976 by Mary Ann Hoberman.

Macmillan Publishing Company: From "Through the Microscope" from *A Short History of Science: Man's Conquest of Nature from Ancient Times to the Atomic Age* by Arthur C. Gregor. Copyright © 1963 by Arthur C. Gregor.

Margaret K. McElderry Books, an imprint of Macmillan Publishing Company: From *Searches in the American Desert* by Sheila Cowing, with photographs and maps by Walter C. Cowing. Copyright © 1989 by Sheila Cowing.

National Wildlife Federation: "Shark Lady" by Bet Hennefrund from *Ranger Rick*, March 1989. Copyright © 1989 by the National Wildlife Federation. From "China's Precious Pandas" by Claire Miller from *Ranger Rick*, July 1989. Copyright © 1989 by the National Wildlife Federation. "Andy Lipkis and the Tree People" by Mark Wexler from *Ranger Rick*, May 1984. Copyright © 1984 by the National Wildlife Federation.

W. W. Norton & Company, Inc.: From *The Winter of the Fisher* by Cameron Langford. Copyright © 1971 by Cameron Langford.

Oxford University Press, Inc.: From *The Sea Around Us, Revised Edition*, by Rachel L. Carson. Copyright © 1950, 1951, 1961 by Rachel L. Carson; copyright renewed © 1979, 1989 by Roger Christie; Golden Press edition copyright © 1958 by Western Publishing Company, Inc.

Prentice-Hall, Inc.: From *order: in life* by Edmund Samuel. Copyright © 1972 by Prentice-Hall, Inc.

Richard Rieu: "The Flattered Flying Fish" by E. V. Rieu. Copyright © 1983.

Simon and Schuster Books For Young Readers, New York: From *Journey Through a Tropical Jungle* by Adrian Forsyth. Copyright © 1988 by Adrian Forsyth.

The Society of Authors as the representative of The Literary Trustees of Walter de la Mare: "Seeds" from *Rhymes and Verses: Collected Poems for Children* by Walter de la Mare. Copyright © 1947 by Henry Holt and Company, Inc.

Doug Stewart: From "These Germs Work Wonders" by Doug Stewart from *Reader's Digest*, January 1991. Copyright © 1991 by Doug Stewart.

Time-Life Books Inc.: From *The Everglades: The American Wilderness* by Archie Carr and The Editors of Time-Life Books. Copyright © 1973 by Time-Life Books Inc.

U.S. News & World Report: Quote by Magic Johnson from "Stunned by Magic" by Tom Callahan from *U.S. News & World Report*, November 18, 1991. Copyright © by U.S. News & World Report Inc.

Franklin Watts, Inc., New York: From *The Black Plague* by Walter Oleksy. Copyright © 1982 by Walter Oleksy.

Wayland Publishers Limited and Franklin Watts, New York: From *The Voyage of the Beagle* by Kate Hyndley. Copyright © 1989 by Wayland Publishers Limited.

v

CONTENTS

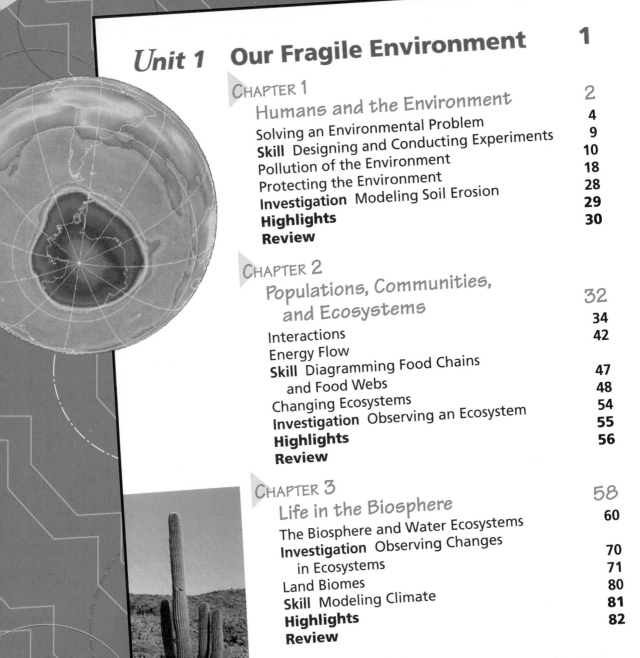

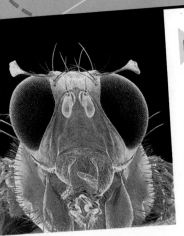

Unit 2 **Studying Living Things** 94

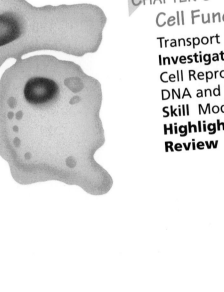

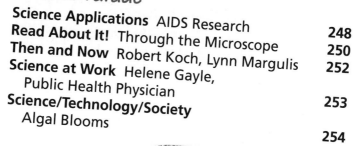

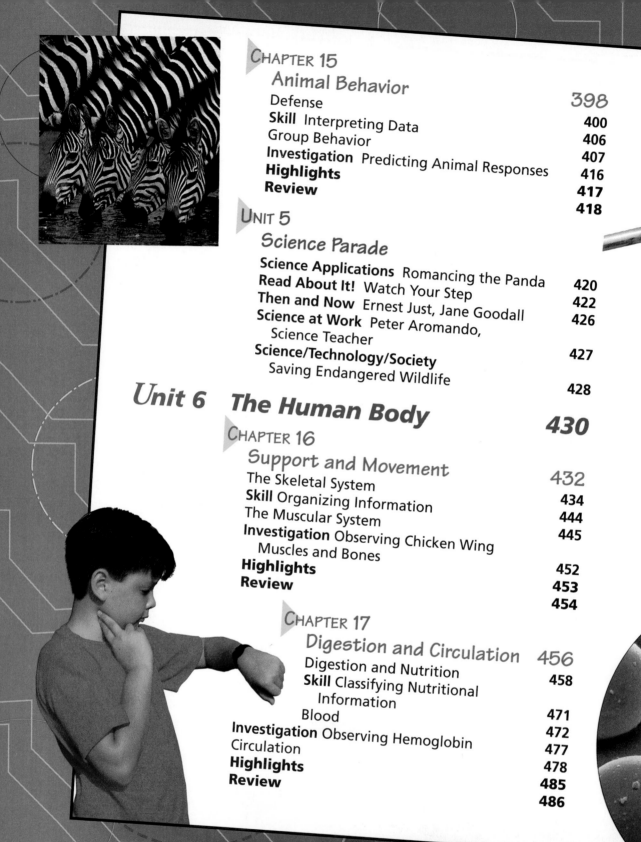

Lennart Nilsson/
Boehringer Ingelheim
International Gmbh.

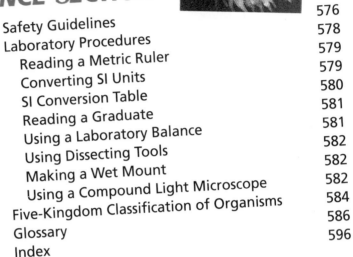

REFERENCE SECTION

SKILL

INVESTIGATION

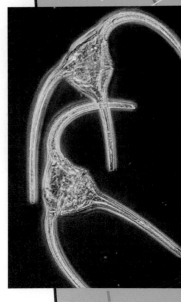

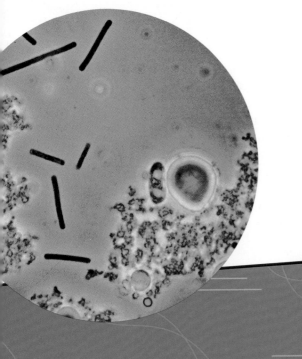

ACTIVITY

DISCOVER BY . . .

About Holt Life Science

Share the Wonder

Have you ever wondered why tropical rain forests are important? How all living things are alike? How you get energy for work and play? These are just a few of the questions that have been asked by life scientists over the years. The answers to these and many of your own questions about life and living things can be found in HOLT LIFE SCIENCE.

Explore the Nature of Science

Have you ever seen a bird catching a bug, a horse eating grass, or an animal taking care of its young? Have you ever grown a vegetable garden? Have you ever eaten yogurt or cheese? If you've ever done any of these things, you have experienced the study of life science. The study of the interactions among living things is what life science is all about. Science is knowledge gained by observing, experimenting, and thinking. Life science is knowledge of living things—their makeup and how they work. It is also knowledge of the relationships between living things and the world around them. You probably have a lot of this knowledge already. HOLT LIFE SCIENCE will help you add to your knowledge through reading, discussion, and activities.

Work Like a Scientist

Much of science is based on observations of events in nature. From observations, you form questions about the things you see around you. By doing experiments, you will gather more information. During your study of life science, you will be working and thinking like a scientist. You will develop and refine certain skills—such as your observation skills. In some cases, you will be asked to make predictions, and through activities you will have a chance to test the accuracy of your predictions. With new information, you will try to answer questions, to think about what you have observed, and to form conclusions. You will also learn how to communicate information to others, as scientists do.

Keep a Journal

One way to organize your ideas is to keep a journal and make frequent entries in it. Like a scientist, you will have a chance to write down your ideas and then, after doing activities and reading about scientific discoveries, go back and revise your journal entries. Remember, it's OK to change your mind after you have additional information. Scientists do this all the time!

Make Discoveries

Life science is an ongoing process of discovery—asking and answering questions about the structure and function of living things. If you choose a career in life science, you may find answers to the questions of students and scientists who lived before you. Through curiosity and imagination, life scientists are able to create a never-ending list of questions about the world in which we live. You can also gain respect and appreciation for all living things.

Prepare yourself for discovery. Through your studies, you will learn many interesting and exciting things. Allow yourself to explore and to discover the ideas, the information, the challenges, and the beauty within the pages of this book. HOLT LIFE SCIENCE can help you understand the world around you and how each living thing is important to the way in which the world works. You will begin to ask new questions and to find new answers. You will discover that learning about life science is fun and helpful.

OUR FRAGILE ENVIRONMENT

Imagine a dense forest five kilometers long and five kilometers wide, containing hundreds of different trees, many of them over 75 meters tall. Living among the trees are thousands of different mammals, reptiles, amphibians, and insects. Now imagine that during this class period, all the trees have been cut or burned to the ground, and all the animals have been killed. You can stop imagining now, because all this is true. Is this to be the future for all of Our Fragile Environment?

Science PARADE

Humans and the Environment

Visible air pollution and water pollution highlight the ability of humans to cause changes in their environment. Solving environmental problems is a challenge people are now facing. Only then will Earth remain a hospitable home to humans and all other forms of life.

From space, Earth looks like a glimmering blue jewel as white puffs of clouds swirl over its blue oceans. Up close, we can see its majestic mountains, rolling plains, and deep canyons. We can view its bubbling streams, reflecting pools, and massive oceans. Earth is a remarkably beautiful, yet fragile, place. On it, we can also see eroded land, polluted waters, and smog. We can watch forests being destroyed and animal and plant species disappearing. Centuries of human activities have helped create these problems. Now, humans have the responsibility to help solve them.

For Your Journal

🖊 List any environmental problems you have observed.

🖊 How do you think the problems can be solved?

🖊 What can individuals do to help solve the problems?

Solving an Environmental Problem

Objectives

List *the processes used to solve a problem scientifically.*

Analyze *problem solving through controlled experimentation.*

Compare and contrast *a hypothesis and a theory.*

You probably have many questions about the world in which you live. Why are plants green? Why is the sky blue? Why do people still pollute the environment? If you ask questions such as these, you are thinking like a scientist. Scientists ask questions about the world in which they live. They also work to find the answers through scientific investigation and observation.

DISCOVER BY Writing

You solve problems every day. Some problems are simple, such as deciding what movie to see. Other problems are more important, such as deciding what courses might prepare you for a certain career. In your journal, describe the steps you might follow when you solve a problem. Are your problem-solving steps like those of a scientist? ✎

Figure 1–1. Asking questions is one way to think like a scientist. What questions might these pictures pose to scientists?

Identifying Problems

There is no list or set of rules that all scientists follow to solve problems. Each problem is different and must be viewed with an open mind. Yet solving a problem scientifically usually involves certain processes, or steps. These processes, called a *scientific method,* are part of the orderly way in which problems are solved.

The first step in using a scientific method is usually to make observations. Once you make observations, you can decide which problem interests you. Suppose a scientist chooses to study coyotes, animals that prey on rabbits and other small wild animals. Coyotes also prey on farm animals. Coyotes cost farmers and ranchers millions of dollars each year. Destroying coyotes reduces the loss of farm animals, but it also disturbs the natural balance of the environment.

A scientist might be concerned about how to protect farm animals without harming coyotes. If it is possible to solve this problem, the solution will help the ranchers and it would help the coyotes.

Figure 1–2. Although coyotes' natural food source is small wild animals, they also kill and eat farm animals.

Once a scientist has identified the problem to be investigated, the next step is to collect more information. Often other scientists are researching similar problems. Scientists share their results so that more can be learned about the problem. Books, magazines, scientific meetings, and computer-databanks can be used to learn about information already available.

Using this information together with new observations, a scientist can then form a question about the problem. A possible question might be stated as follows: "Can coyotes be prevented from killing farm animals without upsetting the natural balance in the wild?"

A scientist's next step is to suggest a possible answer to the question based on known information. This possible answer is called a **hypothesis** (hy PAHTH uh sihs). A scientist studying coyotes, for example, might make the following hypothesis: "Coyotes made slightly ill by eating farm-animal meat treated with chemicals will not kill and eat farm animals in the future."

This hypothesis states an action to be taken and a possible outcome. Coyotes are to be given chemically treated meat that will make them slightly ill. The scientist expects the coyotes to associate becoming ill with eating the treated meat and, therefore, to avoid preying on farm animals.

 ASK YOURSELF

What is a scientific method of solving problems?

Conducting Experiments and Drawing Conclusions

The most important characteristic of a hypothesis is that it must be able to be tested with experiments. An experiment is an organized way of collecting facts, or *data,* through observations.

An experiment often includes the study of two groups. These two groups should be exactly alike, except for one difference, or *variable.* In this case, the hypothesis describes one group of coyotes. This group feeds on a small amount of chemically treated meat that makes them ill. Meat with no chemicals is given to the second group of coyotes. The variable between the two groups in this experiment is the chemical in the meat.

The group given chemically treated meat is the experimental group. The group given untreated meat is the control group.

Figure 1–3. The collection of experimental data involves careful observations during an experiment.

A scientist compares the data collected from the two groups. In the case of the coyotes, the scientist wants to see the effect that eating chemically treated meat has on the experimental group. Remember, the two groups are exactly alike except for the single variable. Therefore, any differences seen between the groups must be caused by the variable. An experiment that uses two groups such as these is called a *controlled experiment*.

Now the scientist must set up the experiment, make observations, and record data. In the following activity, you can practice making observations and collecting data.

DISCOVER BY Doing

Observe an animal, such as a dog, a cat, a bird, a lizard, or a fish, for a period of 15 minutes. Write your data—all the things you observed—in your journal, and report your observations to your classmates. ✏

Just as you observed and recorded data on the animal you studied, scientists make and record their observations. For the coyote problem, scientists studied two groups of wild coyotes in Saskatchewan, Canada. The experimental group was given a small amount of chemically treated sheep meat that would make the coyotes slightly ill for a short time. Untreated meat was put out for the other group. The number of lambs and sheep killed by coyotes in each group was recorded for a period of one year. Data from the experiment is shown in Table 1-1.

Table 1-1	Sheep Killed by Coyotes	
Subjects of Study	**Control Group**	**Experimental Group**
Total Number of Sheep	100 000	100 000
Number of Deaths	10 000	10 000
Deaths Due to Coyotes	400	150

The data shows that the coyotes in the experimental group killed and ate 63 percent fewer sheep than the coyotes in the control group. The data supports the hypothesis. Scientists concluded that feeding coyotes sheep meat that makes them ill reduces their killing and eating of sheep.

Figure 1–4. Controlled experiments lead to meaningful results.

If this experiment had been done without a control group, no comparisons could have been made. The data from the experimental group is meaningful only when compared to the data from the control group. Knowing that 150 sheep were killed by the coyotes given chemically treated meat is of little value by itself. However, when this information is compared to the number of sheep killed by coyotes in the control group, the effect of the variable can be seen.

The results of this experiment were reported in the *Journal of Range Management*. By reading the journal, other scientists can share the information and decide if more investigation should be done. Other studies may be made to test this hypothesis further.

When a hypothesis is supported by the work of many scientists, it may be called a **theory**. A theory is not a fact. A theory is, however, more than a hypothesis. Theories explain why things happen the way they do. Every theory is changed as new data is gathered. A theory can be judged by how well it supports observations and by how much good evidence supports it. Theories are useful because they help us predict what might happen in new situations.

 ASK YOURSELF

Why are controlled experiments needed to support a hypothesis?

SECTION 1   *REVIEW AND APPLICATION*

Reading Critically

1. What are the processes a scientist uses to solve a problem by a scientific method?
2. In what ways are observations important?
3. Why is it important to use a control group in an experiment?

Thinking Critically

4. How would you design a controlled experiment testing the hypothesis "blue light causes plants to grow taller"?
5. Would it be difficult to investigate scientifically the claim of a dog food company that its product was "better" than a competitor's? Explain.

SKILL *Designing and Conducting Experiments*

▶ MATERIALS
- bean seeds ● three flower pots of soil ● water

▼ PROCEDURE

1. State a hypothesis about the effect of light on the length of bean sprouts. Decide on a variable to include in your hypothesis—for example, the effects of different amounts of light or different kinds of light.

2. Using a scientific method, design an experiment that could test your hypothesis. Your design should include your hypothesis, a list of materials you will need, an identification of your experimental and control groups, and your procedure. Be sure that the steps required for the experiment are outlined in your procedure and that all instructions are clear. Your procedure should accurately describe the way in which you will make your observations and the method you will use to record what you find.

3. Have your teacher check your experimental design. Make any necessary changes or revisions in your plan before you begin.

▶ APPLICATION

1. After your teacher has approved your design, set up and conduct your experiment. Make daily observations and record what you see carefully.

2. Organize your observations so that other students can understand what you have seen. You may wish to use a table, chart, or graph for this purpose.

3. Analyze your observations. Be sure to state any relationships you observe.

4. State your conclusion. Be sure to explain how your data supports your conclusion. Also indicate if your hypothesis was proved or disproved by your experiment.

❋ Using What You Have Learned
State a new hypothesis that deals with an aspect of sprout growth other than length (number of leaves, for example). Using the sprouts that you have already grown, collect additional data to test your new hypothesis; then state and explain your conclusion.

Pollution of the Environment

Objectives

Report about ways humans pollute the environment.

Describe several environmental problems.

Relate the causes and effects of specific environmental problems.

More than five billion people live on Earth today! Every second of every day, three people are added to the world's human population. As the population grows, it requires more and more fuel, housing, food, and clothing. Raw materials used to produce products must be taken from the environment. At the same time, the use of these materials produces more waste that is dumped into the environment.

Pollution and Its Causes

Whether drying your hair with a blow dryer or riding on a bus, you are using energy. Every day, people use electricity, gasoline, batteries, and other forms of energy to meet their needs and to make life more comfortable. Look around you. How many devices require the use of energy to operate? How many of these things do you use daily? How would your life be different without them? Would you be willing to give them up? How would your life change if you did? Answering these questions points out our tremendous need for energy to maintain our lifestyles.

Figure 1–5.
Environmental damage results from the pollution of the environment or the destruction of natural resources.

The fuels used to produce energy also produce waste. Making the environment unclean with waste products is **pollution.** Any kind of waste can cause pollution. Other than waste produced by energy use, what other waste products can you identify? How can these waste products pollute the land, water, or air?

Manufacturing, agriculture, mining, and transportation help improve the quality of life for humans. However, these activities are also major sources of pollution. The waste products from these activities contribute to the pollution of the earth's water, air, and land. These waste products may be in the form of chemicals, heat, radiation, or even noise. In what ways have manufacturing, agriculture, mining, and transportation improved your life?

Waste products can pollute the air you breathe and the water you drink. Pollutants can be found in the soil in which food is grown and on which homes are built. Some pollutants can be tasted, smelled, seen, or heard. Other pollutants cannot be detected by human senses. They must be identified through the use of laboratory instruments.

Some waste materials can be broken down by living organisms. Waste materials that are broken down by microorganisms in the soil and water are called *biodegradable.* Other waste materials, such as glass, metal, plastic, and certain chemicals, are non-biodegradable—they remain in the environment forever.

Figure 1–6. Dumping wastes into waterways causes serious damage to the environment.

 ASK YOURSELF

Give some examples of land, water, and air pollution.

Solid Wastes

One of the most difficult problems facing humans today is the disposal of solid wastes. *Garbage, refuse,* and *litter* are some of the terms used to describe solid waste. In the United States, each person produces about 3 kg of refuse a day. That's about 750 000 000 kg of garbage a day—every day! In addition, industry and agriculture produce many times this much solid waste daily. In the next activity, you can see where some of this waste comes from and what could be done to reduce the amount produced.

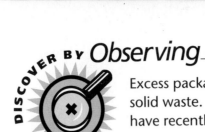

DISCOVER BY Observing

Excess packaging is one of the major sources of solid waste. Look at some packaged items you have recently purchased. How much excess packaging is there on the product? Check some other brands of the same product in the store. How much waste could be eliminated by reducing the excess packaging? ✎

Figure 1–7.
Barges are used in some places to haul garbage out to sea. In other cases, garbage is dumped in the open or buried in a landfill.

Excess packaging contributes to solid waste, and disposing of packaging and other solid waste is a serious problem. Some communities dispose of refuse in an open dump, where the refuse is left on the ground. These open dumps are a health hazard and a threat to the environment. Rain carries materials from the dump into water supplies. Since rats, insects, bacteria, and fungi are found in open dumps, such dumps may be sources of disease.

Other communities solve their disposal problems by dumping solid waste into the ocean. This removes the refuse from sight, but it damages the ocean's environment. The refuse causes changes in the ecosystem that kill or injure living things.

Still other communities dispose of their solid waste in sanitary landfills. Landfills are locations where solid waste is buried with a layer of soil around the waste on all sides. When sanitary landfills are properly made, they are less harmful to the environment than are open dumps or ocean dumping. However, many communities are running out of sites for landfills.

 **ASK YOURSELF**

What is one of the biggest sources of solid waste?

Figure 1-8. Acid rain has damaged many forests and destroyed statues.

Acid Rain

Water is one of the most important needs for life. Water collects on the earth's surface, giving living things the moisture they need to survive. Water that evaporates into the air eventually returns to the earth's surface in the form of rain or other precipitation. Organisms need precipitation for their supply of fresh water. Unfortunately, because of air pollution, rain that falls in some parts of the world is harmful to plants and animals.

Carbon dioxide in the air dissolves in rainwater. Normal amounts of carbon dioxide in the air make rain slightly acidic. Sulfur oxides and nitrogen oxides are other gases that dissolve in water and make rain even more acidic. All three gases are produced when high-sulfur coal or oil is burned. Coal and oil are commonly used as fuel to power factories and automobiles and to heat and cool homes. The polluted rain that results from large amounts of sulfur oxides and nitrogen oxides combining with rainwater is called *acid rain.*

These photographs show the damage acid rain can do to trees and statues. In addition to destroying forests and eating away statues, acid rain kills fish in lakes and streams. In parts of Scandinavia, Canada, and the United States, hundreds of lakes are now nearly lifeless due to acid rain. In the next activity, you can see how acid affects rocks and plants.

DISCOVER BY Doing

Put a few drops of vinegar, which is an acid, on a piece of chalk. Chalk is similar to limestone, a rock used in buildings and statues. Observe and describe what happens. Now put a few small leaves into a bowl of vinegar. Observe the leaves after five or ten minutes. What does the acid do to these plant parts?

In the activity you saw the effect of one type of acid on leaves and chalk. Acid rain can have a similar effect on plants and buildings. Tests have shown that rainfall in some areas is 10 to 30 times more acidic than ordinary rain. In the next activity, you can test rainwater for acid.

DISCOVER BY Doing

You will need a clear cup, a teaspoon, rainwater, and purple grape juice. Grape juice is an indicator—it turns red in an acid and green in a base, the opposite of an acid. Collect one-half cup of rainwater. Add a teaspoon of grape juice. Observe any color change. Does the rainwater you collected contain acid? If it does, what might the source of the acid be? ✐

Acid rain damages the leaves of plants, reducing a plant's ability to make its own food. Acid rain changes the soil so that some plants can no longer survive. When plants die, the animals that need those plants for food also die. Acid rain also damages wood, metal, and stone structures. Automobiles, bridges, and buildings show the effects of acid rain.

▶ ASK YOURSELF

Name two types of dissolved gases that contribute most to the creation of acid rain.

A Hole in the Sky

Ultraviolet waves are invisible light waves emitted by the sun. Exposure to the sun's ultraviolet rays is very harmful to living organisms. Fortunately, high in the earth's atmosphere is a thin layer of gas that absorbs many of the sun's ultraviolet rays. This thin, protective layer of gas is the ozone layer. *Ozone* is a type of oxygen. It is produced when ultraviolet rays strike oxygen.

If the ozone layer did not absorb most of the sun's ultraviolet rays, the rays would reach the earth. Many plants would die. Animals and people would suffer severe sunburn, skin cancer, and blindness. Since all living things are interrelated, severe effects on food chains would result. Over time, Earth would become a dead planet.

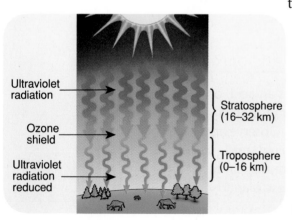

Ultraviolet radiation

Ozone shield

Ultraviolet radiation reduced

Stratosphere (16–32 km)

Troposphere (0–16 km)

This photograph was taken by a satellite over Antarctica. The dark area in the picture is caused by a hole in the ozone layer. Scientists discovered the hole in 1985, and they have observed it growing larger since then.

Scientists have evidence that the hole may have been caused by chemicals called *chlorofluorocarbons* (klawr uh FLOOR uh kahr buhnz), or CFCs. These chemicals are used for coolants in refrigerators and air conditioners and in aerosols. Some industries, such as the microchip industry and makers of plastic foam, also use CFCs. CFCs have been drifting around in the atmosphere since they were developed in the 1920s. They slowly rise into the stratosphere, where they break down the ozone layer.

The only major hole found in the ozone layer so far is the one over Antarctica. However, a smaller hole has been found over the Arctic Ocean. Recent evidence suggests that the ozone is getting very thin or a hole is developing over populated areas. Scientists are very concerned. If people keep using CFCs, more ozone will be destroyed. The United States and more than twenty other countries have agreed to cut use of CFCs by 50 percent before the year 2000.

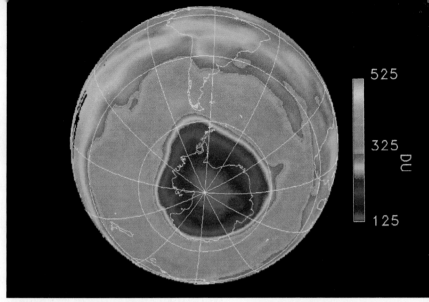

Figure 1–9. The ozone layer over the Arctic may soon have a hole as large as this one over Antarctica.

 ASK YOURSELF

Why do you think it is important that all countries agree to cut down on the use of CFCs?

The Greenhouse Effect

Holes in the ozone layer are not the only problem with Earth's atmosphere. Pollution could cause a global warming of Earth's climate due to the greenhouse effect.

Just what is the greenhouse effect? As the sun shines on the earth, some of its heat is absorbed by the land and by water. Much of the sun's energy is reflected back into the atmosphere. The reflected energy is held in the atmosphere by carbon dioxide, water vapor, and other gases. This warms the atmosphere and helps the earth keep a fairly constant temperature as it rotates.

Figure 1–10. Earth's temperature is balanced by the amount of carbon dioxide and water vapor in the atmosphere.

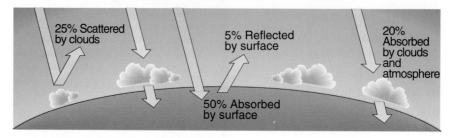

25% Scattered by clouds

5% Reflected by surface

20% Absorbed by clouds and atmosphere

50% Absorbed by surface

However, too much carbon dioxide can make the earth's atmosphere too warm. The atmosphere is like the inside of a closed car on a sunny day. Sunlight goes through the windows of the car and warms up the interior, but the heat cannot escape. The temperature inside the car can get dangerously high.

Carbon dioxide traps the earth's heat just as a closed car traps heat. A greenhouse has the same problem with overheating as a closed car, especially in the summer. For that reason, the heating effect of carbon dioxide in the atmosphere is sometimes called the *greenhouse effect*. The next activity can show you what has happened to the amount of carbon dioxide in the atmosphere.

ACTIVITY

How has the amount of carbon dioxide in the atmosphere changed?

MATERIALS
graph paper, ruler

PROCEDURE
1. Look at the table. It shows the amount of carbon dioxide (CO_2) in the atmosphere in the years between 1960 and 1990. The amounts are given in parts per million (ppm).
2. Make a line graph using the data from the table.

APPLICATION
What happened to the amount of CO_2 in the atmosphere between 1960 and 1990? Predict what might happen to the amount of CO_2 in the future. Estimate the amount of carbon dioxide in the air in the year 2000. Find out if the average temperature of the atmosphere changed between 1960 and 1990.

Table 1: Amount of Carbon Dioxide in the Atmosphere

Year	ppm of CO_2	Year	ppm of CO_2
1960	317	1980	337
1965	320	1985	340
1970	325	1990	342
1975	330		

Figure 1–11. What other problems, in addition to rising carbon dioxide levels, are related to the clearing of the rain forests?

In the activity, you learned how the amount of carbon dioxide has increased. Today, there is 15 percent more carbon dioxide in the atmosphere than there was in 1980. Most of that results from the burning of fossil fuels, such as oil. This picture shows another source of carbon dioxide. As the tropical rain forests are burned, carbon dioxide is released into the atmosphere. Scientists predict that the amount of carbon dioxide will rise an additional 10 percent by the year 2000.

 ASK YOURSELF

What might happen to the earth's climate if the amount of carbon dioxide in the atmosphere keeps increasing?

SECTION 2 *REVIEW AND APPLICATION*

Reading Critically

1. What is the greenhouse effect?
2. Why is the disposal of solid wastes such a problem?
3. What causes acid rain?

Thinking Critically

4. Why is acid rain found downwind of industrial centers?
5. Why should people be concerned about acid rain?
6. In what ways might you be able to reduce the problem of solid waste pollution?

Protecting the Environment

Objectives

Distinguish between renewable and nonrenewable resources.

Explain the need for soil and water conservation.

Describe some wildlife conservation practices.

What do a wooden desk and the pages of a book have in common? Both are products made from a natural resource—timber. Natural resources are materials from the environment that humans use. Timber used to supply wood for a desk and the paper in a book is one natural resource. Soil, water, coal, and oil are also natural resources. Soil and water are needed to grow food. Coal, oil, and gas provide energy.

Natural Resources

Some natural resources are renewable. *Renewable resources* are natural resources that can be replaced by the environment.

Figure 1–12. When forests are cut for lumber, these resources can be replaced by replanting seedlings.

Forests and animals are examples of renewable resources.

Trees are cut down for timber, but in time, new trees will grow. Seedlings can be planted to replace the trees cut down. Over a period of years, the seedlings become fully grown trees. Livestock, fish, and wildlife are also renewable resources. Through reproduction, these resources are replaced.

Figure 1-13. Although fish are a renewable resource, fish farms help to prevent certain species from becoming endangered.

Renewable resources can be replaced only if they are used slowly. Enough fish and wildlife must survive to produce a new generation. Seedlings may take forty years or more to grow into a new forest. If resources are used more quickly than they can be renewed, they will disappear.

Other resources are nonrenewable. *Nonrenewable resources* are not replaced by the environment fast enough to build new supplies. Coal, natural gas, oil, metals, and minerals are nonrenewable resources. These resources were formed under very special conditions. The formation of many nonrenewable resources took millions of years. The supply of each nonrenewable resource is limited. It continues to decrease as the resource is used or destroyed.

To make resources last as long as possible requires conservation. **Conservation** is the careful use of resources. Both renewable and nonrenewable resources should be conserved. Conservation allows resources to be used wisely. With proper conservation, the earth's resources can supply the needs of humans for many years to come. You can read about the work of a conservationist in the feature on page 92.

 ASK YOURSELF

What is conservation of natural resources?

Soil Conservation

Most people think of soil as "dirt," but it is much more than that. Soil is a mixture of both living and nonliving parts. The main part of soil is tiny bits of broken rock. Weather conditions such as wind, rain, heat, and cold produce tiny cracks in the surface of rocks. Water and air get inside these cracks and react with the chemicals in the rock, causing the rock to break down. Water expands as it freezes and breaks the rock into smaller pieces. However, the most important part of soil is composed of material from decaying organisms. Soil also contains water and living organisms.

Figure 1–14. This diagram shows the layers of soil.

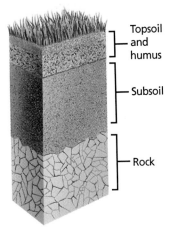

Soil can be divided into several layers as shown in the illustration. *Topsoil* is the uppermost layer of soil. In addition to bits of rock, topsoil contains humus, or dead and decaying organic material, such as leaves. Usually, the darker the soil, the more humus it contains. Humus holds moisture and supplies the nutrients needed for plants to grow. The layer located below the topsoil is the *subsoil*. Subsoil contains much less humus and more rock than topsoil. Because subsoil has fewer nutrients than topsoil, plants do not grow well in it.

Soil forms very slowly. It can take as long as 500 years to make a 2.5-cm layer of soil. The amount of soil varies greatly from one area to another. The soil of a desert may be only a few centimeters thick. In some parts of the western plains of the United States, the topsoil alone is 15 meters thick. Unfortunately, poor land practices cause soil to disappear faster than it forms. The carrying away of topsoil by water, wind, or glaciers is called **erosion.** Runoff from heavy rain causes a great deal of soil erosion. Water runs quickly downhill, carrying soil with it. The soil is carried into streams and rivers and eventually into the sea. In the next activity you can observe soil erosion.

Figure 1–15. Soil erodes when it is not held in place by plants.

ᴅɪꜱᴄᴏᴠᴇʀ ʙʏ *Doing*

You will need four cups of soil, two cups of water, and one cup of pebbles. Make a mound with about two cups of soil. Slowly pour about one cup of water over the mound. Observe and record what happens to the soil.

Now make another mound of soil. Place several pebbles on the mound. Again, pour water over the mound and record your observations. How do the pebbles affect what happens to the soil? ✐

Erosion is a natural process, but it can be slowed down by soil conservation practices. Plants that cover the soil help to slow the runoff of rain. This allows more water to sink into the soil, and less erosion occurs. Some farmers protect their fields by planting a cover crop during the winter months. This crop holds the soil in place and slows erosion. Before spring planting, the cover crop is plowed under to add more humus to the soil.

Figure 1–16. Plowing around a hill slows water flowing downhill and helps prevent erosion.

Each spring as the rainy season begins, farmers plant their crops. Before the plants begin to grow, the soil is exposed to erosion. Contour plowing is a method of protecting the soil until the crops begin to grow. In contour plowing, the land is plowed across a slope rather than up and down it. This form of plowing builds little dams of soil that slow the flow of rainwater.

Soil is also damaged when the same crop is planted in a field every year. This practice results in the same nutrients being removed year after year and never being replaced. To avoid this problem, some farmers change, or rotate, their crops from year to year. For example, crops such as beans and peas can replace nutrients removed from the soil by other crops, such as corn.

In a natural environment, nutrients are returned to the soil as dead plants and animals decay. Farming and harvesting remove plants from the soil. This breaks the natural cycle of nutrients. A common solution to this problem is the use of fertilizers. A fertilizer is a material added to the soil to replace the nutrients used by crops. Some fertilizers are made from animal or plant wastes. Other fertilizers are manufactured from chemicals.

Good soil conservation practices reduce wind and water erosion by keeping the soil covered. Soil conservation also allows more water to be absorbed, reducing runoff. Nutrients removed from the soil by farming are returned by practicing soil conservation.

ASK YOURSELF

How do soil conservation practices reduce soil erosion?

Water Conservation

About 97 percent of the earth's water is in the oceans. This sea water is too salty for land animals and plants to use. They need fresh water. Much of the earth's fresh water is locked in glaciers and the polar ice caps. Part of the supply of fresh water is vapor in the air. Liquid fresh water makes up less than one percent of the earth's water supply. This relatively tiny amount of water must be shared by all life on land.

Almost half of all the water used by humans is used to irrigate plants. Water is also needed to manufacture products and to generate electricity. In addition, each person in the United States uses about 500 L of water each day. What does one person do with so much water each day? Each person drinks only about 2 L of water a day, but he or she also uses water for bathing, food preparation, and watering lawns. People also use water to flush away wastes and to fill swimming pools.

The goal of water conservation is to preserve our supply of clean fresh water. The most obvious method of water conservation is to reduce the amount of wasted water. Allowing faucets to leak and taking long showers are common ways of wasting water. What are some other ways in which water is wasted?

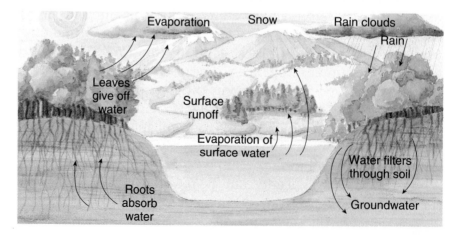

Figure 1–17. Water is continuously removed from and returned to the surface of the earth through the water cycle.

Figure 1-18. Although watersheds seem to have unlimited amounts of water, waste threatens that supply.

Much of the water people use can be recycled for use again and again. Supplies of fresh water must also be restored. Water that is pumped from the ground through wells must be replaced. Swamps and other wetlands allow rainwater to sink back into the ground to replace some of the water being removed. Forests slow runoff and allow water to be absorbed into the ground. A *watershed* is a section of land from which water drains into a river or lake. Conservationists try to protect watersheds whenever possible because watersheds help resupply underground water reserves.

 ASK YOURSELF

How does each person use water every day?

Wildlife Conservation

You probably know that many animals that once lived on Earth are now *extinct,* or completely gone. Dinosaurs and other prehistoric animals became extinct because of natural events. But in the last two hundred years, hundreds of animals have become extinct because of the actions of humans. One example is the passenger pigeon.

In the 1800s, there were billions of passenger pigeons in the United States. Now there are none. What happened to them?

Figure 1–19. A passenger pigeon from a nineteenth-century engraving

Figure 1–20. The last passenger pigeon died in 1914.

Passenger pigeons were hunted commercially for many years. Millions of them were killed every year. They were shipped by train to markets in the eastern United States, where they were sold as food.

Hunters not only shot the passenger pigeons but also attacked the birds in their nests. Trees were cut down, and woods were set on fire to drive the birds out. Without the protective cover of the woods, the birds were easier to find.

After many years, people realized that they were no longer seeing large flocks of passenger pigeons. Every year, fewer and fewer of the birds were found. Finally, on September 1, 1914, the last passenger pigeon, named Martha, died at the Cincinnati Zoo. There will never be another passenger pigeon.

Today, laws in most states protect animals from being hunted to extinction. For example, some animals can be hunted only during certain seasons. Laws often place limits on the number of animals that can be hunted. And *endangered animals,* those identified as being nearly extinct, cannot be hunted at all.

The survival of the American alligator shows that such laws can protect animals. By the 1960s, hunting had endangered the alligator's chances of survival. Then laws were passed making it illegal to hunt alligators. By 1987, the number of alligators had increased, and limited hunting was allowed again in some states.

Figure 1–21. American alligators were once endangered but now are thriving.

Although the American alligator is no longer endangered, it is still threatened in some areas. *Threatened animals* are in danger of becoming extinct if large numbers of them should suddenly die. The next activity can help you find out how threatened and endangered species are protected by laws in your state.

DISCOVER BY *Researching*

Visit the library in your school or community. Find out what animals in your state are threatened or endangered. Make a list of them. Find out if these animals are also threatened or endangered in other states. Make a table that shows this information. Then call an agency in your area that sells hunting licenses. Find out what laws there are to protect the animals on your list. Report on your findings to your classmates.

Figure 1–22. This endangered elephant was killed for its valuable tusks.

In some parts of the world, the hunting of threatened and endangered species continues, even though it is against the law. Endangered cheetahs, Siberian tigers, Asian lions and elephants, mountain zebras, snow leopards, and lowland gorillas are still hunted. What are some other ways in which humans contribute to the extinction or near-extinction of wildlife species?

The place where a population of organisms lives is called its **habitat.** The destruction of habitat is one of the most common causes of extinction. As the human population grows, it expands into new areas. Forest, prairie, and freshwater habitats are replaced by cities of concrete and steel. People build dams to store water. The dam creates a new habitat for some organisms, but it destroys the habitats of others. The prairie habitat disappears as more land is plowed for food crops. When humans change a habitat, some species' food sources are destroyed and their natural shelters are lost. Many animals and plants adapted to one habitat cannot adapt to a new one.

The goal of wildlife conservation is to protect the wild animals and plants that live on the earth. Saving natural habitats is one way to protect wildlife. Wildlife refuges, national parks, national forests, and wilderness areas preserve natural habitats. Conservationists also try to control the size of wildlife populations. Overpopulation can destroy a habitat or result in animals starving. Laws that permit hunting and fishing at certain times of the year can help protect wildlife populations. Good conservation practices reduce the effect of humans on wildlife.

 ASK YOURSELF

Why do you think some people continue to hunt threatened or endangered animals?

A Success Story

Figure 1–23. The California condor, America's largest bird, is nearly extinct.

The California condor is America's largest flying bird. An adult condor may weigh more than 45 kilograms and have a 3-meter wingspan. Because of its large size, the California condor needs a great amount of open space. But because more and more land is used by humans, the condor's range is getting smaller. Today there are fewer than 30 condors. The condor is nearly extinct.

Scientists decided that the only hope for saving condors was to try breeding them in captivity. Wild condors were captured and kept in zoos and wildlife parks. In 1987, the last wild condor was captured. Why do you think scientists caught the few remaining wild condors?

Breeding condors in captivity proved to be difficult. But in May 1988, the first condor chick was hatched at the San Diego Wild Animal Park in California.

Figure 1–24. Molloko was born at the San Diego Wild Animal Park.

The chick was named Molloko, a Native American word for *condor*. Molloko was taken from her nest and raised by a group of scientists. They wanted to be sure that nothing happened to her. To help her feel secure, the scientists played tape-recorded condor sounds to Molloko. They also handled her with a puppet made to look like a condor mother.

Until Molloko hatched, no one was sure the plan for breeding condors would work. Now scientists have begun releasing condors bred in captivity back into the wild. There they may breed and start a new population of wild condors.

Figure 1–25. A hand puppet was used as a "mother" to help feed Molloko.

The story of the California condor may have a happy ending. But for many other species, extinction is likely. Only strictly enforced laws and caring people can save them. In 1854, a great Native American chief named Seattle gave his reason for caring about wildlife:

What is man without the beasts? If all the beasts were gone, man would die from a great loneliness of the spirit. For whatever happens to the beasts soon happens to man.

 ASK YOURSELF

What is the goal of wildlife conservation?

SECTION 3 *REVIEW AND APPLICATION*

Reading Critically
1. What is a renewable resource? List four examples.
2. How does contour plowing help conserve soil?
3. What are some of the factors that can lead to the extinction of a species?

Thinking Critically
4. Why might the use of fertilizers be a good soil conservation practice but a poor water conservation practice?
5. How can people conserve water in their homes?
6. What are some ways you can help protect threatened or endangered species?

INVESTIGATION

Modeling Soil Erosion

▶ MATERIALS

- aluminum cake pans (2) ● potting soil ● ruler ● sod ● dishpan ● liter bottle of water

▼ PROCEDURE

1. Form a hypothesis about how grass might affect the rate of water erosion of soil.

2. Using the materials provided, set up an experiment to test your hypothesis.

3. Identify and describe the variable and control in your experiment.

4. List the procedure you will follow. Remember to establish some method for recording your data.

5. Perform your experiment several times. Record the data for each trial.

▶ ANALYSES AND CONCLUSIONS

Examine your data and draw a conclusion based on the data. Was your hypothesis supported by your data? Does grass affect the rate of water erosion of soil in some way?

▶ APPLICATION

Do you think grass would have a similar effect on wind erosion of soil? Explain your answer.

❋ Discover More

Devise an experiment to test your hypothesis about the effect of grass on the rate of wind erosion of soil.

The Big Idea

Earth's environment provides the air, water, and land needed by diverse forms of life to survive. Humans depend on the resources of the environment to meet their needs. The use of these resources has led to changes in the environment, which can disrupt the natural balance within environmental systems. Human activities have had a great effect on other organisms and on the air, water, land, and climate of the earth. While some human activities cause changes that are destructive to the environment, others help to protect the environment. The preservation of the earth's environment is dependent on the human ability to recognize and solve problems.

For Your Journal

Look back at the ideas you wrote in your journal at the beginning of the chapter. How have your ideas changed? Revise your journal entry to show what you have learned. Be sure to include information about the problems people have created in the environment and the ways we can solve some of these problems.

Connecting Ideas

Copy the following cause and effect chart in your journal.
Complete the chart by identifying the missing cause or effect.

Cause	Effect
Scientists feed a group of coyotes chemically treated sheep meat.	
	The ozone layer is thinning.
	The passenger pigeon became extinct.
Carbon dioxide, sulfur oxides, and nitrogen oxides combine with rain water.	
A California condor is bred in captivity.	

Understanding Vocabulary

Explain how the terms in each set are related.

1. scientific question, hypothesis (5), theory (8)
2. control group (6), hypothesis (5), experiment (6)
3. acid rain (13), hole in ozone (15), pollution (10)
4. water, plants, erosion (20), topsoil (20)
5. wildlife, refuges (26), habitat (25)

Understanding Concepts

MULTIPLE CHOICE

6. In an experiment designed to determine how fertilizer might affect the growth of plants, the amount of fertilizer given the control group would be
   a) zero.
   b) less than the experimental group.
   c) more than the experimental group.
   d) the same as the experimental group.

7. Which of the following is *not* a water conservation method?
   a) watering house plants with rainwater
   b) taking a short shower
   c) fixing a leaky faucet
   d) throwing cooking water down the drain

8. Because raccoons seem to be able to adapt easily to new habitats, they are likely to
   a) become extinct.
   b) be placed on the endangered species list.
   c) thrive.
   d) be found only in one habitat.

9. Copy and complete the following concept map.

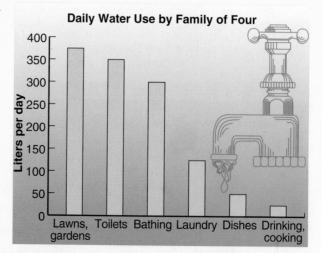

Interpreting Graphics

10. Look at the graph. About how much water is used for cooking every day?

11. If all outdoor use of water were banned during a drought, estimate the amount of water that would be saved in a week.

Daily Water Use by Family of Four

Liters per day

400 — 350 — 300 — 250 — 200 — 150 — 100 — 50 — 0

Lawns, gardens   Toilets   Bathing   Laundry   Dishes   Drinking, cooking

Reviewing Themes

12. Environmental Interactions
Acid rain begins as air pollution. Using acid rain as an example, explain how damage to one part of the environment can affect other parts.

13. Conservation
The evidence indicates that CFCs are destroying the ozone layer. Explain the changes that are likely to result if the use of CFCs is not reduced or eliminated.

Thinking Critically

14. How would you design a controlled experiment to test the effects of softened water on the growth of plants?

15. Why should humans conserve the supply of coal, oil, and natural gas?

16. How does a theory differ from a hypothesis? Give a specific example of a hypothesis that might later be developed into a theory. What conditions are necessary for the establishment of a theory?

17. Is recycling a means of conserving renewable or nonrenewable resources? Justify your answer.

18. Describe a program that you, as an individual, could follow to conserve both water and energy. What specific steps could you and your family take to implement this program?

19. The formation of soil is a long process, whereas the erosion of soil may be quite rapid. Describe how human activities can contribute to both erosion and conservation of soil.

20. If all of a rabbit's natural predators were eliminated in a wildlife preserve, what might happen to the rabbit population? What effect might this have on the preserve habitat? What might be done about the situation?

Discovery Through Reading

Javna, John. *Fifty Simple Things Kids Can Do To Save the Earth.* Andrews and McMeel, 1990. Explains how specific things in your environment are connected to the rest of the world, how using them affects the planet, and how an individual can develop habits and projects that are environmentally sound.

Populations, Communities, and Ecosystems

A warm, tropical sea is one kind of environment. Clown fish and sea anemones are two of the organisms that live in that environment. They share it with other living things, such as moray eels, seahorses, and coral, and with nonliving things, such as sand, rocks, and water. All living and nonliving things in an environment are interconnected in many ways.

The sunlight filters through the water in a warm, tropical sea. Many colorful fish dart about a coral reef. Parrot fish nibble at the coral. A moray eel lurks.

A sea anemone can move, but generally it attaches itself to a surface and stays there. Its cylinder-shaped body is topped

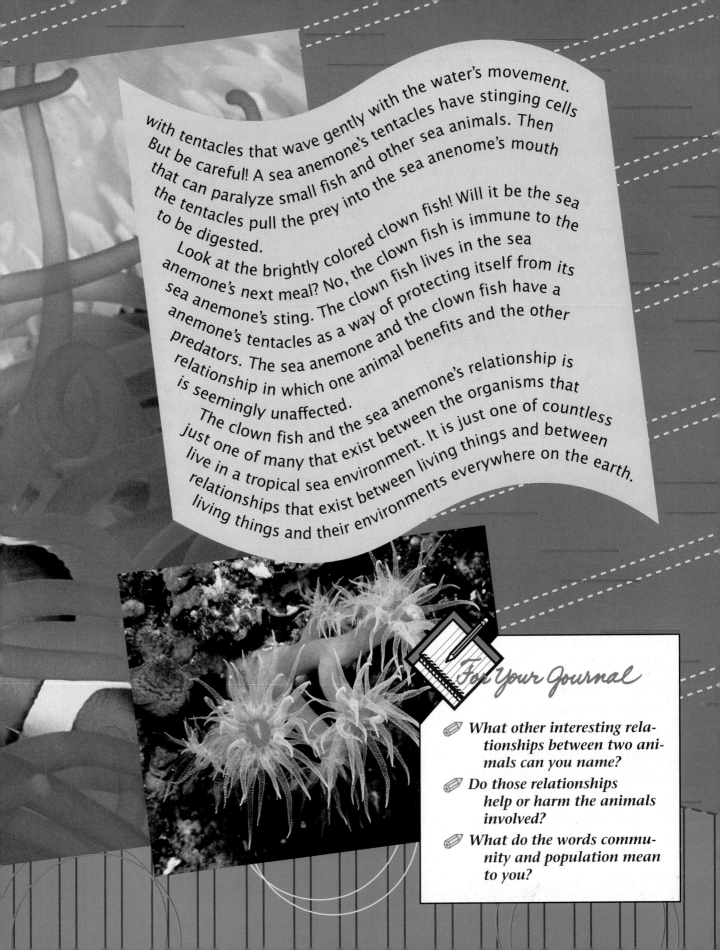

with tentacles that wave gently with the water's movement. But be careful! A sea anemone's tentacles have stinging cells that can paralyze small fish and other sea animals. Then the tentacles pull the prey into the sea anemone's mouth to be digested.

Look at the brightly colored clown fish! Will it be the sea anemone's next meal? No, the clown fish is immune to the sea anemone's sting. The clown fish lives in the sea anemone's tentacles as a way of protecting itself from its predators. The sea anemone and the clown fish have a relationship in which one animal benefits and the other is seemingly unaffected.

The clown fish and the sea anemone's relationship is just one of many that exist between the organisms that live in a tropical sea environment. It is just one of countless relationships that exist between living things and between living things and their environments everywhere on the earth.

For Your Journal

✎ *What other interesting relationships between two animals can you name?*

✎ *Do those relationships help or harm the animals involved?*

✎ *What do the words community and population mean to you?*

Interactions

Objectives

Distinguish among populations, communities, and ecosystems.

Summarize the characteristics of a population.

Compare the relationships between populations.

A single organism has many levels of organization. Cells group together to form tissues; tissues make up an organ; and a group of organs makes up a system. Together, cells, tissues, organs, and systems form an organism. Levels of organization are also found in places where humans, other animals, and plants live.

Discover by Doing

Choose one area near your school or your home. Choose one type of organism that lives in the area. The organism may be a tree, an insect, a bird, or even a human. Through careful observations over several days, estimate the number of individuals of that type of organism living in the area. What factors enable the organisms to live there? How do these organisms affect other kinds of organisms in the area? What name would you give to the group of organisms you observed?

Living Things Interact

Figure 2–1. These grizzly bears are part of a population.

Organisms of the same species living together in a particular place make up a **population.** Each population is described by kind of organism, time, and location. For example, the grizzly bears currently living in Montana make up a population. The grizzly bears that were living in Montana in 1983 made up a different population. All of the populations of animals and plants living together in an area make up a **community.** The populations living in a park near your neighborhood or city form a community. In such a community, you might find populations of mice, oak trees, mosquitoes, mushrooms, and many other kinds of organisms living together.

Each community depends on its environment. Nonliving things in the environment, such as minerals in the soil, are necessary for a community to survive. An **ecosystem** is the combination of all living things in a community and its nonliving, or physical, environment.

Figure 2–2. A forest ecosystem includes not only the trees and animals but also the physical environment in which the organisms live.

Consider this example. The Mississippi River runs from Minnesota to the Gulf of Mexico. The river itself forms one type of ecosystem where fishes and other aquatic organisms live. The banks of the river form another ecosystem. When you move away from the river, you can find many other ecosystems. Each is made up of a community of many populations of animals, plants, and other organisms interacting with its physical environment. The study of the relationships between organisms and their environments is called **ecology.**

 ASK YOURSELF

How are the concepts of a population, a community, and an ecosystem related to each other?

Characteristics of Populations

Organisms are described scientifically according to their characteristics, such as body structures, body coverings, and chemical makeup. Populations are also described according to certain characteristics.

One characteristic of a population is the number of organisms it contains. The number of organisms per unit of living space is the *population density*. Notice that population density is not just the number of organisms in the population.

Describing a population in terms of the number of organisms only can be misleading. The following illustration shows two populations. Each is made up of the same number of oak trees. The trees in Population I are spread over 10 km², while those in Population II occupy only 1 km². Which of these populations has greater population density? Explain why.

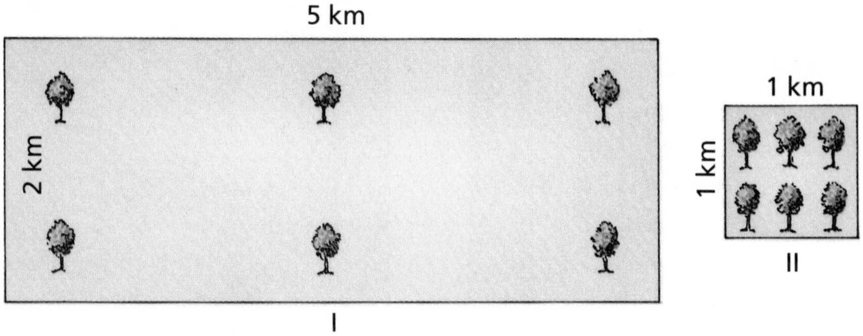

Figure 2–3. Population density is determined by calculating the number of organisms that live in a defined area.

In some cases, a population is spread over a large area, as wildflowers might be spread out over a field. In other cases, a population may be clumped together within a smaller area, with the members of the population spaced at regular intervals. The red-winged blackbirds shown in the photograph are evenly spread throughout the marsh. Why do you suppose they spread out evenly?

Another important characteristic of a population is its size. The size of a population is constantly changing. One way populations increase in size is through reproduction of the organisms. A population's birthrate is the number of offspring produced in a certain amount of time, such as the number of people born in the United States in a ten-year period. The size of populations also changes as organisms die. The death rate is the number of organisms that die in a given amount of time.

Figure 2–4. The population density of these red-winged blackbirds is high.

Birthrate and death rate within a population can be analyzed separately. However, only when they are considered together can their real effect on population size be measured. For example, suppose that a population of rabbits lives in a field measuring 1 km^2. At the beginning of one year, the population included 225 rabbits. At the end of the year, the population had grown to 310 rabbits. Does the increase mean that the birthrate was 85 rabbits a year? Although the population increased by 85 individuals, the birthrate might actually have been 150 rabbits. Perhaps 65 rabbits died.

Consider the same rabbit population the next year. The total population increases from 310 to 400 rabbits. The birthrate for this year is still 150, but the death rate increases to 75. The difference between 150 births and 75 deaths produced a net gain of 75 rabbits. However, the population grew by a total of 90 rabbits. Where did the other 15 rabbits come from?

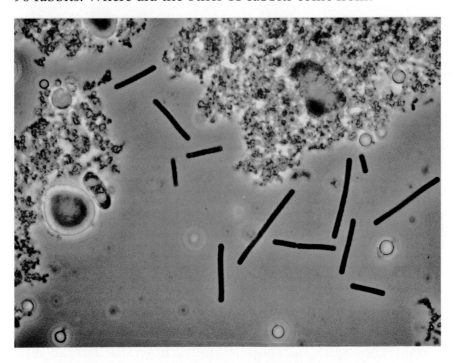

Figure 2–5. The population of the bees can gain or lose members by immigration and emigration, while the population of the bacteria (left) cannot.

A population also changes size when members enter or leave the area. When new members join a population, the population is increased by *immigration*. If members of a population leave, the population is reduced by *emigration*.

From the birthrate, death rate, immigration, and emigration, the total change in the size of a population can be determined. Every ten years, the United States Census Bureau compiles these figures in order to determine the actual size of the human population in this country. In the next activity, you can practice counting populations.

How can you count a population?

MATERIALS
journal, pencil

PROCEDURE
1. Select a large area that has several different populations, such as a park, field, roof garden, and so on.
2. Write a description of the area you are going to observe.
3. Pick out a plant population and an animal population.
4. Describe each of the organisms you have selected and provide a sketch of each organism.
5. Over a period of two or three days, count the members of the populations several times and take an average.

APPLICATION
Did you notice any changes in the populations you chose? What might account for these changes?

 ASK YOURSELF

Why is population density more descriptive than total population?

Biotic Potential

Figure 2–6. The biotic potential of the green turtle is high.

A beehive may contain hundreds of honeybees. Thousands of salmon live in the Great Lakes. Millions of bacteria live in a spoonful of yogurt. Each population reaches a size that can survive in a particular environment. But any population can grow too large for its environment. When that happens, there is not enough space and food for all the members of the population. *Biotic potential* is the number of individuals that could be produced under the best possible conditions.

For a population to reach its biotic potential, each individual must survive and reproduce at the maximum rate. Many animals have the possibility of reproducing in large numbers. A female green turtle, for example, lays about 1800 eggs in her lifetime.

Fortunately, few populations ever reach their full biotic potential. If they did, the earth's resources would be gone by now. A population's biotic potential is controlled by such things as food supply and space. The factors that prevent a population from reaching its full potential are called *limiting factors*.

Figure 2–7. Due to limiting factors, few young turtles survive.

Examples of limiting factors can be seen in the green turtle. As mentioned, a female green turtle lays about 1800 eggs in her lifetime. However, about 1400 of the 1800 eggs never hatch due to genetic defects, injury to the eggs, or their being eaten by predators. Young turtles from the eggs that do hatch find their first hours and days very dangerous. After digging their way to the surface of the beach, the hatchlings must travel as far as 50 m to reach the sea. Along the way, they may be eaten by birds and other animals. If the hatchlings reach the water, they are in danger of being eaten by fishes. More than three-fourths of the hatchlings do not survive their first few months. Of the 1800 possible offspring, only about three turtles survive to reproduce.

While the biotic potential of the green turtle population is high, there are a large number of limiting factors on this population. The combination of the biotic potential and the limiting factors determines the population size that can be supported by the environment. The largest number of individuals that a specific environment can support is called the environment's *carrying capacity*.

DISCOVER BY *Doing*

Graph the data about the human population found in Table 2–1. Study your graph. What patterns do you observe? Predict what the population might be in the year 2000. What factors might affect the accuracy of your prediction? ✐

▼ **ASK YOURSELF**

How does carrying capacity relate to biotic potential and limiting factors?

Table 2-1 **Human Population**

Year	Population
1650	500 million
1850	1 billion
1930	2 billion
1980	4.5 billion

Relationships Between Populations

Some populations have close, permanent relationships with one another. Sometimes both populations benefit from the relationship, or one population may benefit while the other is unaffected. The relationship may help one population while harming the other. The predator–prey relationship is one of the most common between populations. A *predator* is an animal that hunts and eats other animals. The animal eaten by a predator is its *prey*. Lions are predators that eat many different prey. The zebra and the antelope are common prey of the lion.

Figure 2–8. Predator–prey relationships, such as those shown here, are common between populations.

A relationship between two types of organisms in which one organism benefits while the other one is harmed is called *parasitism* (PAR uh syt ihz uhm). One example of this type of relationship is athlete's foot, an infection caused by a fungus that lives on human skin. The warm, moist conditions created by shoes and socks provide the right habitat for the fungus.

In the case of athlete's foot, the fungus is the parasite. A *parasite* is an organism that lives in or on another organism and is harmful to that organism. The infected organism is the *host,* or the organism that is harmed by the parasite. Other examples of parasites are fleas, ticks, lice, and mosquitoes. These animals live off the blood of the host.

Figure 2–9. Parasites cause diseases, such as Dutch elm disease (left) and chestnut blight (right), which can kill trees.

A parasite is different from a predator. The parasite does not need to kill the host organism to benefit. In fact, the death of the host usually results in the death of the parasite also.

A relationship between two types of organisms in which one organism benefits while the other is not affected is called *commensalism* (kuh MEHN suh lihz uhm). In many cases, the host organism makes a home for another organism. Remember the example of the clown fish that lives among the venomous tentacles of the sea anemone? The tentacles do not harm the clown fish; they protect it from predators.

Figure 2–10. A remora cleans the mouth of a moray eel and feeds on the eel's leftovers. What type of relationship is this?

A relationship between two types of organisms in which both organisms benefit is known as *mutualism* (MYOO choo wuhl ihz uhm). For example, mutualism exists between a termite and the protozoan that lives in the termite's gut. Termites eat wood, but they cannot digest it. The protozoan living inside the termite digests the wood for the termite and, at the same time, gets food for itself and a home. Both the termite and the protozoan survive because of their cooperation.

 ASK YOURSELF

How does a parasite differ from a predator?

SECTION 1 *REVIEW AND APPLICATION*

Reading Critically
1. Describe an ecosystem near your school.
2. How is biotic potential different from carrying capacity?

Thinking Critically
3. What kinds of limiting factors might act on a population of tropical fish living in an aquarium?
4. In what ways, do you suppose, could the human population growth rate be slowed?

Energy Flow

Objectives

Define habitat *and* niche.

Distinguish among producers, consumers, and decomposers.

Explain how energy flows through a community.

Every population has a place where it lives in the ecosystem—its *habitat*. Within every ecosystem there are many habitats. For example, a woodland ecosystem has many different populations living in it. Each population lives in a particular habitat within the ecosystem. Bears, squirrels, and foxes hunt for food in the forest. Earthworms live in the soil. Some birds search the forest floor for food. Other birds, such as woodpeckers, spend their time searching for insects higher up, on the trunks of trees. Still higher, hawks might look for food while circling in the sky.

DISCOVER BY *Observing*

Identify the organisms in the illustration shown here. In your journal, design a table with two columns. In column 1, write the name of the organism. In column 2, write the type of food the organism might find within its community. Include all possibilities. Give your table a title. Below the table, write a statement that describes the relationships among the organisms of this community.

Habitats and Niches

Many populations can occupy the same habitat. The earthworm is not alone as it inches its way through the moist soil. Ants, sow bugs, slugs, mushrooms, trees, and other species also live in the soil. The habitat of one species may even be on or in another species. Mosses and mushrooms live on the trunks of trees. Bacteria live in the digestive systems of many soil organisms.

Figure 2–11. Many different types of organisms can be found in the same habitat.

Each population has a particular function in its habitat. The function of a population includes its lifestyle, the food it eats, and the place where it builds its home. The function of an earthworm, for example, is to dig through the soil by eating the material in its path. The earthworm digests this material, which is then released from its body. This material adds nutrients to the soil. An organism's function in the habitat is somewhat like a human's job or profession. The function an organism plays in the habitat is its **niche** (NIHCH).

Although many populations can share a habitat, only one population can occupy a particular niche. An interesting example of this fact is found in Australia. Two very different animals are both called *kangaroo mice*. The two populations live close together in the dry, sandy regions of the continent.

One population is made up of true mice. They get the name *kangaroo mice* because of their long back legs. When they are frightened, they quickly hop away from danger. These hopping mice dig into the dry, sandy soil to make their homes. They eat seeds, berries, and other plant foods.

Figure 2–12. The hopping mouse (left) and the pouched mouse (right) share the same habitat. Why don't they share the same niche?

The second population of kangaroo mice are not really mice at all. They have a pouch like a kangaroo's. They are commonly called *mice* because of their size. These pouched animals eat beetles, cockroaches, termites, spiders, and other small animals. Since these two types of kangaroo mice do not compete for the same niche, they are often found in the same habitat. They have even been found sharing the same burrow.

 ASK YOURSELF

What is the difference between a habitat and a niche?

Food Chains and Webs

Every population within a community needs energy. The source of all this energy is sunlight. Green plants trap light energy and change it into chemical energy during photosynthesis. Because they make their own food, green plants are called **producers.** All communities need food from producers for energy. In addition to green plants, cyanobacteria and a few other types of bacteria are also producers.

Figure 2–13. Scavengers, such as these vultures, do not kill their own food. They eat animals that are already dead.

Organisms that eat other organisms for food are called **consumers.** There are different types of consumers. Animals that eat only plants are *herbivores*. Cattle, deer, and rabbits are examples of herbivores because they eat only grasses. Animals that eat only other animals are *carnivores*. Wolves, lions, and owls are carnivores. Like the lion that hunts and kills a zebra, the wolf and the owl also hunt and kill their food. Animals that eat both animals and plants are *omnivores*. Bears, raccoons, and pigs are omnivores. Humans are also omnivores. *Scavengers* are animals that eat animals they find already dead. Crows, vultures, and hyenas are scavengers.

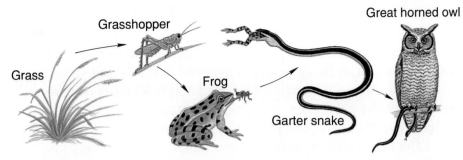

Figure 2–14. A typical food chain is shown here.

Energy is transferred through a community by *food chains*. In a meadow community, for example, grass changes light energy to food energy. The grass uses some of this energy for itself. A grasshopper eats the grass, and, in turn, a meadow frog eats the grasshopper. The frog is eaten by a garter snake that is later eaten by an owl. Energy from sunlight passes through each link in the food chain. Each organism uses some energy to stay alive and stores some energy in its tissues. How might energy be lost from the food chain?

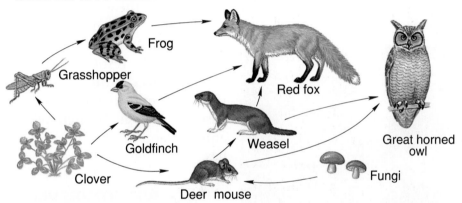

Figure 2–15. Many food chains connect to form a food web.

Many food chains exist within a community. A second meadow food chain might begin with the clover that is eaten by a mouse. The mouse provides food for the weasel that is later eaten by an owl. The owl is part of several food chains. These food chains connect to form a *food web*. The organisms in a food web can eat or be eaten by many other organisms. In the next activity, you can diagram a food web from a community where you live or one that your teacher shows you.

 **ᴅɪꜱᴄᴏᵛᴇʀ ʙʏ** *Doing*

Using producers and consumers from a community near where you live, diagram several interconnecting food chains that form a simple food web. ✎

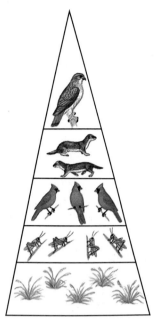

Figure 2–16. This pyramid shows a loss of energy from one consumer level to the next.

A community must have many more producers than consumers because producers provide the energy necessary for both themselves and the consumers. The energy found in the levels of a food web can be shown by a pyramid, such as that shown in the illustration. The base of the pyramid represents the producers. All the producers taken together have the most energy. What happens to the energy as you go higher in the pyramid?

The organisms at each level use much of their energy just to stay alive. Because only about 10 percent of the energy can be passed on to the next level, the number of organisms in each population gets smaller at each higher level of the pyramid. Only a few animals are found at the top of the energy pyramid. These animals, called *top carnivores,* such as lions, tigers, and wolves, exist in only very small populations. They depend on very large populations of other animals and plants to survive.

The materials that make up the organisms in a community—carbon, oxygen, nitrogen, and others—are constantly reused. *Decomposers* return these chemicals to the ecosystem. Bacteria and fungi are common decomposers. Decomposers use some of the materials for food and return the rest to the soil. Green plants can then reuse these nutrients to make new tissue. This tissue is eaten by animals, and the cycle continues.

Figure 2–17. Decomposers, such as these mushrooms, break down dead organisms and return nutrients to the soil.

 **ASK YOURSELF**

How do decomposers complete the energy cycle?

SECTION 2   *REVIEW AND APPLICATION*

Reading Critically

1. How does an organism's habitat differ from its niche? How can several niches exist within a single habitat?
2. What are the relationships that can exist between populations in a community?
3. What is the difference between a producer and a consumer?

Thinking Critically

4. How are scavengers and decomposers similar? How are they different?
5. Describe a food chain in an ecosystem near your school. What organism is at the top of the chain? What organism is at the bottom of the chain?
6. Compare recycling a product, such as paper or aluminum, with the natural recycling of chemicals in the environment. How does recycling benefit the environment?

SKILL Diagramming Food Chains and Food Webs

▼ PROCEDURE

1. As a class, select an ecosystem near your school with which all the class members are familiar. This ecosystem might be a prairie, a woodland, a pond, a city park, or a roof garden.

2. On index cards, write the names of the organisms that you think may be found in this ecosystem. Use one card for each organism. Make cards for as many organisms as you can think of.
3. On additional index cards, write the names of all the nonliving parts of the ecosystem you have chosen. For instance, you could include the sun, wind, water, and so on.
4. Using your cards, arrange the organisms into separate food chains. You should have more than one food chain. Arrange the cards for the nonliving parts of the ecosystem near the organism they affect. Copy your food chains into your journal.
5. Now combine your chains into a food web. Copy your food web into your journal.
6. As a class, make a giant food web by attaching your cards to a bulletin board.

► APPLICATION

1. Identify each organism in the giant food web as a producer, consumer, or decomposer.
2. Identify the herbivores and carnivores in the giant food web.
3. Some communities are very simple. They have only a few kinds of organisms arranged in a simple food web. Other communities are complex; they have many different kinds of organisms arranged in a complex food web. Which type of community is more stable or more resistant to change? Why?

✳ Using What You Have Learned

1. Look at one of the food chains that you developed. What would happen to the other animals in the food chain if the herbivore were eliminated from the community?
2. Find this same herbivore in the giant food web. What would happen to the other animals in the food web if this herbivore were eliminated from the community?

Changing Ecosystems

Objectives

Describe the differences between stable and unstable populations.

Distinguish between pioneer and climax communities.

Compare and contrast primary and secondary succession.

A fire rages through a once majestic forest. But what appears to be the destruction of a woodland ecosystem is not, because the damage is not permanent. Over time, the forces of nature will slowly begin the process of regrowth.

Succession

One illustration shows a forest that has been destroyed by fire. Since the fire, however, many changes have taken place. Grasses, shrubs, and seedling trees have begun to grow. The slow changes that occur when a community recovers from an event such as a fire is called **succession.**

Succession also occurs in areas such as a sand dune or bare rock that has not previously had any life. This succession, called *primary succession,* also occurs on land recently exposed by a melting glacier or on lava fields created by a volcano.

Figure 2–18. Recently burned forests provide areas where succession can take place.

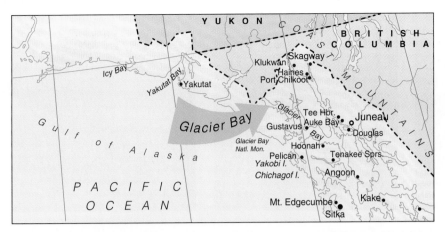

Figure 2–19. Glacier Bay, Alaska, provides an area in which to study primary succession.

One interesting example of primary succession can be seen at Glacier Bay National Park in Alaska. This area contains glaciers that formed during the last Ice Age. Since that time, a slow but steady melting has taken place, exposing land that has been covered by ice for thousands of years. This area gives biologists an unusual and excellent opportunity to study primary succession.

The land closest to the glacier has been exposed for only a matter of days. The slow process of primary succession is just beginning to take place on this barren land. Near the mouth of the bay, the land has been exposed for hundreds of years, and the final stages of succession are taking place. Traveling down the bay is like traveling through time.

Pioneer Stage   The area closest to the melting glacier is, at first, lifeless. The first living things to develop in a previously lifeless area are *pioneer plants*. The habitat of pioneer plants is very harsh. They must be able to survive high winds, bright sun, and heavy rains. There is usually little or no soil during the first stage of primary succession.

Figure 2–20. The pioneer stage of succession can be seen in Glacier Bay.

Pioneer plants grow in crushed rock and gravel and are able to survive without soil. Lichens are perhaps the hardiest pioneer plants. A lichen is a combination of a fungus and a green alga, living together in a unique and mutually beneficial relationship. Lichens attach themselves to bare rock and obtain nutrients directly from the rock. The action of the lichens combined with the action of the weather breaks up the rock. The cracks and depressions in the rock hold dead lichens and bits of rock. These materials are the beginnings of soil.

Figure 2–21. Many stages of succession can be seen in Glacier Bay. The mossy stage of succession (left), the grassy stage of succession (center), and a later stage of succession (right) are shown here.

Mossy Stage

A short distance down Glacier Bay, the next stage of primary succession can be observed. After many years of growth, lichens are slowly replaced by mosses. Mosses grow in the soil formed by the lichens and the rock. Mosses capture bits of dead matter brought in by wind or waves. The decomposing remains of sea creatures and bird droppings enrich the thin soil held in place by the mosses.

Grassy Stage

Slowly the habitat improves, and the mosses are crowded out by new plants. Farther down Glacier Bay, grasses follow the mossy stage of succession. Grasses have spreading roots that hold soil in place better than mosses do. After the grasses are established, flowering plants begin to grow from seeds carried in by the wind and by birds. These flowering plants attract more animals. The spreading layer of roots traps more of the decaying matter, and the soil continues to thicken and improve.

The grassy stage of succession creates a habitat where trees can grow. At Glacier Bay, the seeds of alder trees sprout in the soil that has been enriched by the grasses. The grasses shelter the seedlings from sun and wind. As years pass, a forest of alder trees slowly forms.

Final Stage

The final stage of succession at Glacier Bay comes as spruce trees slowly crowd out the alders. As in the earlier stages, the alders improved the habitat, making it suitable for new populations. Spruce seedlings grow well in the shade of the alders. The alder seedlings, on the other hand, need bright sunlight. Thus, most of the young trees in the forest are now spruce. As time passes, the taller-growing spruce trees eventually shade the older alders, which slowly die out.

 **ASK YOURSELF**

What are the characteristics of each stage of primary succession?

Climax Communities

Spruce trees are part of the *climax community,* or final stage of succession, at Glacier Bay. Unlike earlier stages, a climax community is a stable habitat. Spruce seedlings grow well in the shade of older trees. As the older spruce trees die, they are replaced by seedlings of the same species. The population maintains itself without increasing or decreasing in size.

The earlier stages of succession consist of populations that are less stable. Less stable populations change the habitat as they increase in size. Changes make the habitat less suitable for old populations and more suitable for new populations. As the populations of new organisms grow, older populations die off. One community is replaced by another. Eventually, a climax community forms. The populations of the climax community live in the habitat without changing it. Thus, the climax community remains stable unless it is disturbed by some outside force.

Figure 2–22. The climax community at Glacier Bay is a spruce forest.

Succession is often described in terms of its plant communities. Yet, animals also play an important part in the process. Each community has both animal and plant populations. As more plants grow, changes in the habitat occur. Such changes provide new niches for some animal populations, while the niches of others may die out. The growth and decline of these animal populations help to change the habitat.

Figure 2–23. How is secondary succession, shown here, similar to primary succession?

Natural disasters, such as fires or floods, can disturb a climax community. Disease and insect pests can destroy an entire community of climax species. Humans can also cause major changes in habitats, affecting the climax community. Agriculture and forestry are common practices that destroy climax communities.

Secondary succession is the series of changes in communities that occurs in a disturbed area. This type of succession occurs on land that has not lost its soil. Since soil is already present and does not have to be built up from bare rock, secondary succession takes place more quickly than primary succession.

An example of secondary succession might be a beech-maple forest that has been burned out or been cut down for timber. The pioneer plants in this situation are grasses and weeds, whose seeds are carried to the area by wind and animals. These plants have shallow roots that spread through the soil, protecting it from wind and water erosion. Within a few months, the ground is covered in green. Soon taller plants with deeper roots begin to appear. These plants make it difficult for the pioneer plants to survive. After many years, the grassy field contains tall weeds and shrubs.

The shrub community is a suitable habitat for the seedlings of pioneer trees, such as dogwoods, sumacs, and aspens. These seedlings grow well in bright sunlight. As the trees grow, they capture more and more of the sunlight and available water and nutrients. Gradually, the shrub community is replaced by the pioneer trees.

Figure 2–24. A shrub community, such as that shown here, is one stage of secondary succession.

Beech and maple seedlings grow well in the shade of pioneer trees. The taller beech and maple trees also have deeper root systems, which make them better able to capture water than the pioneer trees. Slowly, the pioneer trees die and are replaced by more beech and maple seedlings. After many years, succession will reestablish a climax community of beech and maple trees. In the next activity, you can find out how rapidly succession occurs in an area devastated by a volcano.

Figure 2–25. In some areas, a beech and maple forest is the climax community.

DISCOVER BY Researching

The eruption of Mount Saint Helens in 1980 was a major disturbance of the ecosystem. Using your library, find books and articles on this or another environmental disaster and on the succession that followed the destruction. Try to find "before" and "after" pictures that you can use. Record your findings in your journal. ✎

▼ ASK YOURSELF

What are the stages of secondary succession?

SECTION 3 REVIEW AND APPLICATION

Reading Critically

1. How is a stable population different from an unstable population?
2. Describe the stages of primary succession.
3. How is secondary succession different from primary succession?

Thinking Critically

4. In which stages of succession are the populations most stable? Why?
5. Suppose a wooden shed was built in a grassy field fifteen years ago. The shed gets knocked down. What would the ground look like where the shed had been standing? When plants start to grow, would this be primary or secondary succession? Why?

INVESTIGATION

*O*bserving an Ecosystem

▶ MATERIALS

- meter stick ● stakes ● string, 4.5 m ● notebook paper ● pencil ● hand trowel
- small jar with cap ● screening material for sifting soil

▼ PROCEDURE

1. **CAUTION: Do not disturb your study area any more than is necessary. Do not grab any small animal with your hands. Identify poisonous plants before you touch anything.**

2. Using the meter stick, mark off an area of study 1 m by 1 m. Use the stakes and string to mark the boundaries of the area as shown.

3. While standing beside the study area, carefully observe your area for animals or animal signs, such as leaves eaten by insects. Write a description of the study area that includes in-formation about the plants and animals or evidence of animals.

4. Dig for insects, small worms, and other living things in your study area. Record names or draw pictures of the types of plants and animals found in your study area.

5. Gather a soil sample by gently digging or scraping a small soil sample from the surface of your study area. Sift the soil through a piece of screening. Collect any animals or animal parts and place them in the jar. Keep the jar cap loose.

6. List all of the animals found on the surface or immediately below the surface of the soil.

▶ ANALYSES AND CONCLUSIONS

1. How does your study area differ from other study areas? What might be the reasons for the differences?

2. How are the study areas the same? Explain.

3. What characteristics do the animals and plants have that help them live in the study area?

4. What are the roles of the plants and animals found in your area? For example, which animals dig into the soil? Which feed on the grasses or weeds?

5. How have humans affected your study area? Has the impact improved or harmed the area? Explain.

6. Draw a possible food chain for the plants and animals found in your area.

▶ APPLICATION

Describe how humans could positively or negatively affect the area you have studied.

※ *Discover More*

Investigate the ways in which humans have both helped and harmed the plants and animals in an environment that is different from your study area.

The Big Idea

Imagine a huge spider web made of many connecting threads. All living and nonliving things in an ecosystem are connected together as if they were all parts of the same spider web. Just as a movement in one part of the web affects the entire web, a change in one part of the ecosystem affects everything in the ecosystem.

There are many different kinds of interactions and relationships that can exist among living things and among living and nonliving things in an ecosystem. These interactions and relationships are like the threads that make up the spider web. Energy and food are two of the threads in the web. The ways in which they are transferred from one organism to another are vital to the survival of the ecosystem.

For Your Journal

Look back at the relationships and definitions you wrote in your journal at the beginning of the chapter. How have your ideas about these concepts changed? Revise your journal entry to show what you have learned. Be sure to include information on the different kinds of relationships and interactions among living things and their environment.

Connecting Ideas

Copy this unfinished concept map into your journal. Complete the concept map by writing the correct term in each blank.

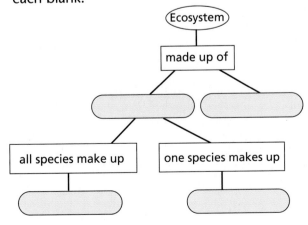

Ecosystem

made up of

all species make up

one species makes up

REVIEW

Understanding Vocabulary

Explain how the words in each set are related to each other.

1. ecology (35), organisms, environment
2. niche (43), habitat (42), ecosystem (35), population (34)
3. succession (48), community (34), primary, secondary

Understanding Concepts

MULTIPLE CHOICE

4. Which term best describes the relationship between a lion and an antelope?
   a) commensalism
   b) predator–prey
   c) mutualism
   d) parasitism

5. Which of the following does *not* contribute to changes in population size?
   a) organization
   b) birthrate
   c) death rate
   d) immigration

6. Because an earthworm and a mushroom both live in the soil, they share the same
   a) niche.
   b) population.
   c) habitat.
   d) species.

7. Which of the following is an omnivore?
   a) wolf
   b) rabbit
   c) vulture
   d) pig

8. During primary succession, lichens appear in the area during the
   a) pioneer stage.
   b) mossy stage.
   c) grassy stage.
   d) final stage.

9. Which term best describes the relationship between a flea and a dog?
   a) commensalism
   b) predator–prey
   c) mutualism
   d) parasitism

SHORT ANSWER

10. Give an example of a food chain involving human beings.

11. What are two types of limiting factors?

12. What type of relationship exists between a family and its pets? Explain your answer.

Interpreting Graphics

13. Look at the following equation. Define the terms and explain how they are related to each other and to the concept of population.

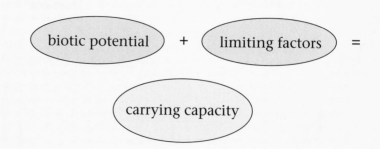

biotic potential + limiting factors =

carrying capacity

Reviewing Themes

14. *Environmental Interactions*
Imagine that you have a small ecosystem sealed off under a plastic bubble. Your ecosystem includes producers, consumers, and decomposers. Why is each group of organisms important to the continued survival of the ecosystem?

15. *Energy*
Describe the three ways in which organisms obtain energy in an ecosystem. Name an organism that plays each energy role.

Thinking Critically

16. Which stage of primary succession is shown in the photograph? How did the area look before this stage? What will it look like in the next stage?

17. Explain how the energy that heats your home came to you from the sun, through a food chain. (Hint: Natural gas and oil are used to heat most homes. These fuels formed from the remains of plants and animals that lived millions of years ago.)

18. How are predators and parasites alike? How are they different?

19. Describe a food web that includes the following organisms: a wolf, wild onions, rabbits, grass, mice, grasshoppers, lizards, snakes, an owl, blackberry plants, a hawk, mushrooms, bacteria, spiders, and oak trees. Identify each type of organism in your food web as a producer, consumer, or decomposer.

20. Explain what would happen to the pond community of the frog if a herbicide accidentally entered the pond.

Discovery Through Reading

Lauber, Patricia. *Summer of Fire: Yellowstone 1988*. Orchard Books, 1991. Describes the season of fire that struck Yellowstone in 1988, and examines the complex ecology that returns plant and animal life to a seemingly barren, ash-covered expanse.

CHAPTER 3

LIFE IN THE BIOSPHERE

Have you ever thought about living on another planet—maybe Mars? It's not out of the question. Humans first landed on the moon nearly 30 years ago. Thirty years from now there will likely be colonies on the moon, and perhaps on Mars as well. What will it be like, living on Mars? How is Mars different from Earth?

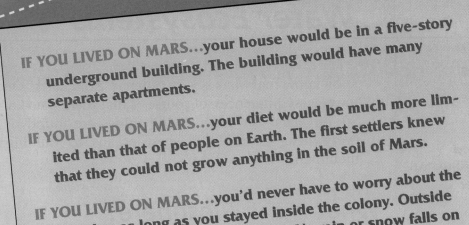

IF YOU LIVED ON MARS…your house would be in a five-story underground building. The building would have many separate apartments.

IF YOU LIVED ON MARS…your diet would be much more limited than that of people on Earth. The first settlers knew that they could not grow anything in the soil of Mars.

IF YOU LIVED ON MARS…you'd never have to worry about the weather as long as you stayed inside the colony. Outside the colony, it is a different story. No rain or snow falls on the surface of Mars. But the weather is almost always cold. The average temperature is a frigid 100 degrees below zero.

IF YOU LIVED ON MARS…you would know that Mars is drier than the driest desert on Earth. No rivers, oceans, ponds, or brooks exist anywhere on Mars. Of course, water is basic to human life. You could not live more than a few days without water.

IF YOU LIVED ON MARS…you'd have to wear a space suit every time you went outside the colony. The airtight space suit, with its connected helmet and boots, has everything you need to keep you alive.

IF YOU LIVED ON MARS…your home would be inside the colony. The colony is the only place on the entire planet where people can live. In fact, it is the only place on Mars where there is any life at all.

from *If You Lived On Mars*
by Melvin Berger

For Your Journal

✎ Why is Earth able to support life, but planets such as Mars are not?

✎ How do conditions for life differ on Earth?

✎ How are organisms adapted to their environments?

The Biosphere and Water Ecosystems

You know that Mars is very different from Earth. One of the biggest differences, of course, is that there is no known life on Mars. On Earth, living things are found nearly everywhere—from the deepest oceans to the highest mountains.

What Is the Biosphere?

Most life requires certain conditions—water, air, and a source of energy. Suitable combinations of these essentials have not yet been found anywhere in our solar system, except on Earth.

Even on Earth, conditions suitable for life are not found everywhere. Life cannot exist high in the upper atmosphere or deep underground. The right conditions exist only in a narrow layer near the surface of the earth.

This narrow layer where life exists is called the **biosphere.** Life exists only within the biosphere because it is the only place where organisms can obtain everything they need to survive.

Figure 3–1. Conditions on planet Earth provide essentials for living things to thrive.

An explorer to Mars leaves the biosphere behind. But an explorer's needs are the same on Mars as they are on Earth. On Earth, the biosphere provides all the things a person needs to survive. On Mars, these things must be provided for the explorer. Mars explorers must take a substitute for the biosphere with them.

Even on Earth, life is not found everywhere. A few organisms live as high as 8 km above sea level, while a few others live 10 km below sea level. However, most life is found from about 110 m above to about 110 m below sea level. The activity that follows can help you visualize the relative size of the biosphere.

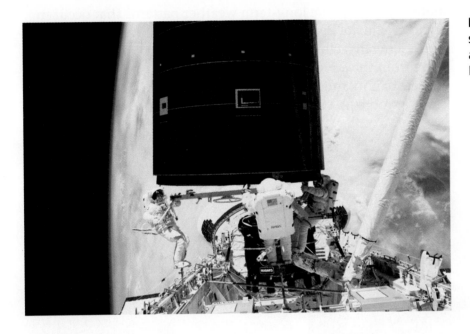

Figure 3–2. How does a spacesuit provide an astronaut with an Earthlike environment?

DISCOVER BY Doing

You will need notebook paper, tape, and a marker or colored pencil. Cut several sheets of notebook paper lengthwise into strips about 5 cm wide. The distance between lines on the strips of paper will represent 0.1 km. Tape enough strips of paper together end to end to represent 10 km. Mark the top line *sea level.* This represents the depth of the oceans. Add other sheets of paper to represent 10 km above sea level. This additional paper represents part of the atmosphere.

Using the marker or colored pencil, color the area of your paper that is between 10 km below sea level and 8 km above sea level. The colored area represents the biosphere. In what percent of the "biosphere" is most life found? ✐

The biosphere is so vast and complex that it would be difficult to study it all at one time. However, the biosphere can be divided into smaller segments to make studying it simpler.

Remember, organisms of the same species living together make up a population, and different populations living together make up a community. Remember too that a community and its physical environment describe an *ecosystem*.

Major land ecosystems are called **biomes.** Each is identified by the dominant plants found there. For example, trees are the dominant plants in a forest biome. What do you think the dominant plants are in a grassland biome?

The factor that most affects the plants in a biome is climate. **Climate** is the long-term weather patterns that occur in an area. Climate depends mostly on temperature and precipitation. Generally, climates become colder as you move farther away from the equator. Coastal climates are usually warmer and wetter than inland climates. And one side of a mountain usually receives more precipitation than the other side.

Figure 3–3. Earth has many different kinds of climates.

Temperature and precipitation also determine the kinds of living things found in a biome. Plants and animals are adapted for certain climates. Cacti, for example, grow well in dry ecosystems, while ferns grow best where it is wet.

The biosphere is made up of six major land biomes plus all the water ecosystems. Look at the map of major land biomes in Figure 3–4. In which biome do you live? What is the climate like where you live?

Figure 3–4. Earth's major biomes

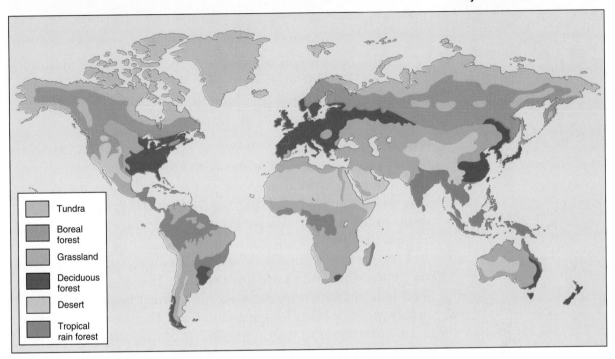

Legend:
- Tundra
- Boreal forest
- Grassland
- Deciduous forest
- Desert
- Tropical rain forest

 ASK YOURSELF

What factors help to determine what biome will be found in a particular area?

Marine Ecosystems

Biomes are defined as major land ecosystems. But nearly three-fourths of the earth is covered by water. While not true biomes, bodies of water are important ecosystems that contain a wide variety of living things. Oceans, seas, lakes, and rivers provide food and other resources.

Although water ecosystems, like biomes, have dominant plants, they are most often identified as freshwater ecosystems or saltwater ecosystems. The water in lakes, ponds, rivers, and streams is fresh, while the water in oceans and seas is salty.

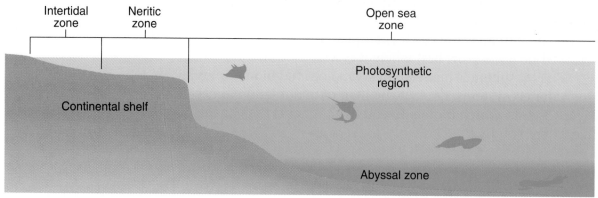

Intertidal zone   Neritic zone   Open sea zone

Continental shelf

Photosynthetic region

Abyssal zone

Figure 3–5. Zones of ocean life

Earth's oceans and seas make up the saltwater, or marine, ecosystems. In these ecosystems, life is found in four zones: the *intertidal zone,* the *neritic* and *open sea zones,* and the *abyss.* Locate each of these zones in the illustration above.

The Intertidal Zone   If you've ever walked along an ocean beach or seen pictures of surfers, you know that the water near the shore is seldom calm. As waves rush in, the beach is covered with water, only to be left high and dry as the water falls back again. Here communities are exposed to air during low tide and covered by water during high tide.

The intertidal zone is closest to the shore, running down the edge of the land to the low-tide line. This zone can be a difficult place in which to live. Organisms must be adapted for life underwater as well as for exposure to air. These organisms must be able to withstand the drying effects of bright sunlight and the pounding of ocean waves. This zone is rich in nutrients that are washed down from shoreline communities. Barnacles, clams, crabs, and seaweed are some of the organisms that live in the intertidal zone.

Figure 3–6. Organisms of the intertidal zone are adapted to harsh conditions.

The Neritic and Open Sea Zones

The neritic zone starts where the intertidal zone ends. Plants and animals of the neritic zone have a more constant environment than those of the intertidal zone. Water always covers the neritic zone. There are no breaking waves, and the temperature is fairly constant. The photographs below show some of the organisms found in the neritic zone.

Figure 3–7. Plankton like those shown here float in the neritic and open sea zones.

Beyond the neritic zone is the open sea zone. The open sea zone begins where the continental shelf drops off. The neritic and open sea zones share many of the same organisms. The most important organisms of these zones are the microscopic plankton, which drift with the ocean currents.

In the upper, or photosynthetic, region of the open sea zone, green plankton make their food as land plants do—by photosynthesis. Plankton are the basis for most of the food chains in the open oceans. Green plankton are also responsible for much of the oxygen in Earth's atmosphere.

Animal-like plankton eat green plankton. They, in turn, are eaten by larger, free-swimming organisms. Through these actions, energy from the sun, converted by green plankton to food energy, is made available for a large and complex web of marine life. Many kinds of free-swimming organisms—mammals, turtles, and fish—live in the neritic and open sea zones.

Figure 3–8. This strange-looking fish is found in the abyss.

The Abyss Below 2000 m, the temperature of the ocean remains just above 0°C. No sunlight penetrates there, so photosynthesis cannot take place. This is the abyss. The abyss, which extends to the ocean bottom, is the largest zone in the marine biome. However, there is little life here. The few animals of the abyss are mostly scavengers, feeding on dead organisms that rain down from above. Jacques Cousteau describes the abyss:

Half of the earth's surface is covered by more than 13,000 feet of water. It is the abyss—vast plains interrupted by volcanoes, rugged mountain ranges, and great scarring fractures. No humans have walked here as they have on the moon. Only a few humans in bathyscaphs have visited it. Our knowledge and understanding of life in this hostile world is derived almost entirely from a limited number of automatic photographs or samples taken almost at random in trawl nets.

 ASK YOURSELF

How is the abyss different from the neritic and open sea zones?

Freshwater Ecosystems

If all the water in the oceans is salty, how could rivers and lakes be fresh water? In his book *Secrets of Rivers and Streams,* naturalist Peter Swensen describes the water cycle.

Fresh pure water is constantly evaporating from the oceans and from the land masses of our earth. When the concentration of this water vapor becomes large enough in the atmosphere, it condenses around wind-blown grit and dust particles. These droplets of water are carried by winds and eventually return to the earth as rain or snow. Some of this precipitation immediately soaks into the soil to become ground water or falls at an elevation cold enough to keep it frozen as glacial ice or snow. The rest of the precipitation fills ponds and lakes or runs into streams.

Taken together, the freshwater ecosystems make up only a tiny fraction of the earth's water ecosystems, but they provide many different environments. Fast-moving bodies of water, such as streams, contain little life. Slow-moving bodies of water, such as rivers, contain more living things.

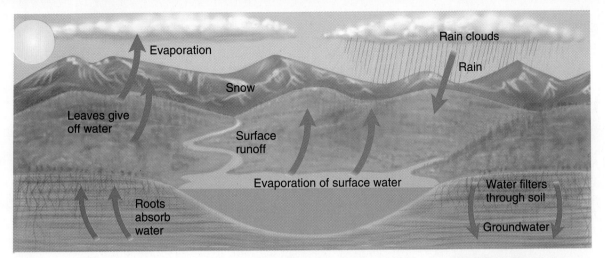

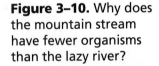

Figure 3–9. All the water on Earth is recycled over and over.

Sunlight and oxygen are the most important limiting factors in quiet waters. Lakes and ponds are rich in nutrients because dead organisms fall to the bottom and decompose. Also soils and sediments washed from their banks remain in the water. Plants near the surface grow well in the nutrient-rich water, but these plants keep the light from reaching the deeper parts of the water. Therefore, photosynthesis occurs only at shallow depths.

Decomposers remove oxygen from the deeper waters and add carbon dioxide. The low oxygen and high carbon dioxide levels make the deeper water less suitable for animal life. Changes in temperature during each spring and fall cause water currents to bring the nutrients toward the surface. With a fresh supply of nutrients, the producers continue to grow, keeping the ecosystem in balance.

Figure 3–10. Why does the mountain stream have fewer organisms than the lazy river?

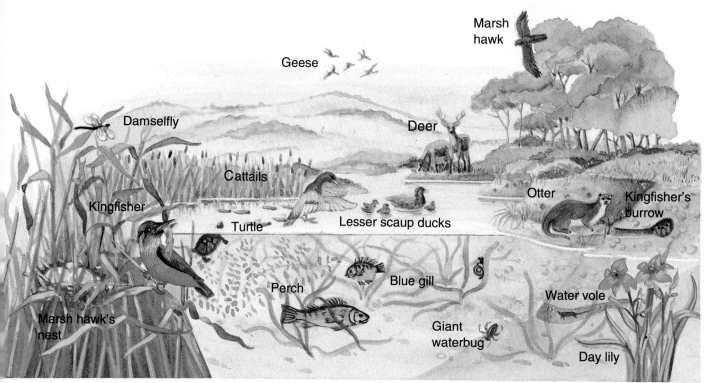

Figure 3–11. A pond is a freshwater ecosystem that contains many different populations.

Fish, turtles, and insects live in and around ponds, feeding on the numerous plants growing there. Birds are also a part of freshwater ecosystems. Even though they don't live in the water, they feed in it and are considered part of the community.

What do you think happens where fresh water empties into the ocean? The next activity can give you an idea.

ACTIVITY

What happens when fresh water and salt water meet?

MATERIALS
glass baking dish, a strip of cardboard that fits across the dish, salt, blue and yellow food coloring

PROCEDURE
1. Fill the baking dish with water. Then place the cardboard across the middle of the dish.
2. Add a teaspoon of salt and a few drops of blue food coloring to the water on one side of the cardboard. Gently stir the blue, salty water.
3. Add a few drops of yellow food coloring to the water on the other side of the cardboard. Gently stir this too.
4. Carefully remove the strip of cardboard. Describe what happens where the waters meet.

APPLICATION
How is the mixing of fresh water and salt water in the dish similar to the mixing of fresh water and salt water where a river enters an ocean?

Where a river and an ocean meet, an *estuary* forms. Estuaries can be mud flats, salt marshes, or swamps, but in all estuaries fresh water and salt water mix, producing water that is called *brackish*. Brackish water is more salty than fresh water but not as salty as the salt water of an ocean.

The unique conditions of an estuary allow many kinds of organisms to live there. The brackish water and changes in tides create many habitats. Both freshwater and saltwater organisms inhabit estuaries.

The waters of estuaries contain many nutrients. Marsh grasses and algae are the main producers for the community. Marine animals—such as clams, crabs, shrimp, and oysters—feed on the producers. Many small fish live in estuaries, and birds, which live near estuaries, feed in the shallow waters.

Figure 3–12. Estuaries often form where slow-moving rivers flow into the ocean.

 ASK YOURSELF

How are stream and pond ecosystems different?

SECTION 1 REVIEW AND APPLICATION

Reading Critically

1. What is the biosphere?
2. How are freshwater ecosystems and saltwater ecosystems alike? How are they different?
3. Why do fast-moving waters, such as streams, have less life than still waters, such as ponds?

Thinking Critically

4. Why is the water of an estuary less salty than the water of an ocean?
5. Why do estuaries support such a wide variety of living things?
6. What do you think would happen if you placed a saltwater organism, such as a starfish, in fresh water?

INVESTIGATION

Observing Changes in Ecosystems

▶ MATERIALS

- container with lid, 5L ● dried straw or grass that has been boiled ● distilled water at room temperature ● wax pencil ● graduate, 10 mL ● medicine dropper ● pond water ● hammer ● nail ● microscope ● slides and coverslips

▼ PROCEDURE

1. Place the previously boiled grass or straw in a clean 5-L container.
2. Fill the container two-thirds full with distilled water. Mark the water level with a wax pencil. On the side of the container, write the names of your lab group members.
3. Add 5 mL of pond water to the container. Using the hammer and nail, punch holes in the lid and loosely place it on your container.
4. Make a slide using the water from your culture. Follow the directions for making a wet mount in the Reference Section on page 582. Observe under both low power and high power of your microscope. Record your observations.
5. Each day for seven days, observe the pond water in the container and record your observations. If the

water level drops, add more distilled water to keep the level at the mark.
6. After two days, make another slide of the water from the container. Observe under low power and high power, and record your observations.

7. Make at least three more slides of the water from the container on different days. Be sure to record your observations, especially any increase in the number of organisms observed or changes in the types of organisms you see.

▶ ANALYSES AND CONCLUSIONS

1. Explain why the water was distilled and the dried grass or straw was boiled before being put into the container.

2. What happened to the water when the pond culture was added to the container?

3. How did the appearance of the water in the container change in the seven days?

▶ APPLICATION

Explain what happened in the container in terms of ecological changes over the length of the investigation.

✳ Discover More

What do you think would happen if 100 mL of ice water or hot water were suddenly added to the container? Find out.

Land Biomes

Ecosystems found in similar climates look alike. For example, the deserts of Australia are similar to those in the southwestern United States. Forests, grasslands, or tundra also look much the same wherever they occur. Scientists disagree about the number of distinct biomes on Earth, but the six most generally agreed upon are described in this section.

Objectives

Identify each biome's dominant plants.

Relate a biome's dominant plants to its climate.

Figure 3–13. The desert in Texas (left) is similar in many ways to the desert in Australia (right).

Tropical Rain Forests

In a hot land near the equator, where winter never comes, a new day is beginning. The climbing sun looks close enough to touch as it turns the sky pink. Out of the mist a vast ocean of leaves appears, slashed with yellow, orange, and violet blossoms. It is the roof of the jungle.

High in the trees, the birds are about to begin their morning chorus. On the branches, monkeys sit motionless, their long tails hanging down behind them, like dark quarter notes dotting the gray dawn. Soon the sun will chase the mist and another day will begin in the mysterious green world below.

The South American rain forest contains tall trees with large leaves, brilliant orchids, and long climbing vines. A tropical rain forest in Africa looks very much the same. The species that make up each of these forests may differ, but the effect is the same—a tropical rain forest

The Jungle, by Helen Borton, describes the mist rising from the canopy, or roof, of the forest. The term *jungle* is often used when describing a tropical rain forest. However, the jungle is only one community within this biome.

During the year, a tropical rain forest receives more than 400 cm of rain, falling nearly every day in short, intense cloudbursts. Temperatures in tropical rain forests stay about the same all year. Locate the tropical rain forests on the world biomes map on page 63. Are there any tropical rain forests in North America?

Scientists hypothesize that more kinds of organisms live in the tropical rain forest biome than in all the other biomes combined. Tropical rain forests probably contain thousands of plants and animals that have not yet been discovered. Tall, flowering trees shade the forest floor and provide homes for many different kinds of animals. Monkeys, apes, and lemurs are common, as are reptiles and amphibians. All flourish in the warm, wet climate.

Figure 3–14. The diversity of life in tropical rain forests is fantastic.

Tropical rain forests are being destroyed at an alarming rate. Farmers need the forest land to grow food crops, but the soil of the rain forest is very poor, so their crops do poorly. After the crops fail, the farmers clear more land, and the cycle continues. Others cut down the trees for the valuable hardwoods they contain—the lumber is often used for making furniture. Once destroyed, it takes a very long time to regrow a tropical rain forest.

Figure 3–15. Over a million hectares of tropical rain forest are destroyed every day.

Tropical rain forests are important to the biosphere. During photosynthesis, forest plants take carbon dioxide from the air and release oxygen. If many trees are cut down, the atmospheric balance between oxygen and carbon dioxide changes. What changes might result from an increase in the amount of carbon dioxide in the atmosphere?

Saving the tropical rain forests is necessary for preserving the biosphere. By doing the next activity, you can make people aware of the importance of saving this biome.

DISCOVER BY *Researching*

Use headlines and pictures from magazines and newspapers to make a poster about saving the tropical rain forests. In your journal, describe efforts being made to save the rain forests. Share your information and your poster with your classmates and with the people of your community. ✐

▼ ASK YOURSELF

Do you think it is important to save the tropical rain forests? Explain your answer.

Deciduous Forests

Red oaks, red maples, and ash trees were everywhere. On the forest floor [were] . . . seedlings . . . trees that like the deep shade. Every autumn the trees lost their leaves. They fell to the ground with dead twigs and branches. All of these things decayed and made a rich layer of stuff called humus. The beeches and sugar maples were the kings of the forest. It is home for many wild animals—for foxes, bobcats, wood turtles, chipmunks, bears, deer, squirrels, mice, porcupines, and many other creatures.

Have you ever sat under a big shade tree on a hot summer day? In his book *How the Forest Grew,* William Jaspersohn mentions the shade of a deciduous forest. Deciduous forests are shady and damp. They receive about 75 cm of precipitation each year. The climate of deciduous forests is temperate—winters are cold and snowy, while summers are warm with occasional rain. Look again at the world biomes map on page 63. Where in North America are deciduous forests located?

Figure 3–16. Fall colors in a deciduous forest

Oak, hickory, beech, maple, and other useful hardwood trees make up the dominant plants of the deciduous forest biome. Deciduous trees lose their leaves each fall.

A wide variety of shrubs and many types of ferns and fungi live on the forest floor. The animals of the deciduous forest include bears, deer, bobcats, snakes, foxes, squirrels, and many birds.

 **ASK YOURSELF**

How do winter and summer differ in the deciduous forest?

Boreal Forests

They opened the bars at the entrance to an old wood-road, and stepped into the cool shadow of the pines. It was quiet in the woods, and dark, except where filtered patches of sun came through to sprinkle the brown needle carpet with flecks of gold. A soft sound, like the hush-sh-sh. . . hush-sh-sh *of waves on a beach, descended from the far green roof of the forest.*

In his book *Lumberjack*, Stephen W. Meader describes the hush of the pine forest. Try to find a stand of pines near your home and listen to the gentle sounds of wind blowing softly through the trees. *Conifers*—trees that produce cones—are the dominant plants of the boreal forest biome, or taiga (TY guh). The leaves of most conifers are modified into needles. Unlike the leaves of deciduous trees, needles are replaced throughout the year, so most conifers are evergreens.

Figure 3–17. The moose is so common in boreal forests that they are sometimes referred to as "spruce-moose" forests.

The climate of the boreal forest is cold and snowy in the winter and cool in the summer. Annual precipitation is only about 60 cm.

Animals of the boreal forest include snowshoe hares, moose, bears, beavers, deer, and many kinds of birds. Look again at the world biomes map on page 63. At what latitudes are most boreal forests found?

 ASK YOURSELF

Do boreal forests receive more or less precipitation than deciduous forests?

Tundra

It was beautiful spring weather, but neither dogs nor humans were aware of it. Each day the sun rose earlier and set later. It was dawn by three in the morning, and twilight lingered till nine at night. The whole long day was a blaze of sunshine. The ghostly winter silence had given way to the great spring murmur of awakening life. This murmur arose from all the land, fraught with the joy of living. It came from the things that lived and moved again, things which had been as dead and which had not moved during the long months of frost.

Remember how you felt on the first warm day last spring? In *The Call of the Wild,* Jack London describes the coming of spring to the Arctic tundra. The tundra biome is found near the earth's poles. Find the tundra on the world biomes map on page 63. Why is there little tundra in the Southern Hemisphere?

Winters on the tundra are long, dark, dry, and very cold—temperatures may drop below −40°C. Precipitation, mostly snow, totals only about 25 cm a year. The ground, which is frozen solid during most of the year, thaws a little on top during the summer. Below this thin layer of soil is the *permafrost,* so called because it never thaws.

Due to the extreme cold, strong winds, and scant precipitation, tundra plants grow only a few centimeters above the ground. Grasses and shrublike trees, mostly willows and birches, are the dominant plants of the community. The grasses and tundra wildflowers must sprout, grow, and reproduce quickly during a summer season that may last only a month or two.

Tundra animals include caribou, musk oxen, and Arctic hares, which feed on the grasses and shrubs. Birds such as puffins and ptarmigans (TAHR muh guhns) live on the tundra, while others, such as Canada geese and Arctic terns, nest there only during the summer. Many insects also inhabit the tundra during the short summer.

Figure 3–18. For all its harshness, the tundra has a large variety of plant and animal life.

 ASK YOURSELF

Why are tundra plants usually only a few centimeters high?

Grasslands

The prairie was glorious, ablaze with wildflowers and overrun with game. The buffalo grass was the best, alive as an ocean, waving in the wind for as far as his eyes could see. It was a sight he knew he would never grow tired of. Everything was immense. The great, cloudless sky. The rolling ocean of grass. Nothing else, no matter where he put his eyes. No road. No trace of ruts for the big wagon to follow. Just sheer, empty space.

Looking at a lawn or the neatly trimmed grass of a park, you may find it hard to imagine a place where a person has to mount a horse just to see over the grasses. But if you have read *Dances With Wolves* by Michael Blake, you know that the prairie, or grassland biome, is dominated by two-meter-tall grasses and similar plants. Corn, wheat, sorghum, and other grains also grow well in the rich prairie soil. Locate the grasslands on the map on page 63. In addition to the prairies of North America, what other continents have large grasslands?

Figure 3–19. Most of the world's major food crops come from the grassland biome.

Winters on the prairies are cold and snowy, while summers are hot and dry. Grasslands receive about 50 cm of precipitation each year, but much of it falls as winter snows. There is generally not enough water for large forests to grow on the grasslands, except near rivers and streams.

Bison and antelope were once common on the American prairies. Today, these populations have been replaced by herds of domestic cattle and sheep. Many small animals, such as ground squirrels, prairie dogs, birds, and insects, also inhabit grassland biomes. In the next activity, you can find out about the root structure of grasses.

MATERIALS

hand lens, several samples of grass, ruler

PROCEDURE

1. Use a hand lens to examine each sample of grass. Observe the roots. How are the roots shaped?
2. Use a ruler to measure several roots.
3. Observe and measure several blades of grass. How do the measurements of the roots and blades compare?
4. Draw each grass sample and label the roots and blades.

APPLICATION

1. What makes grasses especially useful for keeping soil in place?
2. Look up the term *grassroots* in the dictionary. Describe how the structure of the roots of grasses might have led to the use of this term.

ASK YOURSELF

Why are trees unable to grow in the grassland biome?

Deserts

The first explorers found the desert a vast and varied place. They crossed flat white valleys shimmering with heat. Spidery gray creosote bushes dotted the bare ground, each off by itself as if guarding its space. Strange poisonous snakes and insects flourished there. On the horizon, the dark, wrinkled mountains looked scorched. The early explorers rode under giant arches of red rock. They climbed vast blocks of stone so flat they called them mesas—"tables." In other places, the land sank into canyons with rivers twisting at their bottoms.

Sheila Cowing describes the vastness of the American Southwest in *Searches in the American Desert,* but not the climate. Most people think of deserts as dry and hot! But there are two kinds of deserts—hot and cold. Hot deserts have an average temperature of 21°C, while cold deserts have an average temperature of 10°C. All deserts receive less than 25 cm of precipitation during the year.

Deserts are found in many parts of the world. They are often found in the shadows of tall mountains that cut off moisture-laden winds. Find the deserts on the world biomes map on page 63. Which ones are near tall mountains?

While some hot deserts support no life, most deserts support a wide variety of living organisms. The dominant plants of a desert community include cacti, small shrubs, and a few trees. Many desert plants have spinelike leaves, and all have extensive root systems to get any available water. Some desert plants store water to be used during prolonged droughts.

Figure 3-20. What adaptations do desert plants and animals have to keep them from drying out?

Desert animals, like desert plants, are adapted to the dry environment. Some animals rest underground or in the shade during the day, becoming active during the cool desert evening. Lizards, scorpions, snakes, and birds, such as the comical road runner, are common in the deserts of North America.

ASK YOURSELF

Describe the characteristics of a desert.

SECTION 2 *REVIEW AND APPLICATION*

Reading Critically
1. Name and describe the six major land biomes.
2. How are deciduous trees different from coniferous trees?
3. In what ways are the desert and tundra biomes similar?

Thinking Critically
4. In Greek mythology, Boreas was the god of the north wind. How might this be related to the name *boreal forest?*
5. In the grassland biome, you may occasionally see clumps of trees growing near rivers or ponds. Why do grassland trees grow only near bodies of water?
6. Cactus plants are called *succulents*. The first definition of *succulent* in the dictionary is "full of juice; juicy." How does this word apply to cacti?

SKILL

Modeling Climate

▶ **MATERIALS**
- graph paper • pencil

▼ **PROCEDURE**

1. A model of a region's climate can be made into a graph called a *climatogram*. A climatogram is a graph of both temperature and precipitation. Temperature is shown on a line graph, while precipitation is shown on a bar graph. You can use climatograms to relate climate to biomes.
2. The scale across the bottom shows the months of the year.
3. Precipitation is read from the scale along the left side of the graph.
4. Temperature is read from the scale along the right side of the graph.
5. Make a climatogram for each of the three regions described in the table.

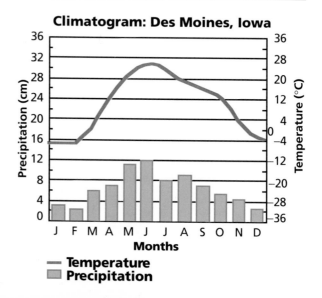

Climatogram: Des Moines, Iowa

— Temperature
▢ Precipitation

TABLE 1: REGIONAL PRECIPITATION AND TEMPERATURE												
Region	Month											
	J	F	M	A	M	J	J	A	S	O	N	D
A Precipitation (cm)	1	1	0	0	2	2	2	1	1	1	1	1
Temperature (°C)	−22	−24	−20	−16	−4	2	4	4	1	−4	−16	−20
B Precipitation (cm)	0	0	1	1	1	0	0	0	1	1	1	0
Temperature (°C)	24	26	27	28	28	27	26	28	28	28	27	24
C Precipitation (cm)	12	14	8	3	2	2	1	1	1	4	10	12
Temperature (°C)	20	18	16	14	12	10	10	12	14	16	18	20

▶ **APPLICATION**

1. Compare the climatograms for regions A and B. How are they similar? How are they different?
2. Describe the climate of region C over the course of a year. What is unusual about this climate compared to the other two? How can you explain this difference?

✳ *Using What You Have Learned*

Obtain data on temperature and precipitation for the past year in the area where you live. Make a climatogram for your area. How do you think the climate influences the biome in which you live?

80   Chapter 3   Life in the Biosphere

The Big Idea

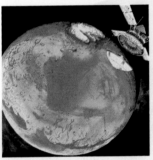

If you lived on Mars, you know that things would be very different than they are on Earth. Mars has no biosphere. On Earth, there is a biosphere between 8 km above and 10 km below sea level. Life exists in the biosphere because it has everything organisms need to survive.

Within the biosphere there are six major land biomes and two water ecosystems. Water ecosystems may be either fresh water or salt water. Within a biome, the climate and the dominant plants are related. The animals as well as the plants are adapted to the environmental conditions of the biome. Tropical rain forests, deciduous forests, boreal forests, tundra, grasslands, and deserts are the major land biomes.

For Your Journal

Look again at the answers you wrote about why planets such as Mars cannot support life and whether all places on Earth can support the same kind of life. How have your ideas changed? How are living things adapted to their environment?

Connecting Ideas

The pictures show six land biomes found on Earth. In your journal, identify the biomes and list three physical characteristics of each. Name some of the living things that are adapted to the environmental conditions in each biome.

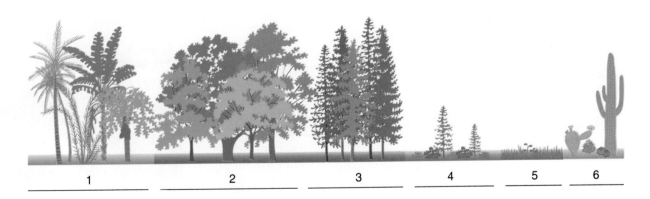

| 1 | 2 | 3 | 4 | 5 | 6 |

REVIEW

Understanding Vocabulary

Explain the relationships between the following sets of terms.

1. biome (62), biosphere (60)
2. biome (62), ecosystem (62)
3. biome (62), climate (62)

Understanding Concepts

4. Write the letters *a* to *f* on a sheet of paper. Next to each letter, fill in the missing information from the table.

BIOME SUMMARY TABLE

Biome	Rainfall	Plants	Animals
Tropical Rain Forest	400 cm	(a)	monkeys
Deciduous Forest	80 cm	(b)	deer
Boreal Forest	(c)	conifers	moose
Tundra	25 cm	(d)	caribou
Grassland	(e)	grasses	prairie dogs
Desert	20 cm	(f)	lizards

MULTIPLE CHOICE

5. Which two biomes are similar in the amount of available water?
   a) tundra and desert
   b) boreal forest and desert
   c) desert and grassland
   d) grassland and deciduous forest

6. Which biomes are similar in temperatures?
   a) tundra and desert
   b) boreal forest and tundra
   c) desert and grassland
   d) grassland and deciduous forest

7. Where are conditions most stable for life?
   a) intertidal zone   b) abyss
   c) neritic zone   d) 2000 m below sea level

SHORT ANSWER

8. Describe the characteristics of a tropical rain forest.

9. Describe the differences among the various forest biomes.

10. Why are temperature and precipitation important factors in determining the dominant plants of a biome?

11. Compare and contrast saltwater and freshwater ecosystems.

12. In what ways are deserts and tundra similar? In what ways are they different?

Interpreting Graphics

13. Look at the plants in the drawing. In which biomes might you find these plants?

14. Look at the trees in the photograph. In what biome are they likely to be found? What factors of that biome influence their growth?

Reviewing Themes

15. Environmental Interactions
Explain the relationship between climate and the kinds of plants and animals that can live in the desert.

16. Systems and Structures
How are the organisms of the intertidal zone adapted to the conditions there?

Thinking Critically

17. Suppose you were hiking on a mountain and found the following pattern of biomes: a grassland at the base of the mountain, deciduous trees higher up, conifers still higher, and grasses and dwarf trees near the mountaintop. Explain why this pattern of vegetation exists.

18. Does the abyss have more in common with quiet freshwater ecosystems or with intertidal saltwater ecosystems? Explain your answer.

19. One of the areas shown below is a normally wet area that is in a drought condition, while the other area shown is a normally dry area that has been flooded by heavy rains. Explain how changes in weather patterns, such as those shown, would affect the plant and animal life in these areas. What would happen if the changes were not temporary conditions but permanent climatic changes?

Discovery Through Reading

George, Jean Craighead. *One Day in the Tropical Rain Forest.* Crowell Jr. Books, 1990. From a colony of army ants to an overview of the importance of rain forests in the biosphere, Ms. George takes readers on a memorable journey through a jungle. This is one of a series of books on nature by the same author that includes *One Day in the Alpine Tundra, One Day in the Prairie, One Day in the Woods,* and many others.

Studying the Rain Forest

Remember the great old jungle movies—the ones with explorers hacking their way through thick vines? You heard cries of ferocious animals and saw giant insects and millions of flowers. You saw all sorts of exotic animals and plants. Jungles, or tropical rain forests, do exhibit a great variety of life. Perhaps half of all plant and animal species on Earth live in the rain forests. For many years, scientists from all over the world have come to study rain-forest life. Now scientists are studying ways to save the tropical forests. Just fifty years ago, one-tenth of Earth's land area was covered by rain forests. Today less than half of these forests remain. Fifty years from now, there may be none.

Exciting Work

Rain forests are so large that scientists have been able to study only small portions of them at a time. Unfortunately, scientists may not have the chance to study the rain forests much longer. Large sections of rain forest are destroyed every day. Some areas are being cleared by lumber companies. In other areas, farmers are clearing rain-forest land for cattle grazing and planting crops.

Poor Soil

When farmers first began clearing the rain forests, they thought the soil would be rich in nutrients, since it supports such thick plant growth. However, farmers soon found out that the soil is too poor to grow crops for more than a year or two.

The trees of the rain forest are able to survive because their roots have certain kinds of fungi growing on them. The fungi use the nutrients in dead leaves, twigs, and fruits that have fallen to the ground. This process helps to decay materials on the forest floor. The nutrients from the decaying plants are absorbed quickly by the roots of the trees. But when the trees are harvested, most of the nutrients are taken away with them. After crops are raised for a few years, all the nutrients in the soil are used up. There is no way to replace lost nutrients, except with expensive fertilizers.

More Problems

To make the soil richer, farmers now use the slash-and-burn method of clearing the land. Trees are cut and burned. Then their ashes are used to supply nutrients for crops. Even with this method, most of the nutrients are used up after a few years, and the farmers abandon the land.

A variety of bromeliads, or air plants, grow on the rooted plants of a rain forest.

There are many kinds of animals in the rain forest, including poison dart frogs, anoles, and bark beetles.

Then the heavy rains that occur every day wash away any remaining soil.

The burning of trees also releases carbon dioxide into the atmosphere. This gas traps heat, adding to other sources of global warming.

Hope for the Future

Scientists who study rain forests know how important they are, both for the organisms living there and for the forests' influence on Earth's climate. If scientists can convince world leaders that rain forests are more valuable for themselves than for lumber or farmland, perhaps the forests and all the species that live there can be preserved for the future. ◆

This farm was once a tropical rain forest.

An Eagle to the Wind

by Nancy Ferrell from *Young World*

Rob Simon first heard the squeal as an urgent call. Though weak, the cry forced attention through the other normal sounds along the ocean shoreline.

"Listen," he said sharply, motioning for his friend Matt Mercer to stop. Squinting his eyes, Rob's head swiveled slowly, trying to pinpoint the direction. "You hear that?" he asked.

"Yes," came the reply, "but what is it?"

"Sounds like something's hurt," answered Rob. Again he listened intently until his eyes focused on a tall, scraggly Sitka spruce. Following the trunk upward with his eyes, Rob put a hand on his friend's shoulder, and pointed. In a crotch two-thirds up the spruce was something that resembled a large haystack, and, Rob felt sure, was the source of the cry.

"Why, it's an eagle's nest," he said. With a slight turn, his eyes gazed around the area from tree to tree and finally out beyond the beach. "Must be some young birds up there," he continued, "but I don't see any eagles around. Do you suppose something's happened to them?" "Don't ask

me," Matt answered. "They're probably flying around getting food."

Rob shook his head doubtfully. "Maybe," he replied, "but usually one of them stays close if there are eaglets in the nest."

"Ah, come on," Matt said, hitching his pack higher on his back. "They'll be all right. Besides, we have to keep going if we want to make Yankee Cove tonight."

"There's plenty of time," Rob answered. "We've got all afternoon, and we can't just leave. What if the birds don't get any food? They'd die for sure." He let his pack drop to the ground and sat down on a boulder. "Let's stay here for a while and just watch. Matt shrugged and sat down with his friend.

For over an hour the two kept watch on the spruce and the surrounding waters, but during that time no mature eagle approached the nest.

Suddenly a raven, attracted by the young squeals, began circling the spruce high above.

"That does it!" Rob cried. He sprang up and ran shouting toward the tree as Matt followed. The startled raven did not fly away, but settled in a nearby tree.

"Something must have happened to the parents," Rob stated. "They'd never let a raven this close to the nest. I'm going up to see."

Matt's mouth dropped open. "But look at that tree!" he exclaimed. "I bet you'd have to have special equipment to get up there."

"I'll have to try," Rob answered. "I have to find out."

His friend stopped him. "Wait a second," Matt said. "I think I read someplace that an eagle won't fly unless his parents teach him. If you take them out of the nest, they may never fly. You ever think of that?"

Staring at the aerie, Rob answered, "No—I didn't." Undecided, he thought for a minute. Finally, he made up his mind. "You could be right," he said, "but we can't just leave them here. Better live eagles who can't fly than no eagles at all. I'm going up."

With that, he began climbing. There were plenty of footholds on the old spruce, but several cracked when pressure was applied, and Rob had to use caution.

At last he reached the base of the aerie and was surprised at the size. The top, or platform, spread nearly six feet across, and Rob realized he had several more feet to go before he could even look into the nest.

Eventually, he stretched over a final bough and peered at the platform. There, staring with fierce dark eyes, stood a month-old eaglet, its brown flight feathers mixed with shedding, light-gray down. The fledgling half-stretched its wings nervously, and Rob caught sight of the talons gripping the nesting material. Scattered nearby and mixed with fluffy down were bits of animal bones or fish bones.

"There's one here," he called down to Matt. But as he finished, his eyes caught the shape of another, smaller eaglet pushed to the rim of the aerie. Rob sensed immediately that the smaller one was dead.

Easing up and over, Rob advanced cautiously upon the young eaglet, who danced backward at his approach—never taking his eyes from the intruder. Rob stopped, fumbled in his pocket, and removed a pair of gloves. Abruptly, he lunged forward and gripped the young legs firmly. The bird fought, pecking Rob's hands, but he was too weak and too young to do any serious damage.

"I've got him," Rob called again. With some effort, he covered the talons with the gloves as best he could. Enclosing the wings with one arm, he drew the bird to his chest and zipped his jacket over the eaglet. In the darkness of this pouch, the bird calmed. Then, using great care, Rob made his way over the edge of the aerie and down the tree. Finally reaching bottom, he gently lifted out the bird, and placed him on the ground.

"The poor guy must be starving to death," he said.

"What kind of eagle is it, do you know?" Matt asked, examining the bird from a distance.

"Probably a bald eagle."

"But his head's not white."

"They don't get white until they're about four or five years old."

"How come you know so much about eagles?"

Rob smiled. "You don't have a biologist for a father without picking up something along the way," he replied.

There was a bird in the nest!

With Matt's help, Rob again tucked the bird in his jacket, "Come on," he said, heading for the water. "Tide's out, and we have to find something for the bird."

Searching the tide flats, the two found sea urchins, the meat of which the bird ate ravenously. With some nourishment, the eagle perked up, and began exploring, pecking among the exposed rocks. At one spot Rob found a dead herring, and, holding his nose, flicked it toward the bird with a stick. But it was only after Rob himself broke the meat and offered it in tiny pieces that the bird would eat. At last the fledgling seemed temporarily satisfied. It was then Matt asked, "What are we going to do with him?" "Well," Rob considered, "we can't leave it here. An animal will get it for sure. I just know something happened to the parents, or they would have returned by now." He shrugged his shoulders. "I guess we'll have to take him with us."

"You mean baby-sit an eagle?" Matt asked, shaking his head but smiling at the same time. "That ought to be a first."

Throughout the following days of camping and hiking, the eaglet rode in Rob's jacket. When free, the bird spent hours preening itself, shedding more down, while the flight feathers strengthened. Occasionally it would instinctively jump skyward in short hops, flexing its wings in preparation for flying. It was at these times that a question darkened Rob's mood—would the eagle be able to fly now that it was gone from its parents?

On the day they were to return home, Matt put the question into words. "You think the little guy will ever fly without someone showing him how? It's for sure I'm not going to run around flapping my arms to show him."

Rob laughed. "That sure would be a picture," he said. Still, his face grew serious as the worry persisted. It was later that afternoon when the two finished their trip and reached Rob's home—a comfortable house away from town. Mrs. Simon, happy at their safe return, fed them while the eaglet perched on the back porch.

"Nice-looking bird," she commented. "You'll have to fix something outside for him. Oh," she added, "better call Fish and Game, too. They'll want to know about the eagle, I'm sure."

Later, after discussing the eaglet with the Fish and Game Department, Rob received permission to take care of the bird. He was assured that the eagle should be able to fly when it was old enough—probably near the end of July or early August.

However, Rob was not entirely convinced.

During the rest of June and July, Rob set aside time each day to exercise the bird on the gravel driveway. The down of babyhood disappeared, and the bird feathered out in full, brown flight dress. The eagle allowed itself to be handled by Rob, who had learned that a heavy glove on his hand made a safe perching place, but the bird

Rob set aside time each day to exercise the eagle.

kept a reserved, often unpredictable attitude.

Each day the eagle would hop about the yard, jumping first one foot, then two feet—three feet into the air, beating its wings trying to gain altitude but never really succeeding. On these occasions, which grew more frequent, Rob wondered if he had done the right thing by taking the bird, and this dark worry haunted him more and more.

"You've got to fly," he would repeat under his breath. "You've got to fly."

On the last day of July, Rob brought the bird outside, perching it on a low shed roof. Again the eagle leaped up and down, stretching its wings desperately, jumping five feet—six feet—eight feet into the air. It was at the zenith of one jump that a gust of wind caught the bird. At first clumsily, the eagle lifted into the air, rocking uncertainly as it hovered, and then with a tremendous beat of its wings, he was flying! Up, up he climbed, testing wind currents, judging wing movements, gaining confidence. Though unpracticed, the eagle demonstrated excellent muscle coordination, giving indication of the strong, proud bird he would become.

Rob laughed from pure joy as he watched. It was as if he, too, were up there. He could almost feel the rushing wind, the smooth, singing freedom of nothing to touch. He didn't want the moment to end.

Eventually, the eagle, wings spread, talons down, dropped to earth a few yards from Rob. They looked at each other for a few moments while the bird rested. But when Rob took a step forward, the eagle turned and was soon airborne again. He circled once, gliding gracefully past Rob, and then headed toward the sea.

Rob watched until the speck disappeared in the sky. Though he sensed a loss inside, he was happy, too. The eagle was free, and somehow, he felt the same way. ◆

The eagle was free, and somehow, he felt the same way.

Rachel Carson
(1907—1964)

Rachel Carson published a best-selling book, *The Sea Around Us*, in 1951. In 1962 she published her most important book, *Silent Spring*, which introduced the public to the dangers of toxic chemicals in the environment. Carson stressed the dangers of the pesticide DDT, which was responsible for the deaths of birds and beneficial insects.

Carson was born in Springdale, Pennsylvania, in 1907. She attended the Pennsylvania College for

Women (now Chatham College), where she was able to combine her two favorite subjects—biology and writing. She also studied genetics and zoology at Johns Hopkins University, where she received an M.A. in 1932, and spent her summers doing research work at the Marine Biological Laboratory in Woods Hole, Massachusetts.

Carson worked as an aquatic biologist for the United States Bureau of Fisheries. In addition to her books, she wrote conservation bulletins for the government. Despite the attempts of the agricultural chemical industry to discredit her and her books, many people believe that Carson was responsible for beginning the environmental protection movement. ◆

Akira Okubo
(1925—)

Akira Okubo has spent many years studying the ecology of the oceans. Through his studies, he has added much to the understanding of topics such as why and how fish live in schools. He has also studied how some large sea animals are able to survive by eating only plankton, and why plankton group together in large patches.

Okubo was born in Tokyo, Japan, in 1925. He studied at the Tokyo Institute of

Technology and at Johns Hopkins University, where he received his doctorate in oceanography in 1963. He has served as a research scientist at the Johns Hopkins Chesapeake Bay Institute and has taught physical ecology at the Marine Science Research Center of the State University of New York at Stony Brook.

Although he is an oceanographer, Okubo has also studied land animals to learn more about marine life. He hopes to form mathematical models that would help him develop laws of the behavior of organisms in their environments. These laws would be similar to the laws of physics that predict the behavior of particles. ◆

DAVID POWLESS, CONSERVATIONIST

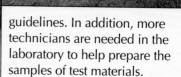

David Powless knows the importance of natural resources. He is a conservationist in Green Bay, Wisconsin.

What kinds of things do you do?

I'm involved in environmental testing. I do complete chemical analyses of soil, air, and water. In addition, I test waste water, ground water, and drinking water for pollutants. I also test hazardous waste and toxic substances and do some air-quality monitoring.

In your job, do you work with other people?

Yes, I work with 23 technicians. These technicians have a variety of degrees in chemistry, biology, geology, and biochemistry. Twelve of these people are Native Americans. My company is owned by the Oneida Tribe of Wisconsin, of which I am a member. This company is a tribal venture based on science and technology. We believe that people are responsible for taking care of the earth and for doing something about the environmental problems.

Are computers and other instruments important in your technical work?

Yes. Computers are used to record and store test information. Computers are also used to prepare reports. Other instruments measure hazardous and toxic substances. With these instruments, I'm able to determine all the pollutants in materials.

What kinds of career opportunities are there in this field?

There are many career opportunities in conservation, especially in chemistry. Environmental groups and agencies are setting very strict guidelines for the levels of pollution to be allowed in manufacturing. Trained technicians are needed to operate the new machines to measure the ultralow levels of pollutants allowed under the guidelines. In addition, more technicians are needed in the laboratory to help prepare the samples of test materials.

What subjects should students study for this type of work?

I would suggest that students study biology, chemistry, geology, and zoology. At the higher levels of science, mathematics is an absolute requirement. Computer skills will also be important.

What is your educational background?

I studied petroleum engineering at the University of Oklahoma and then got a degree in marketing and economics at the University of Illinois. I first worked in the steel industry; then I received a National Science Foundation grant, the first given to a Native American. I used this grant to develop a process to recycle hazardous waste in steel mills.

Do you think people will be able to stop pollution and clean up the earth?

Yes, I do. We are coming to a point of choice. The earth will always be here, but will we always be able to live on it? The choice is up to us. I believe that we can do it. ◆

Discover More

For more information about careers involving conservation and the environment, write to the

National Wildlife Federation
1400 16th St.,
N.W. Washington, DC 20036-2266

A Hole in the Sky

"Make sure you use lots of sunscreen!" Has anyone ever said that to you? We're often told to use sunscreen to filter out the sun's ultraviolet rays. These rays give people a suntan, but they also cause sunburn and some kinds of skin cancer. The sunscreen we use may become even more important in the future because the natural "sunscreen" in our atmosphere is rapidly disappearing.

A Serious Problem

One of the most serious problems the world is facing today is the formation of a "hole" in the ozone layer over Antarctica. Scientists have been aware of this hole for several years, but in recent years it has been growing rapidly.

What is ozone, and why are scientists concerned that the hole is getting larger? Ozone is a relative of oxygen. It forms a layer above the earth and shields the earth from the harmful ultraviolet rays of the sun. A hole in the ozone layer allows more of these rays to reach the earth. If ultraviolet radiation reaches the earth in

The size of the hole in the ozone layer has been increasing recently.

greater amounts, several things could happen.

Ultraviolet Hazards

First, ultraviolet rays are damaging to human skin. More ultraviolet rays would mean more cases of fatal skin cancer—up to 200,000 more cases each year in the United States alone. There could also be 60,000 additional cases each year of cataracts, a clouding of the eye.

Second, there could be increased global warming. A temperature increase of as little as one degree would cause some polar ice to melt.

This melting would raise sea levels and put many coastal areas underwater.

Third, the increased ultraviolet radiation could increase the amount of smog in the atmosphere. Smog is a hazard to human health and to food crops.

What Caused the Hole?

Many scientists believe that chlorofluorocarbons (CFCs), which chemically destroy ozone, are the major cause of the hole. The largest sources of CFCs are Freon leaks from car air conditioners and the burning of foam cups and foam insulation.

The simplest way to stop the loss of ozone is to limit the use of CFCs. Many scientists would like to have the use of CFCs stopped entirely. Acceptable substitutes to CFCs are already available. It is hoped that these substitutes will soon be put in use, since scientists have recently observed a thinning of the ozone layer over heavily populated northern Europe. ◆

In an underground tunnel, workers discover parts of an ancient city. There are many questions to be answered. Archaeologists and other scientists are fascinated by the discoveries they make. Through careful excavation, the ancient Roman city of Pompeii emerges from a slumber of almost 17 centuries. Buried beneath volcanic ash, Pompeii provides a glimpse of what life was like in ancient times. Tools, wooden artifacts, and works of art are discovered. By thorough examination, the scientists are able to find out what the people wore, what they ate, how they lived, and how healthy they were.

CELLS

*W*hat do you think it was like to be the first person to look through a microscope and to discover a previously unknown world of tiny creatures living in a drop of water? The microscope was only the first of many scientific tools that have changed scientists' approach to their work and added to our knowledge of ourselves and the world around us.

he development of the microscope has given us an opportunity to make contact with the hidden world of nature. Yet at first the microscope will probably tell us as little as a glance into a kaleidoscope. What we see may look beautiful and intriguing, but it is also completely incomprehensible. We have discovered an entirely new world.

Though microscopes have been in use for about three hundred years, interpretations of what they reveal are considerably more recent. When a curious amateur constructed the first microscope he could not possibly have understood what he saw. He had no body of knowledge to help him and no reference books to turn to. No one had ever seen what the new lens revealed. Fifty years ago the scanning electron microscope opened up yet another new world, this time smaller by a factor of hundreds than anything seen before—a discovery that made the world seem even richer and more complex. Gradually, however, the structure of nature's hidden world has begun to emerge.

from *Close to Nature: An Exploration of Nature's Microcosm* by Lennart Nilsson

For Your Journal

🖋 Have you ever looked through a microscope? Describe the experience.

🖋 How do microscopes help scientists in their work?

🖋 Why might it be useful to study the microscopic structures that make up living organisms?

Tools for Life Science

Objectives

Identify *the SI units of distance, volume, mass, and temperature.*

Explain *how to use a compound light microscope.*

Compare and contrast *the SI and customary systems of measurement.*

Imagine that you are a member of a team of scientists investigating a new life form. Everything that you discover about this life form will be of interest to your scientific colleagues around the world. As a result, it is important that you begin by making accurate measurements. Studying a living organism requires many different measurements and special tools. Common scientific measuring tools include rulers, balances, and scales. Each type of measuring tool uses one or more special units of measurement.

You are probably already familiar with many of the units of measurement scientists use in their work. The work of a scientist depends on the *quality* of the observations made. For observations to be correct, measurements must be accurate. Scientists all around the world must be able to share the results of their experiments. So it is also important that all scientists agree to use the same units of measure. What do you think would happen if scientists did not use the same measurement system? See for yourself in this activity.

 **R BY** *Doing*

Measure the length and width of your textbook, but do not use any standard unit of measurement. Choose your own unit, such as the length of a paper clip, a pencil, or an index card. Your teacher will record all the measurements on the board. In your journal, explain why standard units of measurement are important. ✐

A System of Measurement

In 1960, scientists from around the world met and decided that a common system of measurement was needed to allow them to communicate more easily with one another. The scientists agreed to use a system of measurement called the *International System of Units (SI).*

Measure Up   In SI, length or distance is measured in meters. A **meter** (m) is about as long as a baseball bat or as high as a doorknob. The meter can be divided into smaller units. One advantage of working in SI is that all the units are based on the number 10. Prefixes are used to show larger and smaller amounts. For example, the prefix *kilo-* means "one thousand." A *kilometer* (km) is equal to 1000 meters. The prefix *deci-* means "one-tenth." A *decimeter* (dm) is one-tenth of a meter (0.1 m). To convert measurements in SI units, you need only to multiply or divide by 10. When would you multiply? Divide? The next activity will give you practice measuring in SI units.

Figure 4–1. Common measurements are usually used in the construction of most homes. For example, most doorknobs are about 1 meter from the floor.

DISCOVER BY *Calculating*

Estimate the height in SI units of several objects in your classroom. Record your estimates in a chart. Then measure the height of the same objects in centimeters, decimeters, and meters. What is the relationship among these units? ✎

In your work in this course, you will need to measure quantities such as length, area, volume, mass, and temperature. The following table shows some common units of measure for these quantities.

Table 4-1 **Some Common SI Units**

Measurement	Common Unit	Symbol
Length	kilometer	km
	meter	m
	centimeter	cm
	millimeter	mm
Area	square kilometer	km^2
	hectare	ha
	square meter	m^2
	square centimeter	cm^2
Volume	liter	L
	milliliter	mL
	cubic meter	m^3
	cubic centimeter	cm^3
Mass	kilogram	kg
	gram	g
	milligram	mg
Temperature	Kelvin	K
	degrees Celsius	°C

Notice three of the units in Table 4–1 that do not have prefixes—the meter, the liter, and the gram. You might think of the SI system as being "built" from these three basic units of measure—the **meter** for distance, the **liter** for capacity or volume, and the **gram** for mass.

More Than One   There is something different about the row of the table for volume. What is the difference? This row uses *two* of the basic units, the liter and the meter. The two units are used because there are two different ways to find volume. In the next activity you will explore these two ways of finding volume.

ACTIVITY

How can you measure volume?

MATERIALS
metric ruler, chiton, cardboard (two sheets), 100-mL graduate

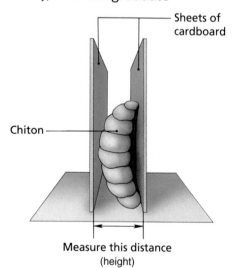

Sheets of cardboard

Chiton

Measure this distance
(height)

PROCEDURE
Part A
1. Using a metric ruler, measure and record the length of the chiton (KYT uhn). Record your data in cm.
2. Find the height and width of the chiton by placing it between two pieces of cardboard as shown.

3. Calculate the volume of the chiton by multiplying the length, the width, and the height. Record your answer in cm^3.

Part B
1. Fill the graduate with 50 mL of water. Record the initial volume of the water as 50 mL.
2. Carefully place the chiton in the graduate and record the new volume.
3. Subtract the initial volume of water from the final volume. The difference is the volume of the chiton in mL. Record this volume.

APPLICATION
1. In SI, 1 cm^3 is equal to 1 mL. Compare your two measurements of the volume. Which do you think was the more accurate? Why?
2. Repeat the activity using a small metal cube. Estimate the volume before you measure and compute. Was your estimate more accurate in Part A or in Part B?

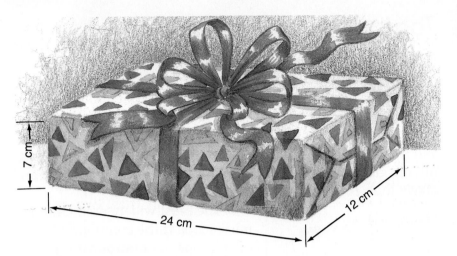

Figure 4–2. The box has a length of 24 cm, a width of 12 cm, and a height of 7 cm. The volume is found by multiplying the dimensions together. What is the volume of the box in cm³?

7 cm

24 cm

12 cm

It often happens that you will have a choice of units in recording scientific data. One example is volume. A second example is temperature. The table shows two different units for temperature, *Kelvin* and degrees *Celsius*. The SI unit for measuring temperature is Kelvin (K). Water freezes at 273 K and boils at 373 K. Because the Kelvin scale is difficult to use in the laboratory, the Celsius (°C) scale is used more often. On the Celsius scale, water freezes at 0°C and boils at 100°C. Normal body temperature is about 37°C.

Another factor in choosing the best unit to use is the size of the object. The gram—approximately equal to the mass of one paper clip—is too small for most everyday uses. In measuring mass, the kilogram is most useful for ordinary objects.

SI is both logical and flexible. Units can be found to measure the extremely large distances between objects in space, as well as the very small distances involved in organisms too small to be seen without the aid of a microscope.

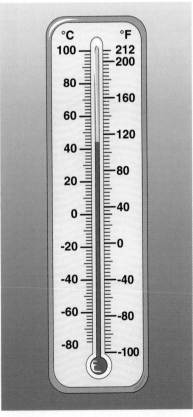

°C   °F
100 — 212
     — 200
80 —
     — 160
60 —
     — 120
40 —
     — 80
20 —
0 — 40
-20 — 0
-40 — -40
-60 — -80
-80 — -100

Figure 4–3. The convenience of the Celsius scale can be easily seen when the boiling point and freezing point of water are compared. At what temperature does water boil and freeze on each of these scales?

▼ **ASK YOURSELF**

Why is it an advantage for scientists in the United States to use SI measurement?

Tools of a Life Scientist

Besides accurate measurements, you and your fellow scientists want to make careful and detailed observations of the new life form. What tools could you use to improve your ability to observe? One very important tool of the life scientist is the *microscope*. Microscopes allow scientists to see small organisms, cells, and parts of cells that otherwise are too small to be seen. A magnifying glass is the simplest kind of microscope. It can make an object look 2 to 20 times its actual size.

The most widely used kind of microscope is the compound light microscope. A **compound light microscope** uses two magnifying lenses to form an image. The lens closest to the object being studied is the *objective lens*. Most compound light microscopes have a high-power objective lens and a low-power objective lens. The lens

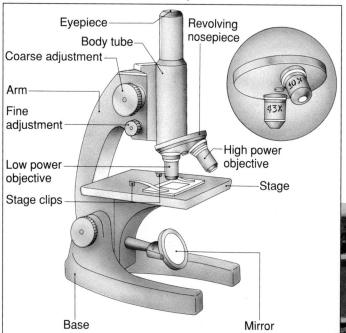

Figure 4–4. A compound light microscope is primarily two lenses mounted inside a tube. The compound light microscope is an important tool used by scientists.

Figure 4–5. A microscope is a delicate instrument and must be handled carefully. When carried, a microscope should be supported as shown.

closest to the eye is the *ocular lens,* or eyepiece. The objective and the eyepiece are at opposite ends of the body tube.

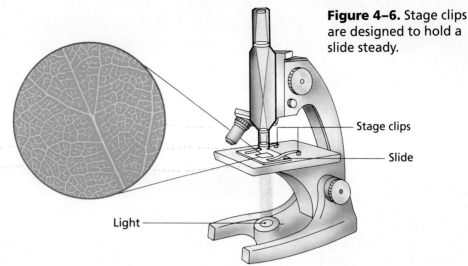

Figure 4–6. Stage clips are designed to hold a slide steady.

Stage clips

Slide

Light

The body tube of the microscope is fastened to the arm. The arm is supported by a heavy base. Always carry a microscope upright by the arm, with your other hand under the base.

A microscope works somewhat like a pair of binoculars. You turn an adjustment knob to bring the object into focus. However, most microscopes have two adjustment knobs. The coarse adjustment is used only when the low-power objective is in place. The fine adjustment is used for final focusing under low power, and it is the only focusing adjustment that is used when the high-power objective is in place.

Over several hundred years, microscopes have been improved. The most powerful microscopes, electron microscopes, can magnify objects hundreds of thousands of times. Electron microscopes have made it possible for scientists to study extremely small organisms, viruses, molecules, and even atoms.

Figure 4–7. This dust mite has been magnified by an electron microscope.

 ASK YOURSELF

Why are microscopes important tools for life scientists?

SECTION 1 *REVIEW AND APPLICATION*

Reading Critically

1. What are the SI units of measure for distance, volume, mass, and temperature?

2. How does using a prefix change the value of a basic unit in SI?

Thinking Critically

3. What types of information about plants and animals would be unavailable to a scientist who did not have a microscope?

4. When measuring an object, why is it important to use both a number and a unit for measurement?

5. How is a magnifying glass like a microscope? How is it different?

INVESTIGATION

Using the Microscope

▶ **MATERIALS**
- compound light microscope ● prepared slides of an insect leg and an insect wing

▼ **PROCEDURE**

Part A

1. **CAUTION: Do not position the microscope or its mirror in direct sunlight. Direct sunlight may damage your eyes.**
2. Check to see that the low-power objective is in place and ready for use. Place the prepared slide of the insect leg on the stage and secure the slide with the stage clips.
3. Predict what the insect leg will look like under the microscope.
4. Use the coarse adjustment knob and then the fine adjustment knob to focus on the leg.
5. Make a drawing of what you see in the low-power field of view. Compare your observations with your predictions.
6. While looking at the leg, move the slide slowly to the left. Record your observations. Repeat, moving the slide upward.

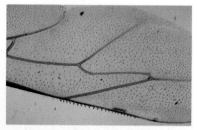

Part B

1. Remove the slide of the insect leg and replace it with the slide of the insect wing.
2. Using the low power, focus on the wing. Make a drawing of the low-power field of view.
3. Move the slide so that the part you want to see under high power is in the center of the field of view.
4. While looking at the slide from the side and at eye level, slowly rotate the revolving nosepiece until the high-power objective is in place.
5. Using only the fine adjustment, focus on the wing. Make a drawing of what you see.

▶ **ANALYSES AND CONCLUSIONS**

1. How does the movement of the image viewed with the microscope compare with the actual movement of the slide?
2. How does the appearance of the insect wing differ under low power and high power?
3. How does examining an insect leg and wing help scientists to better understand the insect?

▶ **APPLICATION**

The amount of magnification on a microscope is shown by numbers marked on the eyepiece and the objective lens. If your eyepiece is marked 10 X and your low-power objective is also marked 10 X, the object you view under low power will be magnified 100 times.

Examine your high-power objective to find out the magnification under high power. How is this number computed?

 Discover More

Make a list of five objects you would like to view under a microscope. Predict what you might discover about each object. How could you confirm your predictions?

Discovery of Cells

Objectives

Identify *the contributions of Schleiden, Schwann, and Virchow to the development of the cell theory.*

Summarize *the three parts of the cell theory.*

Evaluate *the importance of the cell theory.*

As part of your investigation of the new life form, you decide that you must make two kinds of observations. You first make observations on a macroscopic scale. A macroscopic object is large enough to be observed with the unaided eye. You record data about how and where the life form lives, what it eats, how it moves, and how it interacts with others of its own kind.

For a better understanding of your new discovery, you also make microscopic observations. In this way, you look at the fundamental structure of the life form and decide if it is similar to any on Earth. Your observations will be added to those made by other scientists.

DISCOVER BY Doing

Observe a leaf with your unaided eye. Then observe the leaf by using a magnifying glass. How many times does the magnifying glass enlarge the leaf? What can you see with the magnifying glass that you could not see before? Which of your observations surprised you? Discuss with a classmate the best way to record your observations. ✎

Early Observations

Have you ever wondered how the first microscope was made? Did the inventor just put two magnifying glasses together?

Actually, that is close to what really happened. At the end of the sixteenth century, a Dutch lens maker, Zacharias Janssen, put two magnifying lenses together in a tube and invented the compound microscope. About eighty years later, another Dutch scientist, Anton van Leeuwenhoek, made many simple microscopes, each with a single lens. Leeuwenhoek used lenses that could magnify objects up to 270 times their actual size. He made careful records of the things he observed with the microscope. You can read more about Leeuwenhoek on page 250 at the end of Unit 3.

Magnifying an object helps you discover more about its structure. During the 1660s, an English scientist, Robert Hooke, constructed a microscope to study thin slices of cork. Cork is the material under the bark of certain trees. The open spaces Hooke discovered in the cork reminded him of the tiny rooms in which monks lived. So Hooke gave the tiny spaces in the cork tissue the same name, **cells.**

Scientists now know that Hooke was not looking at living cells. He was looking at the walls that had once surrounded the living cells of the cork. However, Hooke's observations opened the door to many new scientific studies.

Figure 4–8. Cork examined under a modern microscope (right) looks very similar to what Hooke saw (left).

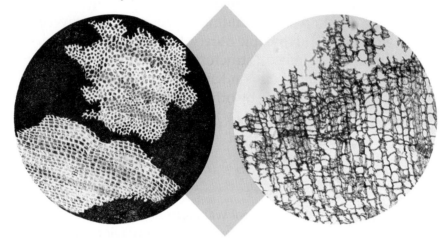

ASK YOURSELF

Why do you think the early scientists were interested in making microscopic observations?

Cell Theory

Nearly 200 years after Hooke's discovery, a German biologist, Matthias Schleiden, was using a microscope to study living plants. Over the years, he looked at the parts of many different plants and found that they were alike in many ways. All the plant parts that Schleiden studied contained cells. His observations led him to conclude that all plants are made of cells.

About the same time, Theodor Schwann, another German biologist, studied many different animals. His many years of observations led him to conclude that all animals are made of cells. The work of Schleiden, Schwann, and many other scientists was studied by Rudolf Virchow, another German scientist. Virchow concluded from his research that all new cells come from other living cells.

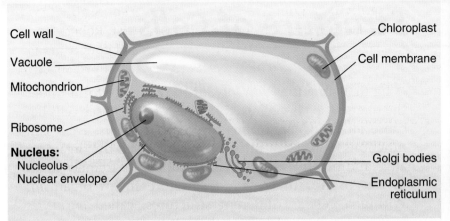

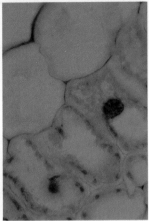

Figure 4–9. The photograph and diagram above show the variety of structures found in plant cells.

The work of Schleiden, Schwann, and Virchow led to one of the most important theories in biology—the cell theory. The **cell theory** is composed of three basic statements.

1. All organisms are made of cells.
2. Cells are the basic units of structure and function for all living things.
3. All cells come from other cells.

The cell theory is important because it shows that all living things share similar structures. This similarity suggests that all organisms are related. In addition to providing answers, however, the cell theory also led to more questions. For example, what is the structure of a cell? How do cells work? How do they reproduce? Although many of these questions have been answered, many others are still being studied by life scientists.

 ASK YOURSELF

Why would a scientist studying a newly discovered living organism expect to see cells under a microscope?

SECTION 2 *REVIEW AND APPLICATION*

Reading Critically

1. State the three parts of the cell theory.
2. Which statements in the cell theory were a result of Schleiden's, Schwann's, or Virchow's work?

Thinking Critically

3. Why was the microscope an important discovery?
4. Why is it important for scientists to study the structure of a cell?
5. A book is made up of chapters, paragraphs, sentences, and letters. What is the basic unit of structure of a book? Use two parts of the cell theory to create a "book theory" similar to the cell theory. Which part of the cell theory were you not able to use? Why?

Structure of Cells

The first time that you look at the new life form under the microscope is very exciting. You expect to see cells because all living organisms are made of cells. Each cell functions along with the other cells to keep the organism alive and healthy. But the cell, although extremely small in size, is not the smallest structure in an organism. Within a cell are different structures, called *organelles* (AWR guh nehlz), that help the cell to carry out its functions. **Organelles** are structures within a cell that have certain jobs to do for the cell, much like each organ in your body has a job to do. So organelles can be thought of as the "organs" of the cell.

The parts of the cell make it possible for each cell to grow, to release energy from food, to get rid of wastes, and to divide to create new cells. You might think of a cell as a small factory. Different people in the factory are all busy at different jobs. But working together, they create a smooth-running and efficient team.

DISCOVER BY Researching

Using reference materials in a library, look for pictures of cells. Make sketches of cells with interesting shapes to share with the class. Do all cells have the same shape? What types of structures might all cells have in common? ✐

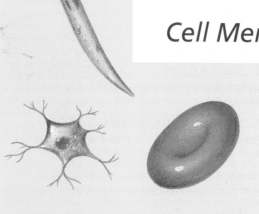

Cell Membrane and Nucleus

Although cells have a wide variety of shapes, sizes, and colors, every cell has an outer covering, much as skin covers the outside of your body. This covering that surrounds the cell is called the **cell membrane.** The structure of the cell membrane allows certain materials to pass through it. Like a security guard at a factory, the cell membrane controls everything that goes into and comes out of the cell.

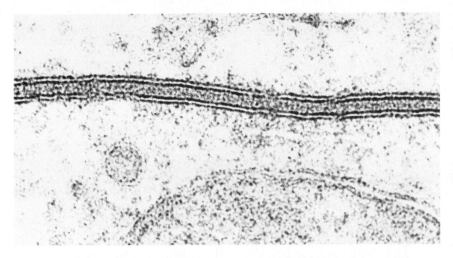

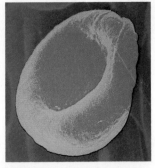

Figure 4–10. An entire red blood cell (right) shows the three-dimensional structure of the cell membrane. The electron micrograph (left) shows the layers of the cell membrane.

The cell membrane has tiny openings that let water enter the cell. Other openings let other small molecules pass through the membrane but usually keep large molecules out.

As you look inside a cell, the most prominent object you see is usually the nucleus. The **nucleus** controls the cell's activities, much as your brain controls the activities of your body or the owner of a factory controls its operations. The instructions for all of the cell's activities are in the material that is found inside the nucleus. Around the nucleus is a special covering called the *nuclear envelope*. Like the cell membrane, this envelope controls the movement of materials into and out of the nucleus.

 ASK YOURSELF

How are the cell membrane and the cell nucleus important to the activities of the cell?

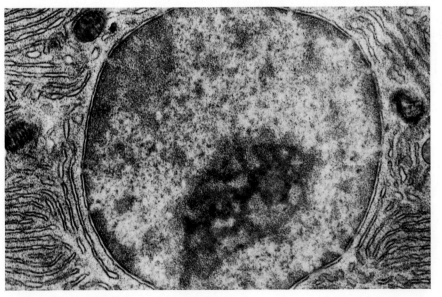

Figure 4–11. The nucleus of a cell is surrounded by the nuclear envelope. How is the nucleus of a cell similar to the human brain?

Figure 4–12. This illustration shows the variety of structures found in an animal cell.

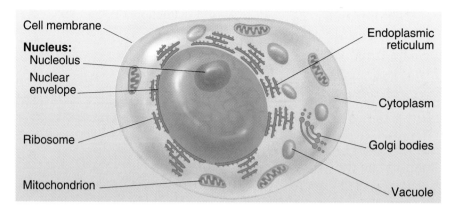

Cell membrane

Nucleus:
Nucleolus
Nuclear envelope

Ribosome

Mitochondrion

Endoplasmic reticulum

Cytoplasm

Golgi bodies

Vacuole

Other Cell Parts

The material between the cell membrane and the nucleus is called *cytoplasm* (SY tuh plaz uhm). The largest organelle in the cytoplasm is the *endoplasmic reticulum* (ehn duh PLAZ mihk rih TIHK juh luhm). This organelle is a network of tubelike canals that run through the cytoplasm.

The endoplasmic reticulum is used much like streets or roads. One of its jobs is to transport materials from place to place in the cell. Some of these materials will be sent to the *Golgi bodies*. The function of the Golgi bodies is to "package" the materials before they are sent to destinations in and out of the cell.

Most of the cell and much of the body are made of proteins. The organelles that make proteins are the *ribosomes* (RY buh sohmz). Ribosomes are like small machines that take raw materials from the cytoplasm to make the proteins. The ribosomes are found throughout the cytoplasm. Many are also attached to some parts of the endoplasmic reticulum. Since ribosomes function to produce materials for the cell, what advantage occurs when they are attached to the endoplasmic reticulum?

Figure 4–13. The ribbonlike structures in the electron micrograph (left) are the endoplasmic reticulum. In the illustration (right), ribosomes can be seen covering the endoplasmic reticulum.

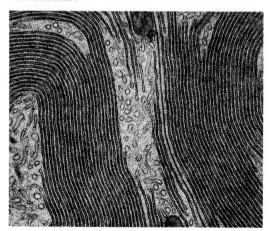

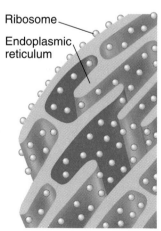

Ribosome

Endoplasmic reticulum

The organelles that release energy are *mitochondria* (myt uh KAHN dree uh). For this reason, the mitochondria are often called the "powerhouses" of the cell. Food molecules are broken down inside these organelles, and energy is released. This energy is used to keep the cell working properly. Energy from trillions of mitochondria in billions of cells keeps the body working properly.

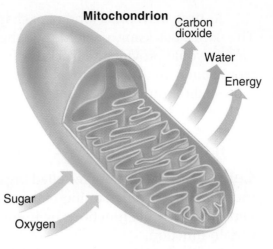

Mitochondrion

Carbon dioxide

Water

Energy

Sugar

Oxygen

Figure 4–14. Mitochondria, as seen here, are responsible for releasing all the energy the cell needs to function.

Another organelle found in cells is the *vacuole* (VAK yu ohl). Vacuoles are sacs filled with liquid. Some vacuoles store water or food. Other vacuoles are used to store waste material until it can be removed from the cell. In a plant cell, there is a large central vacuole that makes up most of the cell. This vacuole is filled with water. By comparison, the vacuoles found in animal cells are much smaller than those found in plant cells.

There are two structures that are found only in plant cells. *Chloroplasts* (KLAWR uh plasts) are large, green, oval-shaped organelles in which photosynthesis takes place. Chloroplasts contain chlorophyll, which, together with light energy, enables the plant to make food. The *cell wall* is the other structure found in plant cells but not in animal cells. The cell wall is found outside the cell membrane, where it helps support the plant and protect the cell. Unlike the cell membrane, the cell wall does not stop materials from entering the cell. The cell wall is made of cellulose, which is nonliving. Why would Hooke see the cell walls of cork cells even after the cells had died?

Chloroplast

Figure 4–15. Chloroplasts convert light energy into chemical energy.

> **ASK YOURSELF**

Why do cells have so many organelles?

Levels of Organization

Before manufacturing became widespread in the nation, many people lived on farms far away from other farms. Each family raised its own food, made its own clothes, and built its own shelter.

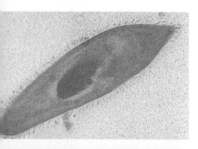

Figure 4–16. A single-celled organism must perform all life functions. In a multicellular organism, cells specialize to perform certain functions.

As time went on, a kind of exchange system developed. The skilled blacksmith would shoe the farmer's horse. The farmer would provide the blacksmith with food. The carpenter would build an office for the doctor, and the doctor would provide medical care for the carpenter's family. As years passed, each person became a specialist. Then factories began to manufacture goods. Each worker in a factory had a specific job. Together the workers produced a finished product.

Cells Organize   Single-celled organisms are similar to the independent farming families. Like each family, each cell must perform all the life functions by itself. Each part of the cell has a certain function.

Most organisms are not made of one cell; they are made of many cells. The organisms are *multicellular.* Each cell in a multicellular organism has all of the basic cell parts. However, each of these cells has also become specialized. Each has a function that benefits the other cells. In this way, the cells of a multicellular organism depend on one another for survival just as workers in a factory depend on one another to produce a finished product.

The specialized cells of multicellular organisms are organized into different levels. The cell itself is at the first level of organization. Most cells have a certain size and shape and certain organelles related to the cell's purpose. For example, in humans there are two types of blood cells and three types of muscle cells. There are also covering cells, nerve cells, bone cells, and fat cells. In plants, there are cells specialized for absorbing water from the soil. There are also cells in plants for covering and protecting, transporting, and growing.

Tissues, Organs, and Systems   The second level of organization is tissues. **Tissues** are groups of similar cells that perform a specific function. Blood cells form blood tissue. One of the functions of blood tissue is to carry oxygen to the body cells. Muscle cells form muscle tissue. Muscle tissue is responsible for movement.

Plant cells also form specialized tissues. Covering cells on leaves are called *epidermal tissue.* Its function is to protect the cells beneath it.

Figure 4–17. The shapes of cells vary with their functions. Shown here are muscle cells (left), blood cells (center), and plant transport cells (right).

 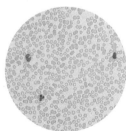 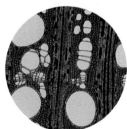

Organs represent the third level of organization. An **organ** is a group of different tissues working together to perform a certain task. The human heart is one such organ. The heart is made up of muscle, nerve, covering, and blood tissues. Together these tissues have a unique function: they pump blood through the body. Other examples of human organs are the eye, the stomach, and the lung.

A leaf is an example of a plant organ. A leaf contains several tissues. Each performs a certain task. Together, as an organ, the tissues function to make food. Roots and stems are other plant organs.

A group of organs working together is known as a **system.** Systems are the fourth level of organization. In humans there are many systems. These include the digestive, respiratory, nervous, circulatory, and reproductive systems. In plants, systems include transport and reproductive systems.

The highest level of organization is the *organism.* A single living organism is the combination of all its systems, organs, tissues, and cells.

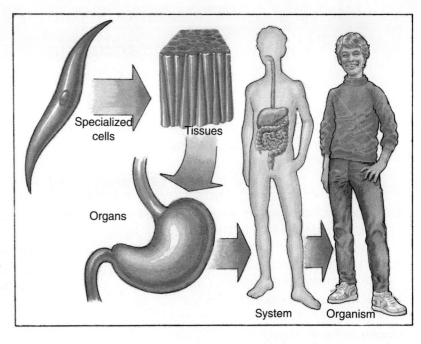

Figure 4–18.
Multicellular organisms exhibit different levels of organization. This results in division of labor, or sharing of functions, within the organism.

 ASK YOURSELF

What is the advantage of specialized cells?

SECTION 3 *REVIEW AND APPLICATION*

Reading Critically

1. How does the structure of the cell membrane help it perform its function?
2. What does the nucleus do for the cell?
3. Make a chart showing each organelle. Next to the organelle, describe its function in a cell.

Thinking Critically

4. Why would a muscle cell have more mitochondria than a skin cell?
5. The chloroplasts in plant cells help the plant make its own food from light energy. How do animals get their food?
6. How might the presence of different organelles help plant cells and animal cells perform different functions? How would this help the organism?

SKILL Comparing Plant and Animal Cells

▶ **MATERIALS**
- slide ● coverslip ● medicine dropper ● *Elodea* leaf ● compound microscope
- prepared slide of human cheek cells

▼ **PROCEDURE**

1. Place a drop of water on a clean slide. Place an *Elodea* leaf in the drop of water and cover with a coverslip.
2. Observe the *Elodea* leaf under low power of the microscope. Observe one cell and make a sketch of what you see.
3. Switch to high power. Again observe one cell and sketch what you see.

4. Obtain a prepared slide of human cheek cells from your teacher. Place the slide on the microscope and focus on one cell under low power. Sketch what you see.
5. Switch to high power and again sketch what you see.

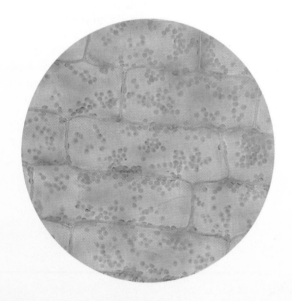 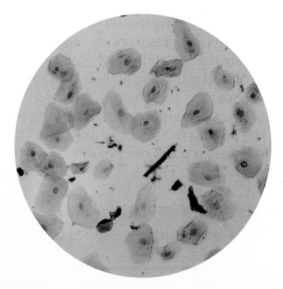

▶ **APPLICATION**

Which cells that you observed were animal cells? Plant cells? How can you tell the difference through your observations?

What structures were you able to observe in each of the cells you viewed?

Make a table in which to write a comparison between the plant and animal cells you observed. Give your table a title and remember to title the columns.

 Using What You Have Learned

Obtain some other prepared slides of plant and animal tissues. Examine these slides under a microscope. Add to your table any new observations you were able to make.

*H*IGHLIGHTS

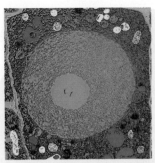

The Big Idea

Life scientists study plants and animals using a wide variety of tools and measurements. They look at an organism and its external structure, how it interacts with other parts of its environment, and—on a microscopic scale—how its internal structures work together.

By using a common system of measurement, SI, scientists around the world can share their findings. In the same way that the cell theory was the result of the work of several different people, scientists today can read and study each other's findings to gain new knowledge about living organisms.

In fact, scientists working in different countries many thousands of kilometers apart are actually part of a common "team." Just as the structures that make up plant and animal cells all coordinate activities, life scientists work together as they gain new knowledge of the living world.

Review the questions you answered in your journal before you read the chapter. How would you answer the questions now? Add an entry to your journal that shows what you have learned from the chapter. What new ideas do you have about why microscopic studies are important to understanding living organisms?

Connecting Ideas

Copy this unfinished concept map into your journal. Complete the concept map by writing the correct term in each blank.

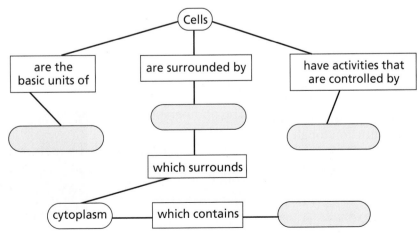

REVIEW

Understanding Vocabulary

Choose the word or phrase that does not belong in each group. Then explain why that term is different from the others. What do the other terms have in common that makes them alike?

1. length, kilogram, volume, mass
2. millimeter, magnifying glass, compound light microscope (102), telescope, binoculars
3. mitochondria (111), cell membrane (108), cell theory (107), cell wall (111), nucleus (109)
4. tissues (112), organ (113), microscope (102), system (113)

Understanding Concepts

MULTIPLE CHOICE

5. Scientists have agreed to use a common system of measurement
   a) because SI is the most accurate.
   b) so that they can communicate more easily with each other.
   c) so that all scientific experiments will have the same results.
   d) because SI is based on the number 10.

6. The early scientist Matthias Schleiden used a microscope to discover that all living plants
   a) come from other living plants.
   b) can make their own food.
   c) contain cells.
   d) include chloroplasts.

7. Which statement is *not* included in the cell theory?
   a) Cells cannot be studied without using a microscope.
   b) All cells are from other cells.
   c) Cells are the basic units of structure and function for all living things.
   d) All organisms are made of cells.

8. Which of the following activities would occur in a plant cell but *not* in an animal cell?
   a) Materials are transported from one place to another.
   b) Food molecules are broken down and energy is released.
   c) Waste material is stored until it can be removed.
   d) Food is manufactured using light energy.

SHORT ANSWER

9. List reasons for and against using SI in the United States.
10. The cell membrane controls the passage of materials into and out of the cell. Give examples of materials that would be entering and leaving a cell.

Interpreting Graphics

11. The photograph shows a slice of stem from a plant. How is this photograph like the cork that Hooke saw under his microscope? How is it different?

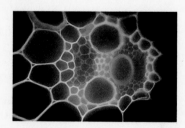

Reviewing Themes

12. *Systems and Structures*
Animals are studied on both the microscopic and macroscopic scales. What information about an animal *cannot* be collected during microscopic investigations?

13. *Systems and Structures*
How would the other parts of a cell be affected if the vacuoles were no longer able to store water and food?

Thinking Critically

14. Write three equations to show the relationships between inches and feet, feet and yards, and inches and yards. Then write three equations showing the relationships among centimeters, meters, and kilometers. Explain how your equations show an advantage built into SI.

15. A scientist measured a plant cell and found that it was about 50 micrometers (μm) across. Estimate the number of these plant cells that would fit across your thumbnail. Use 1 cm for the width of your thumbnail and the fact that 1 m = 1 000 000 μm.

16. The photograph below is an electron micrograph of a Mediterranean fruit fly magnified 21 times. How have both the compound light microscope and the electron microscope added to our knowledge of living things?

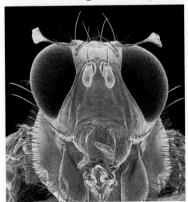

17. Why does a cell's survival depend on the movement of certain materials through its cell membrane?

18. Unicellular living organisms are made of only one cell; multicellular organisms have many cells. Compare the function of a cell in a unicellular organism with the function of a cell in a multicellular organism.

19. Make a three-dimensional model of an animal cell or a plant cell, showing the structures described in this chapter. Use modeling clay or gelatin to form the cytoplasm of the cell. Use modeling clay to form the organelles.

20. Use an example other than a pioneer community to compare division of labor in society with levels of organization in a multicellular organism.

Discovery Through Reading

Bender, Lionel. *Through the Microscope: Atoms and Cells.* Gloucester Press, 1990. Text and photographs introduce microscopic plant and animal life, viruses, microspores, and other forms of life that can be viewed only through a microscope.

CELL FUNCTION

The great detective Sherlock Holmes has a puzzle to solve. He thinks that the dancing men are part of a secret code. But how can he decipher the code and find out what the message means? Like Sherlock Holmes, scientists are always trying to solve mysteries. For example, understanding what happens inside cells was once a scientific puzzle. Today, even though much is known about the function of cells, there are puzzles that scientists are still working to solve.

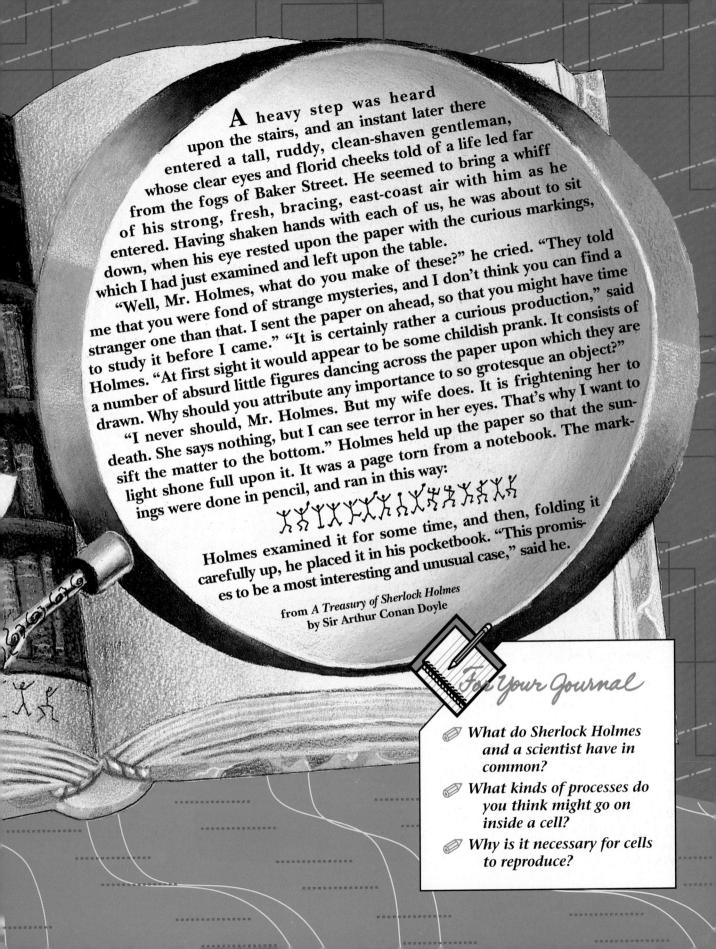

A heavy step was heard upon the stairs, and an instant later there entered a tall, ruddy, clean-shaven gentleman, whose clear eyes and florid cheeks told of a life led far from the fogs of Baker Street. He seemed to bring a whiff of his strong, fresh, bracing, east-coast air with him as he entered. Having shaken hands with each of us, he was about to sit down, when his eye rested upon the paper with the curious markings, which I had just examined and left upon the table.

"Well, Mr. Holmes, what do you make of these?" he cried. "They told me that you were fond of strange mysteries, and I don't think you can find a stranger one than that. I sent the paper on ahead, so that you might have time to study it before I came." "It is certainly rather a curious production," said Holmes. "At first sight it would appear to be some childish prank. It consists of a number of absurd little figures dancing across the paper upon which they are drawn. Why should you attribute any importance to so grotesque an object?"

"I never should, Mr. Holmes. But my wife does. It is frightening her to death. She says nothing, but I can see terror in her eyes. That's why I want to sift the matter to the bottom." Holmes held up the paper so that the sunlight shone full upon it. It was a page torn from a notebook. The markings were done in pencil, and ran in this way:

Holmes examined it for some time, and then, folding it carefully up, he placed it in his pocketbook. "This promises to be a most interesting and unusual case," said he.

from *A Treasury of Sherlock Holmes*
by Sir Arthur Conan Doyle

For Your Journal

- ✎ *What do Sherlock Holmes and a scientist have in common?*
- ✎ *What kinds of processes do you think might go on inside a cell?*
- ✎ *Why is it necessary for cells to reproduce?*

Transport in Cells

The first thing Sherlock Holmes noticed about the dancing men message was that the figures could be grouped into words. He thought each figure holding a flag could be the last letter in a word.

In the same way that letters combine to make words, atoms combine to make molecules. But unlike the letters in the message, atoms and molecules have the ability to move from one place to another! In fact, molecules have two different ways in which they can move into and out of cells.

Diffusion and Active Transport

All living things are made of matter. Matter is anything that has mass and takes up space. Just as buildings are made of individual bricks, matter is made from small particles called *atoms*. Atoms are considered the smallest building blocks of matter. When two or more atoms join together, molecules are formed. Atoms and molecules are always moving. When moving molecules hit each other, they bounce and change direction like rubber balls that are moving around in a box.

Think about what happens when a drop of red ink is added to water. The molecules of ink are bunched together as the drop enters the water. An area with many molecules of a substance is called an area of *high concentration*. As molecules of water and ink bump into each other, the ink begins to mix with the water. After a while, the ink and water molecules are mixed equally throughout the beaker, and the water turns red.

Figure 5–1. Diffusion is movement from an area of high concentration to an area of low concentration. The molecules of ink will continue to diffuse throughout the water until the water is an even red color.

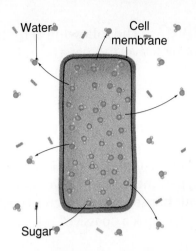

Water   Cell
membrane

Sugar

Figure 5–2. The movement of water molecules into and out of a cell is controlled by the concentration of water inside the cell. If the water concentration is higher inside the cell, water will move out. If the concentration of water is higher outside the cell, water will move in.

The ink spreads through the water by the process of diffusion (dih FYOO zhuhn). **Diffusion** is the movement of molecules from an area of high concentration to an area of low concentration. The movement of many molecules within cells is the result of diffusion. The following activity shows you one example of diffusion. What other examples can you think of?

DISCOVER BY *Observing*

Open a bottle of perfume in one corner of the classroom. How does the fragrance diffuse through the classroom? Which students were able to smell the fragrance first? Later? Explain your observations.

Diffusion is a process used by cells to take in materials from their environment. To function properly, animal cells need food molecules for energy. Plant cells make their own food, but they need water and minerals for the food-making process to take place. Waste molecules must be removed to keep the cells healthy. Some materials, like the food made in plant cells, need to be moved to other parts of the organism. Many of these molecules enter and leave cells by diffusion.

The Gate to the Cell   The cell membrane controls the movement of molecules into and out of the cell. This membrane lets only certain molecules pass through its openings. This is similar to the way in which a kitchen strainer lets water pass through but not the spaghetti you are draining. A membrane that lets molecules pass through it is called a *permeable* (PUR mee uh buhl) *membrane*. Cell membranes are selectively permeable because only certain molecules can pass through.

Passive or Active? The following activity will introduce you to the concept of active and passive transport.

When you are using a car or a bus for transportation, you are not using your own energy. Your role as a rider is passive. Diffusion in a cell is also a type of **passive transport.** This means that no energy is used by the cell during diffusion. However, diffusion is not the only way molecules move across the cell membrane. For instance, salt molecules move from your bloodstream into your cells until almost all the salt molecules are out of the bloodstream. To do this, the salt molecules must move from an area of low concentration to an area of high concentration. The movement of molecules across the membrane from an area of low concentration to an area of high concentration is called **active transport**. During active transport, the cell uses energy to move the molecules across the cell membrane.

Figure 5–3. Canoeing can be much like passive and active transport. If the canoe is being carried along by the current (left), no energy is being used, as in passive transport. However, if the canoe is being moved against the current (right), the people paddling would have to use energy, as in active transport.

▼ ASK YOURSELF

Why is diffusion not a form of active transport?

ACTIVITY

MATERIALS
agar-filled Petri dish, plastic straw, scissors, filter paper, empty Petri dish, medicine dropper, scoop, food substances to be compared, water

PROCEDURE
1. Using a straw, make five holes in the agar in one Petri dish, forming an X-shaped pattern as shown. Put one hole in the middle and the other four holes at the ends of the X.
2. Cut the filter paper so that it fits into the top half of a Petri dish, and put it into a dish. Mark circles on the filter paper that are just like the holes in the agar.
3. Using the medicine dropper, moisten the filter paper with 10 to 15 drops of water.
4. With the scoop, put a small amount of one of the food substances into each hole in the agar. Add three drops of water to each hole.
5. Put the same amount and kind of food substance on the circles on the filter paper. Do not add more water to the filter paper.

6. Watch the agar and the filter paper for two or three minutes. Keep track of the time it takes the food substance to reach the edge of each dish. Record your observations.
7. Cover both Petri dishes and observe them after 24 hours. Record your observations.

APPLICATION
1. How does the rate of diffusion through the agar compare with the rate through the moistened filter paper?
2. Compare your results with those of your classmates. Which food substance diffuses the fastest? Which food substance diffuses the farthest?
3. In your experiment, were the results the same for both the agar and the filter paper? Did all groups have the same results? Explain your results.

Osmosis

The activity demonstrates how molecules are moved during diffusion. Another movement of molecules into and out of a cell is called *osmosis* (ahs MOH sihs). **Osmosis** is a type of diffusion that occurs only when water moves across a selectively permeable membrane. During osmosis, water molecules move from an area of high concentration to an area of low concentration just as in diffusion. Like diffusion, osmosis is a passive process. The movement of water into and out of cells is so important to living things that the diffusion of water through a membrane is given this special term—osmosis.

Figure 5–4. As the concentration of water outside a cell decreases, water diffuses out of the cell. As a result, cell membranes collapse. A lack of water will cause a plant to wilt. However, if water is added to the soil, it diffuses into the plant's cells, and the plant stands up straight.

Have you ever seen a wilted plant straighten up after water has been added to the soil? The movement of water molecules into the plant causes it to straighten.

Living cells get most of their water by the process of osmosis. Osmosis keeps plants from wilting. Usually, there is more water in the soil than there is in the cells of a plant. This concentration of water causes plant cells to swell as water flows across the membrane and into the cells. Eventually, the cytoplasm is pushed tightly against the cell walls, causing the stems and leaves to stand straight.

Osmosis also occurs if there is more water in the roots of a plant than there is in the soil around them. Water will leave the cells of the roots and cause the roots to shrivel. The plant may die because of this loss of water. This can happen if the soil has too many nutrient molecules as a result of too much fertilizer. If there are more nutrients in the soil than in the roots, the water will leave the roots.

 ASK YOURSELF

Compare osmosis with diffusion.

SECTION 1 *REVIEW AND APPLICATION*

Reading Critically

1. Why is the cell membrane described as being selectively permeable?
2. What is active transport?
3. Define *diffusion* and give an example of it from your everyday life.

Thinking Critically

4. What might happen if the concentration of water inside a cell is greater than the concentration of water outside the cell? Explain why this would occur.
5. What causes the stems and leaves of a plant to stand up straight?

INVESTIGATION

*O*bserving Osmosis

▶ MATERIALS

- water ● *Elodea* leaf ● microscope slide ● coverslip ● microscope ● salt water
- medicine droppers (2) ● paper towel

▼ PROCEDURE

1. Using tap water, make a wet mount of an *Elodea* (el uh DEE uh) leaf. Observe the *Elodea* leaf under the microscope with low power. Draw what you see.

2. Switch to high power. Pay careful attention to the location of the chloroplasts within the cells. Draw what you see. Label the cell wall, the cytoplasm, the chloroplasts, and the vacuole.

Show the location of the cell membrane, even though it is not visible.

3. What might happen if you put salt water on the leaf? Discuss your predictions with a classmate before you continue.

4. Without moving the slide, place two or three drops of salt water along the right edge of the coverslip. Hold a torn edge of the paper towel along the left edge of the coverslip. The paper towel will absorb some of the tap water, allowing the salt water to flow beneath the coverslip.

5. Observe the *Elodea* leaf for 3–4 minutes. Describe the changes that occurred. How could you make a record of the results of the experiment?

▶ ANALYSES AND CONCLUSIONS

1. Compare the change that took place in the *Elodea* leaf with what you predicted might happen.
2. Explain how this change occurred.

▶ APPLICATION

Blood cells contain approximately one percent salt. Explain what would happen to the nearby blood cells if pure water were injected into the bloodstream.

✳ *Discover More*

Continue the investigation by flooding the slide with fresh water. Compare the leaf with the drawings you made at the beginning of the investigation.

Cell Reproduction

Objectives

Define *the term mitosis.*

Explain *why interphase is not a part of mitosis.*

Summarize *the four steps that occur during cell division.*

Think back to the dancing men puzzle in the Sherlock Holmes story. Until Holmes collected more messages showing the dancing men, he did not have enough information to solve the puzzle of the code. Holmes wanted to understand how the code in the message worked so that he could write a message of his own.

Many scientists have worked on the puzzles involved in cell function. By studying a great number of cells under the microscope, they have been able to describe the key functions of cells. One very important function of cells is reproduction.

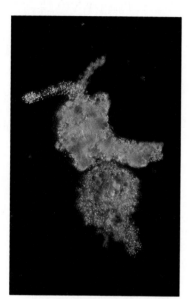

Figure 5–5. All living things reproduce. Single-celled organisms, such as this amoeba, reproduce by dividing into two new cells.

Methods of Cell Reproduction

Look at your hand. Each square centimeter of skin is made of more than 150 000 cells. Most of these cells will be gone by tomorrow. During the next 24 hours, two complete generations of skin cells will live, reproduce, and disappear. What do you suppose will happen to the old skin cells on your hand during the next 24 hours?

In other parts of your body, other cells are also reproducing rapidly. Red blood cells are made at the rate of about 100 million cells a minute. If you have a broken bone, new bone cells are being made so that the broken bone can heal. Since you are getting taller, your bones are growing in length by adding new cells at the ends of the bones. There is much cellular activity taking place in your body.

Cells reproduce by dividing into two cells. The original cell is called the *parent* cell. The two new cells that are formed are called *daughter* cells.

Cells reproduce in several different ways. One type of cell division occurs when a single-celled organism splits into two organisms. Another type of cell division occurs in multicellular organisms when cells, such as skin cells, make exact duplicates of themselves. A third type of cell division occurs when sex cells, or reproductive cells, are produced.

 ASK YOURSELF

What is the purpose of cell division?

Mitosis

The type of cell division by which two daughter cells are formed is called *mitosis* (my TOH sihs). **Mitosis** produces daughter cells that are exactly the same as the parent cell. Mitosis is actually the process by which the cell nucleus duplicates. After the nuclear material is duplicated, the rest of the cell simply divides in two.

Mitosis is a continuous process. However, the events of mitosis are easier to understand if the process is broken down into steps, or phases. There are four phases in mitosis. They are *prophase, metaphase, anaphase,* and *telophase.* In the following activity, you will learn more about these four terms.

Figure 5–6. Through mitosis, an embryo steadily increases the number of its cells.

 **DISCOVER BY** *Researching*

Notice that the names of the four stages of mitosis—prophase, metaphase, anaphase, and telophase—all end with *phase.* Use a dictionary to analyze the meanings of the four different prefixes. Then, as you study the process of mitosis, check to see whether the meaning of each word is a good description of what is happening inside the cell at that stage. ✏

Preparing for Mitosis   Before mitosis can begin, several events have to occur in the nucleus. First, the hereditary material in the cell must be duplicated. *Hereditary* means "passed on from parents to offspring." The hereditary material, called **DNA,** is found within the nucleus. DNA makes up threadlike structures called **chromosomes** (KROH muh sohmz). Since the DNA duplicates, the chromosomes also duplicate. The duplicated chromosomes are joined together at a point called the *centromere.* Identical hereditary instructions are carried on the two chromosomes.

The time before mitosis, when the DNA and chromosomes duplicate, is called *interphase. Inter* means "between." Interphase is not a part of mitosis; it is the time between the end of one mitosis and the beginning of the next. Most of a cell's life is spent in interphase. However, interphase is a time of much activity. During interphase, the cell performs all life activities except mitosis. During this time, a cell stores the extra energy that mitosis will require.

Prophase   The first phase of mitosis is called *prophase.* Prophase begins as the membrane around the nucleus breaks apart. The chromosomes inside the nucleus begin to twist and thicken. As the nuclear envelope breaks apart, the chromosomes begin to move toward the center of the cell. As mitosis continues, thin tubes begin to form between organelles called *centrioles.* These tubes are called *spindle fibers.*

Metaphase   During metaphase, the spindle fibers seem to push and pull the duplicate chromosomes until they are arranged in a line across the middle of the cell. The centromere of each pair of chromosomes is attached to a spindle fiber.

Anaphase   During anaphase, the spindle fibers shorten and pull each chromosome pair apart at the centromere. The spindle fibers continue to shorten, pulling the chromosomes through the cytoplasm.

Telophase   During telophase, the last phase of mitosis, the cells complete their division. As the separate chromosomes reach the opposite ends of the cell, telophase begins. The nuclear envelope forms again as the chromosomes untwist and become longer and thinner. At the end of telophase, the new cells separate. In animal cells, the cell membrane pinches together, dividing the cytoplasm in two. The number of

Figure 5–7. Mitosis takes place in four stages—prophase, metaphase, anaphase, and telophase. Each stage blends into the next stage. Interphase is not part of mitosis. However, all life activities of the cell, except mitosis, take place during interphase.

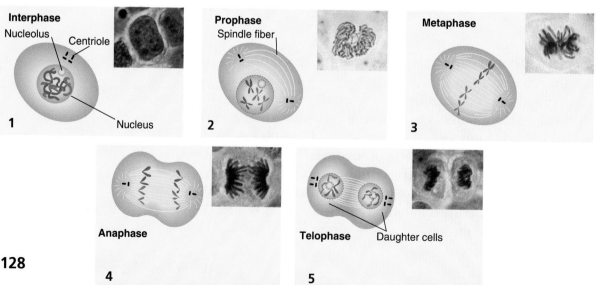

Interphase
Nucleolus
Centriole
Nucleus
1

Prophase
Spindle fiber
2

Metaphase
3

Anaphase
4

Telophase   Daughter cells
5

128

chromosomes in each daughter cell is the same as it was in the parent cell. The daughter cells now enter interphase.

The stages of cell mitosis can be seen under a microscope. In the next activity, you will have an opportunity to see for yourself what happens during mitosis.

DISCOVER BY *Problem Solving*

Obtain from your teacher a set of slides showing the stages of mitosis. Before looking at them, ask a classmate to arrange the slides so they are not in order. Use a microscope to examine each slide. Place the slides in the correct sequence. What problem-solving strategies did you use to order the slides? On a separate sheet of paper, sketch and name each stage of mitosis. ✎

The division of cytoplasm in plant cells is different from the division in animal cells. The thick cell wall of a plant cell is too stiff to pinch together. Instead, a structure called the *cell plate* forms between the daughter cells. The cell plate begins forming in the middle of the cell and moves outward until the daughter cells are separated from each other.

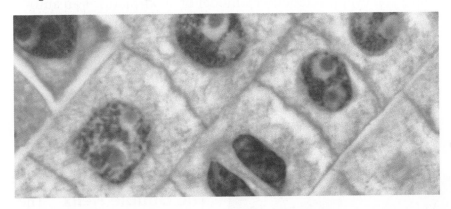

Figure 5–8. In both plant and animal cells, the cell membrane pinches off in the middle to form two new cells. However, in plant cells, a cell plate forms to divide the two new cells.

◢ ASK YOURSELF

Describe what happens in a cell during mitosis.

SECTION 2 *REVIEW AND APPLICATION*

Reading Critically
1. Why is interphase not a part of mitosis?
2. How is mitosis different in plant cells and animal cells?

Thinking Critically
3. How would a chemical that prevented spindle fibers from forming affect the cell division of an organism?
4. What would happen if DNA did not duplicate?

DNA and Cell Energy

Objectives

Summarize the process of DNA replication.

Compare respiration and fermentation.

Describe the process and purpose of photosynthesis.

With a moderate amount of effort, Sherlock Holmes unraveled the puzzle of the dancing men. The message read, "AM HERE ABE SLANEY." Holmes then used his knowledge of the dancing men code to trap Abe Slaney.

One of the most exciting scientific puzzles to be solved in recent times has been that of understanding how the genetic material in cells is passed on from one generation to the next. The scientists who unraveled this puzzle faced problems considerably more difficult than the ones Sherlock Holmes faced, but in many ways, they approached the problems in the same way.

DNA and Replication

Have you ever put together a model of an airplane or an automobile? Would you be able to put all the tiny pieces of the model together correctly if you lost the instructions? Scientists often have to work with no instructions. For example, scientists had tried to figure out the structure and function of DNA. They knew that DNA was contained in chromosomes and that both DNA and chromosomes duplicated before mitosis. But how did DNA work? How was it made? These are questions that puzzled scientists for many years.

Figure 5–9. Watson, Crick, and their DNA model

The Double Helix The shape and makeup of the DNA molecule were finally worked out by James Watson, an American biologist, and Francis Crick, an English physicist. Watson and Crick made a three-dimensional model of what they thought the molecule was like. Their model was based on information collected by many different scientists over many years. Watson and Crick used the pieces of information like the pieces of a puzzle to build their model and to prove the structure of DNA. Now scientists know that DNA molecules are very long and thin. Each molecule is made of two chains formed from many small parts. The chains are arranged side by side and are connected by other smaller molecules. Together the chains form a molecule shaped like a ladder that has been twisted into a spiral. This twisted ladder is called a *double helix.* Watson later said of the discovery, "It seemed almost unbelievable that the DNA structure was solved, that the answer was incredibly exciting, and that our names would be associated with the double helix . . . ."

DISCOVER BY *Researching*

Using poster board, draw and label a time line showing the events of DNA research since DNA was first removed from a cell nucleus in 1869. Use reference materials from the library to include the most recent discoveries.

The Letters in the Code   To fully understand how the DNA molecule is formed, scientists looked at its internal structure. The pieces that form the DNA molecule are made of molecules called *nucleotides* (NOO klee uh tydz). Each nucleotide is made of three smaller molecules: a sugar, a phosphate, and a nitrogen base. There are four different nitrogen bases that may be found in DNA: adenine (AD uh neen), guanine (GWAH neen), cytosine (SY tuh seen), and thymine (THY meen). These bases are often abbreviated A, G, C, and T, respectively. The bases form pairs, but each base can pair with only one other. Adenine (A) always pairs with thymine (T), and guanine (G) always pairs with cytosine (C).

The sugars and the phosphates that accompany the bases form the sides of the DNA ladder. The bases of the nucleotides form the rungs. Each kind of DNA has a different sequence of bases. The sequence of the base pairs forms the hereditary code.

If we compare this code to our own language, we see that the hereditary language of DNA is a language built on only a four-letter alphabet—A, T, G, and C. But by using those letters in different combinations and in different lengths, an endless number of DNA "words" and "sentences" can be written. These can be grouped into endless numbers of different DNA "stories." Each of us received our DNA combination from our

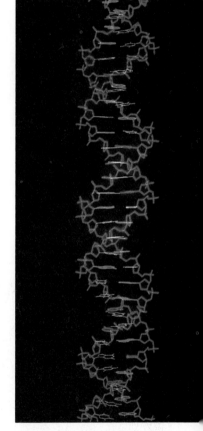

Figure 5–10. A DNA molecule is composed of four different kinds of nucleotides. Each nucleotide contains a sugar called *deoxyribose,* a phosphate molecule, and a nitrogen base.

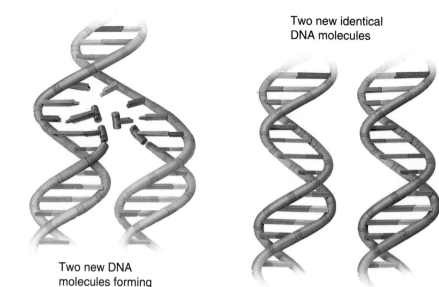

Two new identical
DNA molecules

Base pairs in DNA
molecule separating

Two new DNA
molecules forming

Figure 5–11. Before mitosis can take place, the hereditary code must be duplicated so that both new cells are identical. The code is duplicated when the DNA replicates.

parents. But because we have a unique combination of DNA, our DNA "story"—our own inherited characteristics—are like no other individual's DNA story—past, present, or future.

An Exact Copy

Have you ever seen a copy machine duplicate the information on a piece of paper? DNA can make exact duplicates of itself in a manner that can be compared to the duplication of information by a copy machine. **Replication** is the process by which DNA makes copies of itself. Replication is necessary to provide each daughter cell with a complete and exact copy of the DNA in the parent cell.

This is really the most important idea in cellular reproduction. In order for your cells to be replaced, exact duplicates must be made. Any mistake in the DNA instructions could be, and usually is, fatal for the daughter cells.

Replication begins when a DNA molecule begins to come apart at one end. The paired bases separate, and the separation spreads along the DNA molecule as if the molecule were being "unzipped." As the chains come apart, each side of the original DNA becomes half of a new DNA molecule. New nucleotides move into position to take the place of those that separated. Remember, each base may pair with only one other kind. C will always pair with G, and A will always pair with T. As the new bases come together with the bases in each half of the old molecule, the base-pair pattern of the original DNA forms again. The two new molecules of DNA have exactly the same base-pair order as that found in the old molecule. Why is this important?

Although replication usually results in exact copies of DNA, changes, or mutations, in DNA can occur accidentally. In recent years, scientists have learned how to make deliberate changes in DNA. DNA from one cell can be combined with the DNA in another cell. The DNA that is formed from this combination is called *recombinant* (ree KAHM bih nuhnt) *DNA*.

Recombinant DNA can turn bacteria into chemical factories that produce important substances. For example, people whose bodies cannot produce enough of a chemical called insulin suffer from a disease known as diabetes. Many of these people must take daily shots of insulin to stay healthy. Insulin is made from the organs of cows and pigs. Recombinant DNA research has helped develop insulin-manufacturing bacteria. It is easier to produce large quantities of insulin made in this way, and animals do not have to be used. Another advantage of this technology is that it produces human insulin, so diabetics who cannot tolerate animal insulin can use recombinant insulin safely.

In the future, recombinant DNA may also be used to replace incomplete or undesirable DNA instructions in some plants and animals. Scientists are now using recombinant DNA to make food crops that are more nutritious and to make farm animals that are more resistant to disease.

 ASK YOURSELF

How does DNA replicate itself?

Figure 5–12. Models of the DNA molecule are used in classrooms and laboratories to help students and scientists understand the molecule.

Figure 5–13. Nutrient molecules and oxygen enter a cell and are used for cellular respiration, which takes place in the mitochondria. Nearly all the energy you require to function is released during cellular respiration. The heat your body gives off is also the result of cellular respiration. The cell gives off carbon dioxide and water as waste.

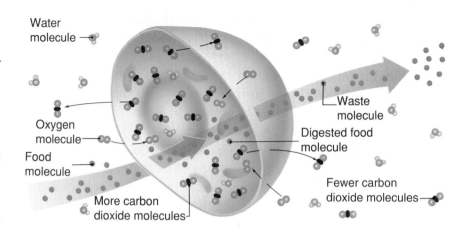

Water molecule

Oxygen molecule

Food molecule

More carbon dioxide molecules

Waste molecule

Digested food molecule

Fewer carbon dioxide molecules

Respiration and Fermentation

You need energy for your body to work properly. Running, walking, breathing, and sleeping are just a few of the things you do that require energy. Your body receives the energy it needs from the food you eat. After your body breaks down food into nutrients, the nutrient molecules enter the cells. Sugar is one of the nutrients that cells use as a source of energy. The sugar molecules are broken down in the cells, and energy is released. Some of this energy is used right away, and some of it is stored.

When Cells Need Oxygen . . . In the cell, oxygen combines with the sugar molecules, forming carbon dioxide and water. This process is called **cellular respiration.** Respiration releases a great deal of energy. Respiration takes place in a series of small steps inside the mitochondria. Each step gives off a tiny bit of energy. Some of this energy escapes as heat, but most of the energy is used to do work in the cell.

Figure 5–14. Just as each rower in a racing shell provides a little of the energy needed to propel the shell through the water, each of the mitochondria in a cell provides a little of the energy needed by the cell.

. . . And When Cells Don't Need Oxygen

Have you ever seen bread being made? The person making it takes advantage of the fermentation process, which causes the bread to rise. Like cellular respiration, **fermentation** (fur muhn TAY shuhn) is a process that gives off energy as nutrients (mostly sugars) are broken down. Fermentation, however, does not use any oxygen.

Many organisms get their energy from fermentation. For example, yeast is used to ferment sugar. This process releases carbon dioxide and alcohol as waste products. Therefore, yeast cells can be used to produce the alcohol in some beverages and the carbon dioxide that makes bread rise. Some bacteria also use fermentation to get energy. One of these kinds of bacteria gives off lactic acid instead of carbon dioxide and alcohol. The tangy yogurt you might enjoy for a snack is made by the action of these bacteria on the sugar in milk. Follow the directions in the next activity to observe the process of fermentation in action.

Figure 5–15. When yeast (above) is used to ferment sugar, carbon dioxide and alcohol are released as waste products. The carbon dioxide causes the dough (left) to rise. The alcohol evaporates as the bread is baked.

DISCOVER BY Doing

Open a package of yeast and put it into a beaker. Add a small amount of warm water to the yeast. Gently swirl the yeast and water mixture inside the beaker. Carefully observe the beaker. What happens to the yeast? Add a small amount of sugar to the beaker. Observe carefully for several minutes. What changes do you see? What causes these changes? ✎

When There Isn't Enough Oxygen Fermentation and respiration are similar because both processes can be used by cells to release energy. Have you ever exercised so much that your muscles seemed to burn? Exercise requires large amounts of energy. Normally, your muscles receive all the energy they need from respiration. However, respiration requires oxygen. During heavy exercise, oxygen cannot be brought to your muscle cells fast enough to keep up with the demand.

When oxygen is in short supply, muscle cells have the ability to release energy from sugar through fermentation. Fermentation releases energy from sugar molecules without the use of oxygen. However, less energy is released from a sugar molecule during fermentation than during respiration. In addition, during fermentation muscle cells form lactic acid instead of the carbon dioxide and water formed during respiration. The lactic acid is what causes the "burning" sensation in the muscles.

You may notice a stiffness and soreness in your muscles the day after heavy exercise. This stiffness and soreness will go away as your body gets rid of the lactic acid. Fermentation in muscle cells will stop after exercise ends and enough oxygen becomes available for the cells to again begin respiration. The table shows a comparison of respiration and fermentation.

Figure 5–16. Fatigue after exercise is the result of lactic acid formation.

Table 5-1	A Comparison of Respiration and Fermentation	
Respiration	**Fermentation**	
Uses oxygen	Does not use oxygen	
Releases much energy	Releases little energy	
Produces carbon dioxide and water	Produces carbon dioxide and alcohol or acid	

 **ASK YOURSELF**

Why do cells need energy?

Photosynthesis

Can your body make its own food? Of course not. Plants, however, can. The process by which plants make food is called *photosynthesis*. Photosynthesis requires light, carbon dioxide, water, and chlorophyll. Chlorophyll traps energy from sunlight and uses it to break down water molecules into atoms of hydrogen and oxygen. The oxygen is given off as a byproduct that can be used by organisms during respiration. The rest of the energy is used to combine the hydrogen atoms with the carbon dioxide. Eventually, sugar is formed and stored by the plant until it is needed.

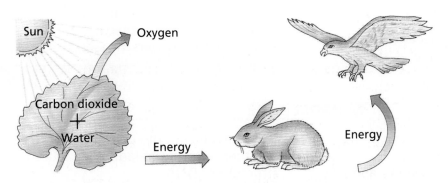

Figure 5–17. Photosynthesis converts solar energy into chemical energy, which is stored by plants. When an animal eats a plant, the animal uses that energy. When one animal eats another animal, energy that was originally solar energy is transferred.

The sugar made during photosynthesis may be used by the plant as food. The sugar can also be converted into starches, fats, and other compounds and stored or used to build new plant structures. An animal that eats a plant takes in the stored molecules and uses them for food. The animal uses the energy from these molecules to do work. Any extra energy is stored in the animal's body. As one animal eats another animal, the energy that came to the plants from the sun is passed on again and again. In this way, all the energy that is used by all the living things on Earth can be traced back to sunlight.

 ASK YOURSELF

What would happen to the animals on Earth if all the plants suddenly disappeared?

SECTION 3 *REVIEW AND APPLICATION*

Reading Critically

1. How is a DNA molecule like a ladder?
2. Describe the pairing of bases in a DNA molecule.
3. How are respiration and fermentation different from each other?

Thinking Critically

4. What do you suppose would happen if DNA replication stopped before a complete copy of the DNA material had been made?
5. If a cell contained fewer mitochondria than normal, what do you suppose would happen to the cell? Explain.
6. Someday space travel may take people to other solar systems. Why is finding a planet inhabited only by plants much more likely than finding a planet inhabited only by animals?

SKILL

Modeling DNA

▶ **MATERIALS**
- 25-cm pieces of fine wire (2) ● miniature marshmallows (6) ● jelly beans (8) ● four different-colored pipe cleaners cut into thirds (6) ● modeling clay

wire = sugar-phosphate backbones marshmallows = sugar groups jelly beans = phosphate groups pipe cleaners = bases

▼ **PROCEDURE**

1. Lay two pieces of wire side by side on your desk.
2. String miniature marshmallows and jelly beans on the wires as shown. Lay the two chains side by side with the larger ends of the jelly beans facing out as shown.
3. Now attach the bases to the sugar groups on each chain. Copy the table and record your color code for each base. Remember, adenine pairs with thymine; cytosine pairs with guanine.

TABLE 1	
Base	**Pipe Cleaner Color**
Adenine	
Thymine	
Cytosine	
Guanine	

4. Twist the ends of the paired bases together to form the ladderlike structure of DNA.
5. Make a modeling-clay base and carefully stick one end of your DNA molecule into the base. Start at the bottom of the model and gently twist it into the characteristic spiral.

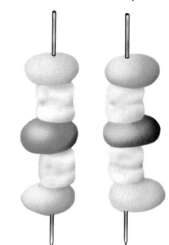

▶ **APPLICATION**
1. If there are 5000 molecules of adenine in one DNA molecule, how many molecules of thymine will there be?
2. Must DNA molecules have the same number of adenine bases as cytosine bases? Explain your answer.

✳ *Using What You Have Learned*
1. Why does using a model often make it easier to understand the structure and function of an object or a substance?
2. Because of the base pairings, DNA replicates exactly. However, sometimes a change, or mutation, occurs. Using your model as a DNA molecule, how do you think the molecule might be changed?

The Big Idea

In the same way that the symbols in a coded message work together to convey the message, the cells in a living organism work together to keep the organism healthy and functioning. But cells cannot perform their function in a plant or animal unless each individual cell can function properly on its own. Cells carry out a wide variety of tasks, with the types of activity depending on the role of the specific cell.

Examples of important cell functions include the transportation of materials by diffusion and osmosis, the release of energy through respiration and fermentation, and the manufacture of food in plant cells through photosynthesis. Reproduction of cells and replication of DNA are other important cell functions.

For Your Journal

Look back at the ideas you wrote in your journal at the beginning of the chapter. Summarize the cell processes you have studied. Do all of the processes involve energy? Why must cells reproduce? How is a scientist studying DNA like a detective trying to break a secret code?

Connecting Ideas

Copy this unfinished concept map into your journal. Complete the concept map by writing the correct term in each blank.

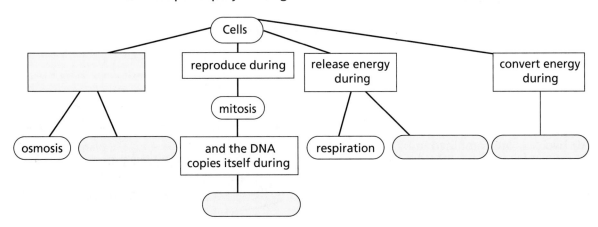

REVIEW

Understanding Vocabulary

In the problems below, the terms in this chapter have been put into different categories. For each term, give a reason justifying its membership in the given category.

1. **Category: reproduction**
   DNA (127), mitosis (127), chromosomes (127), replication (132)
2. **Category: transportation**
   active (122), passive (122), osmosis (123), diffusion (121)
3. **Category: uses energy**
   photosynthesis (136), active transport (122), respiration (134)
4. **Category: releases energy**
   respiration (134), fermentation (135)

Understanding Concepts

MULTIPLE CHOICE

5. Osmosis is a type of diffusion in which
   a) the cell manufactures its own food.
   b) all types of molecules are allowed to pass through the cell membrane.
   c) water is moved through the cell membrane.
   d) the cell uses energy.

6. Diffusion is *not* a form of active transport because
   a) no energy is required from the cell.
   b) the cell membrane allows only some types of molecules to pass through.
   c) oxygen is not required for diffusion.
   d) food molecules are not created.

7. During mitosis
   a) the cell prepares for the next cell division.
   b) the cell's hereditary instructions are modified.
   c) two new daughter cells are formed.
   d) two cells are combined into one.

8. The hereditary instructions of a cell are formed by
   a) a combination of different chromosomes.
   b) mitosis.
   c) scientific research involving DNA.
   d) a particular sequence of bases in the DNA molecule.

9. Which of the following does *not* describe a difference between respiration and fermentation?
   a) Respiration produces water.
   b) Respiration releases energy, but fermentation does not.
   c) Respiration requires oxygen.
   d) Fermentation gives off less energy.

Interpreting Graphics

10. The diagram shows a simplified picture of the carbon cycle. What role does photosynthesis play in this cycle?

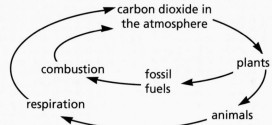

Reviewing Themes

11. Systems and Structures
Compare the events of mitosis with the activities of a group of students putting on a school play.

12. Energy
What functions can still occur in a cell in the absence of energy?

SHORT ANSWER

13. Why does a cell's survival depend on the movement of certain materials through its cell membrane?

14. How is a selectively permeable membrane different from a permeable membrane?

15. Give reasons why this statement is false: Nothing much happens in a cell during interphase because the most important events all occur during mitosis.

16. Why are daughter cells exactly the same in structure and function as their parent cell?

17. How is division of cytoplasm in plant cells different from division in animal cells?

Thinking Critically

18. Explain how external respiration, or breathing, is related to respiration in the cell.

19. How might recombinant DNA research help scientists find a way to make farm animals more resistant to disease?

20. Why would a cell that uses fermentation require more food molecules than a cell that uses respiration?

21. All organisms are controlled by the DNA in their cells. How does the structure of a DNA molecule help to account for the great variety of life that exists on Earth?

22. The photograph below shows a model of a DNA molecule. Explain why its shape is described as a "double helix." Then give other examples from life science of things that have the shape of a helix or a spiral.

Discovery Through Reading

Bender, Lionel. *Through the Microscope: Frontiers of Medicine.* Gloucester Press, 1991. Examines the advances in medical science made possible by the use of microscopes, from the first studies of plants and insects in the seventeenth century to the discovery of bacteria and viruses two centuries later.

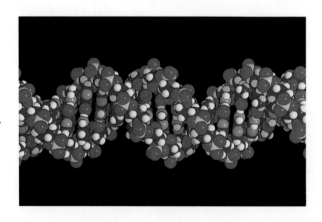

Natural Selection and Heredity

Charles Darwin was born in Shrewsbury, England, on February 12, 1809—the same day as Abraham Lincoln. Everyone thought he would become a doctor, like his father and grandfather, but Darwin preferred collecting stones and minerals and watching birds. He entered Cambridge University to become a minister, and although he passed his courses, the only thing Darwin really enjoyed was observing nature.

In 1831 an invitation was sent to Charles Darwin. He was invited to join HMS Beagle on a surveying trip around the world as the ship's naturalist. The offer came from Captain Robert Fitzroy who was to lead the expedition. They were to make detailed maps of the coast of

Galápagos Islands

South America and to take scientific measurements around the world, which would make maps more accurate.

Darwin was very excited about the voyage. He set off immediately to say goodbye to his family and to pack his clothes and equipment. He took with him a microscope, a compass, a book on taxidermy, binoculars, a magnifying glass and jars of spirits for preserving specimens.

At last, on December 27, 1831, the Beagle set sail from Plymouth. Darwin liked life on board ship but was very seasick. He suffered from this illness throughout the whole trip and had to lie in his hammock eating only raisins when the weather was bad.

As they sailed closer to the South American coast, the sea grew calmer and Darwin was able to go on deck and see the strange birds and sea creatures around the ship.

from *The Voyage of the Beagle*
by Kate Hyndley

For Your Journal

- Why are there so many different kinds of plants and animals?
- Why do organisms change over time?
- How are these changes passed on?

A Great Adventure

Although Charles Darwin was never a good student, he was an avid reader. One of the books that influenced him the most was a book on geology by Charles Lyell. In this book, *Principles of Geology,* Lyell presented many new ideas about the earth, and he hypothesized that the earth was constantly changing. Although most of the scientists of the day disagreed with Lyell, Darwin thought there was some merit to Lyell's hypothesis.

Darwin's Voyage

Lyell said that not only does the earth change, but there are also two kinds of change—fast change, such as that caused by earthquakes and volcanoes, and slow change, such as that caused by erosion.

The evidence for fast change was all around Darwin. While visiting the city of Concepción, Chile, in South America, he got to see firsthand the effects of an earthquake. Here's how the effects are described in *The Voyage of the Beagle:*

> *The most remarkable effect of this earthquake was the permanent elevation of the land.... There can be no doubt that the land around the Bay of Concepcion was upraised two or three feet.*

Figure 6–1. Lyell's hypotheses seem reasonable; perhaps the earth is older than most people believe.

It became obvious to Darwin that the earth does indeed change. Volcanoes created mountains where there had been none before, and earthquakes altered the shape of the land. But the slow changes that Lyell spoke of were not so so obvious. Most people of the 1800s believed that the earth was only a few thousand years old. Slow change, such as Lyell described, could have taken place only if the earth were much older.

One of Darwin's most important discoveries was made while digging in the gravel near Punta Alta in Argentina. He uncovered fossilized bones of large animals that had probably died out a long time before. Darwin wondered why these animals died out, since at one time they must have existed in large numbers.

One day, Darwin observed fossils of seashells at the top of a cliff. He wondered how the shells of sea organisms could have gotten there. Darwin reasoned that the layer of rock containing the shells must have been under the sea at one time. Perhaps some force of nature had lifted the rock to its present position. Maybe this was evidence of the slow change that Lyell had written about.

Wherever the *Beagle* anchored, Darwin went ashore to explore. In Brazil, he was amazed by the beauty of the tropical rain forests. Flowers and brightly colored birds were everywhere. To Darwin, the huge trees covered with vines and the lush vegetation were like a paradise. He spent many hours studying plants and animals no scientists had seen before.

Figure 6–2. The earth does indeed change. Here is an example, just as Lyell said.

Figure 6–3. These rocks must have been under the sea at one time.

Figure 6–4. There are plants and animals here that scientists have never seen before.

As the *Beagle* sailed around the coast of South America, Darwin found many fossils. One skull of particular interest was of an extinct animal known as a *toxodon*. It reminded Darwin of the capybara he had seen in the Brazilian forests.

Darwin continued to wonder why so many animals had become extinct. Darwin was further puzzled after finding the fossils of a horse near Buenos Aires, Argentina. When the Spanish colonists arrived in South America in 1535, there were no living horses on the continent.

Figure 6–5. These animals greatly resemble this ancient skull I have found.

In September 1835 the Beagle *left Lima to sail west across the Pacific Ocean. The party intended to return to England by sailing around the Cape of Good Hope at the southern tip of Africa. Their first port of call was a group of islands 1000 kilometers west of Peru called the Galápagos Islands.*

At first sight, the islands were not very attractive—black volcanic rocks sticking up from the sea with only a few scorched bushes growing on them. Darwin soon realized that the islands had once been underwater volcanoes, which had now risen above sea level.

DISCOVER BY *Researching*

Locate the Galápagos Islands on a globe or map. Note the position of the islands in relation to the United States. Near what country in South America are they located? Use an encyclopedia or atlas to find out more about the islands. What are their elevations? What is their climate? Are they inhabited by people? What kinds of plants and animals live there? Write a paragraph summarizing your findings. ✎

Figure 6–6. The Galápagos Islands support a wide variety of living things such as the iguana (top) and the blue-footed booby (bottom).

Figure 6–7. These animals seem completely unafraid of humans.

On the beaches lived black lizards that were about one meter long. They could swim well but never stayed in the sea for long. These sea-living lizards ate seaweed. The land-living lizards were more brightly colored and fed on cactus. They were very gentle animals and did not attack or bite anything, even when provoked.

The word galápagos *means giant tortoise in Spanish, and these huge animals thrived on the islands. Some were so large that it took eight men to lift one of them. Darwin walked inland to a spring where he watched the tortoises drink greedily. Sometimes they would drink only once a month. He also observed that on each island the tortoises had different shell markings.*

Figure 6–8. The tortoises are different on each of these islands!

Darwin noticed that nearly all living things show differences among members of the same species. Some differences seem to have little importance. But was that true of all differences? And why would there be so many differences among tortoises living on islands within sight of each other? Darwin began to notice that the birds, too, differed from one island to another.

Darwin counted thirteen species of finches on the islands. Each species had a different kind of beak. He was puzzled as to why these variations occurred.

Figure 6–9. There seem to be variations among all the animals living on these islands.

Darwin saw the different finches as evidence that, like the earth itself, organisms change over time. He reasoned that several kinds of finches had come to the islands from the mainland of South America. Variations in their beaks allowed some finches to feed more effectively than other finches in this new environment. Therefore, certain finches were likely to survive, while others died out. Those that survived reproduced and slowly established themselves on certain islands.

Figure 6–10. Finches with certain kinds of beaks will survive better on this island.

ASK YOURSELF

What observations did Darwin make about the tortoises and finches of the Galápagos Islands?

Darwin's Hypothesis

When he returned home, Darwin began the long process of classifying all of the specimens he had collected. He also began to sort out his ideas on how organisms can change over time. Darwin knew that farmers selected and bred plants and animals with certain desirable traits. However, he didn't know what could cause such selection in nature. Again, something Darwin read influenced him.

Figure 6–11. I know that species do change. Now I must determine how.

In *An Essay on the Principle of Population,* Thomas Malthus explained that the human population increases when food and other necessities are sufficient. The population decreases when necessities are in short supply. He suggested that people would overrun the earth if starvation, disease, crime, and war did not keep their numbers down. Malthus called this the struggle for existence.

The struggle for existence was the key to Darwin's own hypothesis. He realized that all organisms must compete for food, water, and other necessities in order to survive. Only the organisms that are better able to compete will survive. Those that are not will die. Now Darwin had the mechanism for **evolution,** or the change in organisms over time. In the next activity, you can see how changes may occur.

Figure 6–12. These tortoises must have evolved over time.

ACTIVITY

What is the role of variation in evolution?

MATERIALS

scissors, red paper, blue paper, 2 paper cups

PROCEDURE

1. Cut 20 small squares of red paper to represent 20 female tortoises. Cut 20 small squares of blue paper to represent 20 male tortoises.
2. Write the words *long neck* on two male and on two female "tortoises." Place all the female tortoises in one cup and all the male tortoises in the other.
3. Suppose your tortoises live on an island. The tortoises are all healthy and full grown, and they begin to pair up for mating.
4. Without looking, select one male tortoise and one female tortoise from the cups. Check to see if either of the tortoises you selected has a long neck.
5. Record the results of the mating. Each mated pair of tortoises has two offspring. All offspring will have short necks unless both of their parents have long necks. Continue selecting

male and female tortoises and recording the results until all the tortoises have been mated.

6. Return the tortoises to their containers and repeat the entire process five times. Of the 100 matings, how many were between two long-necked tortoises? How many long-necked offspring resulted?
7. Suppose there is a drought and the only food remaining is shrubs. Half the tortoises die. Remove 10 tortoises from each of the paper cups, but don't remove any with long necks.
8. Repeat steps 4 and 5, recording 100 more tortoise matings. In the second round of matings, how many were between pairs of long-necked tortoises? How many long-necked offspring resulted?

APPLICATION

Suppose the four long-necked tortoises did not mate with one another. If the food problem on the island got worse, what would happen to the tortoise population?

 ASK YOURSELF

What did Darwin hypothesize about changes in animals?

SECTION 1 REVIEW AND APPLICATION

Reading Critically

1. Give several examples of variations that Darwin observed.
2. What did Darwin mean when he said that animals change over time?

Thinking Critically

3. How might a variation in the size of a bird's beak help the species survive?
4. Create a concept map relating Darwin's observations and inferences.
5. If long-necked tortoises mated only with other long-necked tortoises, how would the population be affected during long periods of food shortage?

A Great Theory

Objectives

Describe what is meant by the "survival of the fittest."

Compare Darwin's and Lamarck's theories of evolution.

List the evidence that supported Darwin's theory.

You know that a library is a building where books are kept. An archive is a place where photographs, public records, and other important historical materials are kept. Anyone curious about the past can go to an archive to get information. Because the earth contains a record of history, it may be compared to an archive.

The Origin of Life

Scientists before and after Charles Darwin have hypothesized about the origin of life on Earth and how living things have changed over time. Many have made observations and done experiments that provide possible solutions to these problems.

Fossil Evidence A great deal of information about the history of the earth can be found in the form of fossils in rocks. **Fossils** are traces of once-living organisms. A leaf, a footprint, a bone, a shell, a skeleton, or even a complete organism could be a fossil. Have you ever made a footprint in wet mud? When the mud dries, the print remains. Some fossils look something like a print you might leave in the mud.

Fossils formed in many ways. Sometimes insects were trapped in plant sap that later hardened. This hardened sap is called *amber*. Some larger animals were trapped and preserved in tar. Other fossils formed when animals were buried in mud. As years passed, the bodies of these animals were covered with layers of soil and dead plants. As these layers became deeper and heavier,

Figure 6–13. The fossils shown here—a trilobite (top), an ammonite (center), and an insect in amber (right)—are a record of Earth's past.

pressure built up. The pressure eventually changed the layers into rock. The organisms trapped by this process left their shapes in the rock. Sometimes, however, parts of an organism were replaced by minerals and the organism became *petrified*, literally turned into stone. The petrified forests in Arizona are examples of this type of fossil. The toxodon skull that Darwin found in the Galápagos Islands was a fossil too.

Scientists have learned much about the early history of the earth by knowing the ages of certain fossils. The age of a fossil can be found by using a chemical process to date it. Another way to date fossils is to examine the rock in which the fossils are found. The older the rock, the older the life forms found in it must be. The layers of the Grand Canyon trace the passage of time from the present back through 2 billion years of Earth's history. Fossils of reptiles, insects, fishes, and plants have been found in various layers of rock. The layer in which a fossil is found indicates the "age" in which the organism lived. Fossils

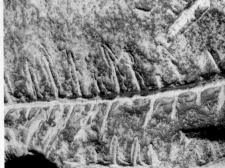

Figure 6–14. The Grand Canyon contains a great deal of Earth's history.

are Earth's archives. They tell about the animals and plants that lived when Earth was younger.

The fossil record does not provide all the pieces to the puzzle of evolution. Sometimes there are "gaps" in the fossil record. Fossils have not been found for every organism that has ever lived on Earth. Since no humans were present, it is impossible for us to know for certain how life originated. However, by studying the fossil record, it is possible to see how life on Earth has changed.

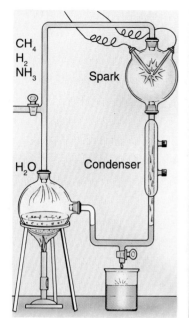

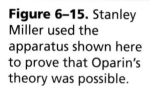

Figure 6–15. Stanley Miller used the apparatus shown here to prove that Oparin's theory was possible.

Chemical Evidence

No one can ever know for certain how life first appeared on Earth. However, Aleksandr Oparin (1894–1980), a Russian biologist, proposed a theory to explain how life began. Oparin's theory states that life developed slowly when the molecules in the oceans were heated by the sun and electrified by lightning. The most abundant chemicals at that time were water vapor, methane, hydrogen, and ammonia.

In 1952, an American scientist, Stanley Miller, demonstrated Oparin's theory by making the instrument shown in the picture. Miller combined water vapor with methane, ammonia, and hydrogen and then passed an electric spark through the mixture. He was able to make the basic molecules necessary for life. However, Miller was never able to create "life."

No one knows exactly how these molecules changed to form cells, but the process took many millions of years. The earliest cells were probably one-celled organisms that needed no oxygen. They could make their own food from the chemicals around them. Later, primitive bacteria developed. These organisms were able to produce food through the process of photosynthesis. Since photosynthesis releases oxygen, the earth's atmosphere slowly changed and became able to support life that needed oxygen for survival. Most of these changes occurred very slowly over a long time. Gradually, more and varied organisms developed.

 ASK YOURSELF

How are fossils used to study the past life of the earth?

Developing the Theory

In 1809, well before the studies of Charles Darwin, Jean Baptiste Lamarck, a French scientist, presented his theory of evolution. Lamarck's theory stated that a change in a structure followed a change in the job. Lamarck said that an animal might acquire a new trait because of an environmental need. In addition, the animal would then pass on the newly acquired trait to the next generation. For example, Lamarck thought that giraffes grew longer necks because they needed to reach leaves growing higher in the trees.

Figure 6–16. Giraffes were the subject of Lamarck's theory. Lamarck believed that giraffes stretched their necks in order to reach higher leaves. The longer neck would then be passed on to offspring.

Both Lamarck and Darwin theorized that evolution occurs. However, they each proposed a different explanation for how evolution occurs. Lamarck theorized that organisms developed new structures to replace old structures that were not being used. He also believed that acquired traits were passed directly to the organisms' offspring so that new generations could immediately benefit from such changes. This could be compared to saying that if your mother had dyed her hair blue, you would have been born with blue hair!

Darwin agreed with Lamarck that the job comes first. However, Darwin said that some organisms in a group are better able to perform the new job because of some difference in their physical characteristics. These organisms are more likely to survive and reproduce. Since only these organisms reproduce, gradually all organisms in the group have the characteristic.

Darwin's theory of evolution is based on natural selection. **Natural selection** is the process by which those organisms best suited to their environment will survive and reproduce. A change in an organism that makes it better able to survive in its environment is called an *adaptation*.

Figure 6–17. The flamingo has adaptations that enable it to feed in a unique manner. It feeds upside down. In the process its upper jaw moves instead of its lower jaw. How does your jaw move?

Figure 6–18. This animal was not fast enough to escape its predator.

As Darwin was exploring the forests of Brazil, he observed animals in a struggle for survival. Those that were best suited to their environments lived, while those that were not died. Darwin hypothesized that this struggle to survive might be true for all living organisms. Later, this struggle was termed "the survival of the fittest."

Since there is only so much food, water, and space for living, only the best-adapted individuals from each species survive. These selected few, therefore, are those that will reproduce, passing on their adaptations to their offspring. Those individuals not well adapted to their environment do not survive, so undesirable characteristics are slowly eliminated from the population. In the next activity, you can see how undesirable variations may disappear completely.

DISCOVER BY *Doing*

You will need a paper punch, four different colors of construction paper, a piece of multicolored cloth, and a watch with a second hand. With the paper punch, make 25 dots from each color of paper. Place the piece of multicolored cloth on your desk. Mix up the dots and spread them randomly over the cloth. The colors of the dots represent differences in one animal species.

Ask a classmate to represent an animal predator that eats the dot "animals." He or she should pick up as many dots as possible—one dot at a time—in 20 seconds. Count how many dots of each color your classmate picked up. Which dots were picked up most often? Which dots were picked up least often? Why were some dots not picked up as often as others? How might color variations help an animal survive? ✐

Darwin worked for many years organizing his ideas into a theory. In 1859, he published *On the Origin of Species by Means of Natural Selection*. In this book, Darwin outlined, in great detail, the way organisms evolve through natural selection. The main points of Darwin's theory follow. As you study them, notice how they differ from the theory of Lamarck.

The Theory of Natural Selection

1. **Each species produces many more offspring than can survive and reproduce.** For example, female fishes lay enormous numbers of eggs. If all of these eggs hatched and the young survived, the waters would be overrun with fish. Even giraffes sometimes produce more offspring than can survive.

2. **The overproduction of offspring leads to a struggle for survival.** The individuals of a species must compete for the necessities of life. All giraffes must compete for food, water, and space to live in their environment. Some will not be successful and will die before they have young of their own.

3. **All organisms of the same species are somewhat different from one another.** Except for identical twins, individuals are not exactly alike in all of their traits. Some giraffes may be stronger or run faster than others. Some may have longer necks than others.

4. **Individuals with certain traits have a better chance of surviving and reproducing.** This is the "survival of the fittest" idea. Giraffes with long necks are better able to survive because they can eat leaves from tall trees, which are plentiful. Giraffes with short necks cannot.

5. **Those organisms that survive and reproduce pass their traits on to their offspring.** These offspring also have a better chance of survival. Giraffes with long necks survive and reproduce. They pass the trait for long necks on to their offspring, which also survive and reproduce because they too can eat leaves from tall trees.

 **ASK YOURSELF**

What is natural selection?

Support for Darwin's Theory

It was 20 years after the *Beagle* returned to England before Darwin began to write out his theory. As he was writing his book, he received an essay in the mail from Alfred Wallace. Wallace, also a naturalist, worked in the jungles of Indonesia. He and Darwin had been writing to each other about their research and ideas. In the essay, Wallace stated the main points of the theory of evolution that Darwin had worked on for so long. However, both men had questions about the way traits are passed from one generation to the next.

Figure 6–19. Mendel's work added meaning to Darwin's theory.

The answer involves heredity—the way traits are passed from parents to offspring. Little was known about heredity before 1900. In that year, the work of Gregor Mendel was found. Mendel had published his research on heredity in garden pea plants in 1865, but scientists did not realize the importance of his findings at that time. You will learn more about Mendel and heredity in the next section.

 ASK YOURSELF

Who else had ideas about evolution that were similar to Darwin's?

SECTION 2 *REVIEW AND APPLICATION*

Reading Critically
1. List the main points of Darwin's theory of natural selection.
2. Why is life a struggle for most organisms?

Thinking Critically
3. What conditions could cause a population to evolve?
4. Compare and contrast Darwin's and Lamarck's theories of evolution.

SKILL

Controlling Variables

▶ **MATERIALS**
You will decide what materials you will need.

▼ **PROCEDURE**

1. Here is the hypothesis you will test in this activity: *When seeds of tall marigolds are planted in sand, the plants will not grow as tall as those from seeds of tall marigolds planted in soil.*

2. Design an experiment that will test this hypothesis. Decide what your *manipulated variables* will be. These will be the things that you will vary. For example,

   you will probably vary the kind of material in which the seeds are planted.

3. Decide what your *controlled variables* will be. These are things you will keep the same. For example, you will need to give each pot the same amount of water.

4. Decide what materials you will need. Then try your experiment.

5. Chart your plant-growth information.

▶ **APPLICATION**
Was the hypothesis correct? What did you find out about seeds of tall marigolds? What inherited trait was affected by the environment?

✳ *Using What You Have Learned*
Compare your experiment with those designed by your classmates. Find out whether you controlled the same variables. Discuss how any differences in the variables that were controlled may have affected the results of each experiment.

Mendel and Heredity

Gregor Mendel was an Austrian monk who lived from 1822 to 1884. He is best known for his studies in heredity. What made Mendel's work especially valuable were the scientific methods he used in his studies. He chose his subjects carefully, used many samples, and kept accurate records of all his experiments. After he had carefully recorded all his results, he checked them using mathematical principles.

The Story of Gregor Mendel

Figure 6–20. Mendel studied the inheritance of traits in garden peas.

As a young boy, Mendel had been interested in the plants that grew on the family farm. He later became the gardener for the abbey in which he lived, and he studied the pea plants he grew in the abbey garden.

Mendel experimented with the heredity of certain traits found in peas. Look at the chart of these traits. Mendel studied each trait separately and discovered certain patterns in the way traits are inherited in peas.

Figure 6–21. Here are seven traits that Mendel studied.

Seed Shape	Seed Color	Flower Color	Pod Shape	Pod Color	Flower Position	Stem Length
Dominant						
Round 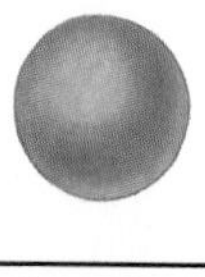	Yellow 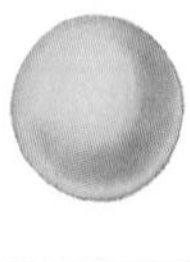	Purple 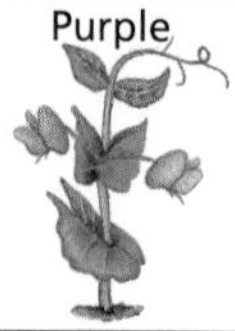	Inflated 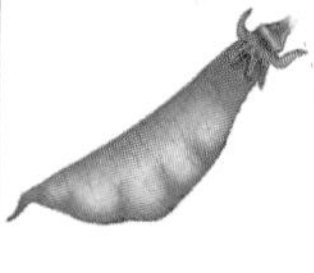	Green 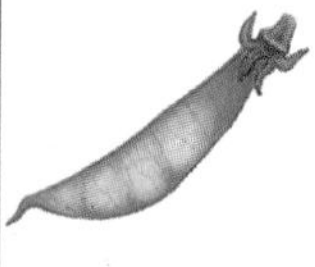	Side 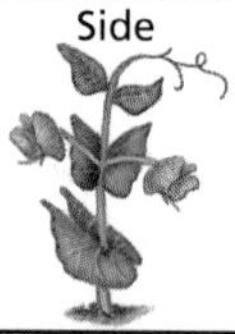	Tall 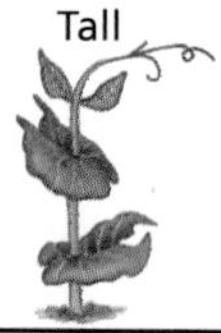
Recessive						
Wrinkled 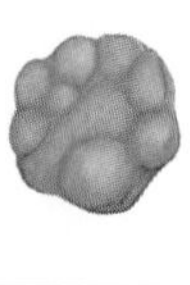	Green 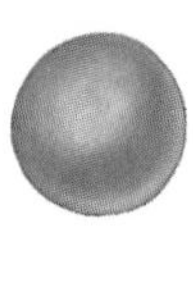	White 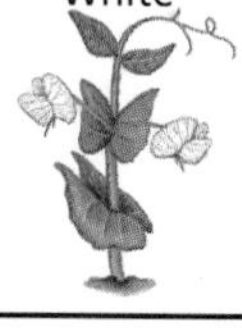	Constricted 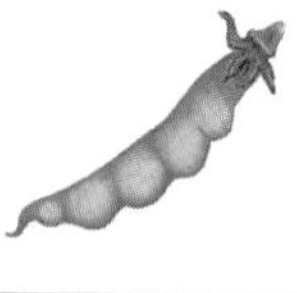	Yellow 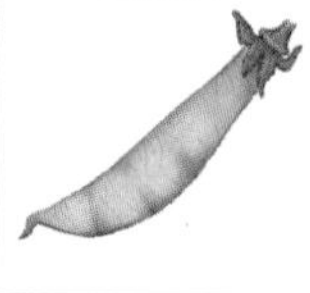	End 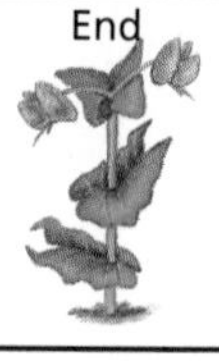	Short 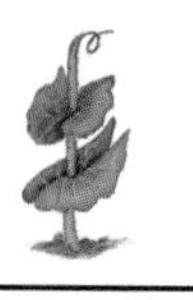

Figure 6–22. Mendel crossed tall plants with short plants.

Mendel chose peas because they had obvious traits and usually produced seeds that grew into plants that were similar to the parents. For example, tall plants produced seeds that grew into tall plants. These plants were called *pure* tall.

For each characteristic, Mendel would cross two different traits. In one experiment, he fertilized pure short plants with pollen from pure tall plants. He planted the resulting seeds and counted the types of offspring that grew. He was surprised to find that all the offspring were tall. He got the same results when he fertilized pure tall plants with pollen from pure short plants. Any cross between a pure tall plant and a pure short plant produced only tall plants.

Mendel next crossed the tall offspring—those produced by the cross of the pure tall and pure short plants—with each other. The seeds from this second cross produced mostly tall plants, but one-fourth of the plants were short!

Figure 6–23. Three-fourths of the plants were tall like the parent plants.

 ASK YOURSELF

Why do you think Mendel was surprised by the one out of four short plants?

Dominant and Recessive

Mendel found similar results when he crossed plants that were pure for round seeds with those that were pure for wrinkled seeds. All offspring of the pure parents showed only the trait of the parent with round seeds. But if these plants, called *hybrids,* were crossed, one-fourth of the offspring were like the grandparent with wrinkled seeds, and three-fourths were like the grandparent with round seeds.

Figure 6–24. Mendel grew several generations of pea plants.

It appeared to Mendel that for each characteristic in peas, one trait was stronger than the other. He called the stronger one the **dominant** trait and the "hidden" one the **recessive** trait. When two pure plants were crossed, all the offspring showed the dominant trait. When two hybrid plants were crossed, three-fourths of the offspring showed the dominant trait and one-fourth showed the recessive trait. Look at the diagram. Even though three-fourths of the offspring of a cross between two hybrid plants show the dominant trait, what fraction are pure and what fraction are hybrid?

Figure 6–25. Mendel's hybrid plants were plants with one recessive factor and one dominant factor.

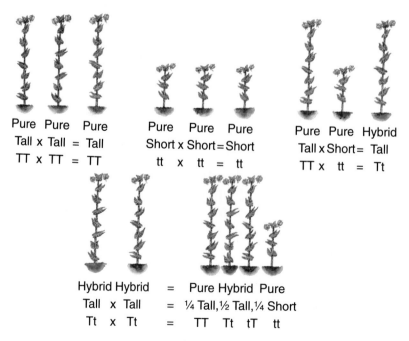

Pure Pure Pure
Tall x Tall = Tall
TT x TT = TT

Pure Pure Pure
Short x Short = Short
tt x tt = tt

Pure Pure Hybrid
Tall x Short = Tall
TT x tt = Tt

Hybrid Hybrid = Pure Hybrid Pure
Tall x Tall = ¼ Tall, ½ Tall, ¼ Short
Tt x Tt = TT Tt tT tt

Mendel also did experiments on flower color with his pea plants. Some results are shown in the following table. Study the table, and determine what factors each set of parents had. Mendel's ideas about dominant and recessive factors may help.

| Test | Flower Colors in Offspring | | Factors in Parent |
	Purple	White	
One	0	12	_____
Two	12	0	_____ or _____
Three	9	3	_____ and _____

Figure 6–26. Offspring inherit recessive traits only if both parents contribute recessive factors.

Has anyone ever told you that you look just like your mother, your sister, or another close relative? The game of "who do you look like" probably began shortly after you were born.

Figure 6–27. It is common for family members to look alike.

Mendel's ideas about factors also apply to human heredity. Your traits are controlled by the factors, now called **genes,** that you inherited from your parents. In the next activity, you can find out which human traits are dominant and which traits are recessive.

ACTIVITY

*A*re your traits dominant or recessive?

MATERIALS

mirror

PROCEDURE

1. Look carefully at the following chart.
2. Use a mirror, or have a classmate observe these traits for you. Copy the data table to record your findings. For each of these traits, you have inherited one factor from your mother and one from your father. Two factors control each trait.

APPLICATION

Having six toes is a dominant characteristic of cats. Why do you think there aren't more cats with six toes?

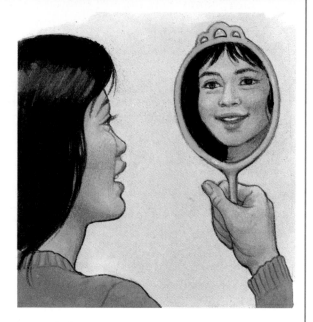

TABLE 1: HUMAN CHARACTERISTICS			
Characteristic	**Dominant**	**Recessive**	**Your Trait**
Hair	Curly	Straight	
Hair whorl	Clockwise	Counterclockwise	
Hair color	Brown, black	Blonde, red	
Hair at forehead	"Widow's peak"	None	
Eyelashes	Long	Short	
Dimples	Yes	No	
Nose	Turned up	Not	
Ear lobes	Free	Attached	
Hair on middle section of finger	Yes	No	
Freckles	Yes	No	
Eye color	Dark	Blue, gray, green	
Tongue	Can roll	Can't roll	

 **ASK YOURSELF**

What is the difference between dominant and recessive traits?

Reproduction and Heredity

Throughout his studies and experiments, Mendel called the characteristics he studied "factors" rather than "genes." He did not know what the factors were or exactly where they were found in living things. When modern scientists discovered those things, Mendel's work became even more impressive.

In 1902, Walter Sutton, a professor at Columbia University, was doing research based on Mendel's work. From this research, Sutton wrote a paper stating a new theory of genetics. The **chromosome theory** says that Mendel's factors are the same as genes and that genes are located on the chromosomes. The chromosome theory also states that traits are passed to offspring by the chromosomes. Moreover, each reproductive cell, or **gamete,** contains these chromosomes, which are located in the nucleus of the cell.

Mendel knew that factors, or genes, separate when gametes are formed. Sutton made it clear that chromosomes, carrying the genes, separated. When fertilization occurs, gametes from each parent join to form a new cell. This new cell divides many times by mitosis to form a complete organism made of billions of cells.

If each gamete had the full number of chromosomes, after fertilization the new cell would have twice that number. In each following generation, the number of chromosomes would double. Since this is not the case, there must be some process to reduce the number of chromosomes in gametes to half. This process is meiosis (my OH sihs). **Meiosis** is the type of cell division by which gametes are formed. Meiosis results in the formation of four daughter cells, each with one-half the number of chromosomes found in the parent cell. Table 6–1 summarizes the differences between meiosis and mitosis.

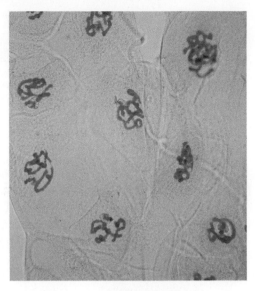

Figure 6–28. The darkly stained objects in this photograph are the chromosomes of a fruit fly.

Figure 6–29. The phases of meiosis are similar to those of mitosis. However, in meiosis there are two cell divisions.

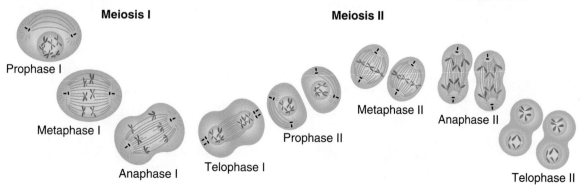

Meiosis I

Prophase I

Metaphase I

Anaphase I

Telophase I

Prophase II

Meiosis II

Metaphase II

Anaphase II

Telophase II

Table 6-1

A Comparison of Mitosis and Meiosis	
Mitosis	**Meiosis**
One cell division	Two cell divisions
Two daughter cells	Four daughter cells
Daughter cells have the same number of chromosomes as the parent cells	Daughter cells have half the number of chromosomes as the parent (one member of each pair)

ASK YOURSELF

What is meiosis?

The Hereditary Code

By 1953, it was clear to biologists that chromosomes were made up of large molecules of DNA. At that time, James Watson and Francis Crick cracked the code of DNA. Have you ever sent a message to someone in code? A code contains information that only certain people know. DNA contains the code of hereditary information. Instead of using letters, as in a written code, the hereditary code is in the nucleotides of DNA.

Parts of the hereditary code—the DNA molecule—are the same in all organisms, including humans. However, no two people, except identical twins, have exactly the same DNA. The order of nucleotides in a DNA molecule controls what a trait will be. The next activity allows you to practice sending and receiving information in code. It may also help you to understand how DNA can be the hereditary code for so many different traits.

Figure 6–30. DNA determines the hereditary code in all organisms.

DISCOVER BY Doing

Shown here is the international Morse code that represents the 26 letters of the alphabet. Write a short message in Morse code. Trade messages with a classmate. Then try to read your classmate's message. How is sending messages in Morse code similar to a gene sending a message in the hereditary code? How is it different?

INTERNATIONAL MORSE CODE

A	·-	N	-·
B	-···	O	---
C	-·-·	P	·--·
D	-··	Q	--·-
E	·	R	·-·
F	··-·	S	···
G	--·	T	-
H	····	U	··-
I	··	V	···-
J	·---	W	·--
K	-·-	X	-··-
L	·-··	Y	-·--
M	--	Z	--··

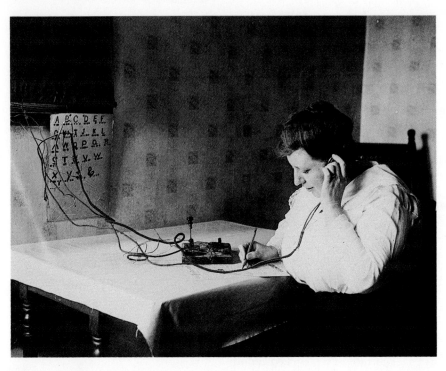

 ASK YOURSELF

How does the hereditary code work?

SECTION 3 *REVIEW AND APPLICATION*

Reading Critically

1. What are genes? What are chromosomes?
2. Why was Mendel's method of studying peas so important to the study of genetics?
3. Why is it necessary for gametes to have only half as many chromosomes as other cells?

Thinking Critically

4. If you crossed a pea plant that was hybrid tall with one that was short, what would the results be?
5. What do you think would happen if chromosomes did not separate during meiosis?
6. If DNA is the same in all organisms, what accounts for the great variety of living things?

INVESTIGATION

Investigating Genetic Probabilities

▶ **MATERIALS**
- pencil ● paper ● coins

▼ **PROCEDURE**

1. Make a table like Table 2.
2. Flip two coins at once to determine the gene pair for each trait. Heads means a dominant gene; tails means a recessive gene.
3. In your table record the gene types and appearance

of the trait. For example, suppose from the first coin flips you got one heads and one tails—one dominant gene and one recessive gene. Using the trait of fur color, you would write B (for brown) and b (for green) in

the gene type column. The appearance is brown fur.
4. Repeat steps 2 and 3 for each trait.
5. Draw the creature that resulted from this investigation.

TABLE 1: TRAITS		
Trait	Dominant	Recessive
Fur	Brown (B)	Green (b)
Tail	Straight (S)	Curly (s)
Horns	Absent (A)	Present (a)
Ears	Pointed (P)	Rounded (p)
Teeth	Double Row (D)	Single Row (d)
Leg Shape	Wide (W)	Narrow (w)
Leg Length	Long (L)	Short (l)
Tongue	Hairy (H)	Smooth (h)
Eye Color	Red (R)	White (r)

TABLE 2: TEST RESULTS		
Trait	Gene type	Appearance
Fur		
Tail		
Horns		
Ears		
Teeth		
Leg Shape		
Leg Length		
Tongue		
Eye Color		

▶ **ANALYSES AND CONCLUSIONS**

1. What combination of factors (gene type) produced a creature with the dominant form of each trait?
2. What gene type produced a creature with the recessive form of each trait?

▶ **APPLICATION**

Assume that you have two creatures with the gene types listed in your table. Make a drawing to show the possible offspring of a cross between the two creatures. Make a different drawing for each trait.

✳ ***Discover More***

For each trait, determine the possibility of an offspring inheriting recessive traits if each parent had opposite gene types of those listed in Table 2.

The Big Idea

After he observed variations among members of the same species, Charles Darwin inferred that these variations allowed some members of a species to have an advantage over other members of the species. These advantageous variations would then be "selected" by nature. Darwin further hypothesized that living things evolve, or change over time. He reasoned that evolution is the result of natural selection and that variations must be passed on to future generations.

Mendel's discoveries gave new meaning to Darwin's theory of natural selection. Mendel discovered that inheritance of variations, or traits, is controlled by factors that we now call genes. The inheritance of genetic traits can be predicted.

For Your Journal

Think again about how organisms change over time and about how these changes are passed from generation to generation. How have your ideas changed?

Connecting Ideas

After many years of study and thought, Darwin formed a theory about evolution by means of natural selection. The major points of Darwin's theory are shown in the illustrations. Use your own words to explain these points.

Understanding Vocabulary

For each set of terms, identify how the following terms are related to the first term.

1. evolution (150): natural selection (155), extinction
2. traits: recessive (162), dominant (162), pea plants
3. chromosome theory (165): genes (163), chromosomes
4. gamete (165): meiosis (165), number of chromosomes

Understanding Concepts

MULTIPLE CHOICE

5. Which of the following combinations would result in inheriting the recessive trait of red hair (b) over the dominant trait of brown hair (B)?
   a) BB **b)** Bb **c)** bB **d)** bb

6. Which of these statements does _not_ support the theory of natural selection?
   a) All members of a species are alike.
   b) Most species produce more offspring than will survive.
   c) A parent may pass on a trait that will give an offspring a better chance of surviving.
   d) Individuals of a species struggle to survive in their environment.

7. To have an offspring with the recessive trait of blue eyes, what must be true of the parents?
   a) Both parents must have brown eyes.
   b) One parent must have blue eyes.
   c) Both parents must have a recessive gene for blue eyes.
   d) One parent must have a recessive gene for blue eyes.

8. Which of the following statements is _not_ part of the chromosome theory?
   a) Genes are located on the chromosomes.
   b) Changes in a species result from natural selection.
   c) Traits are passed on from parent to offspring.
   d) Mendel's factors are the same as genes.

9. If one parent has two recessive genes and the other parent has one dominant and one recessive gene, what is the chance that an offspring will inherit the dominant trait?
   a) one in four **b)** two in four
   c) three in four **d)** four in four

SHORT ANSWER

10. Explain how natural selection depends on genetic principles.

11. Explain why gametes must be formed through the process of meiosis.

Interpreting Graphics

12. Look at the figure below. If the plant on the left is pure tall, could an offspring of these parents be short with wrinkled seeds? Explain.

13. Look at the figure below. If the tall, smooth plant is hybrid for height and seed type, how many chances out of four does an offspring have of being tall and wrinkled? Why?

Reviewing Themes

14. *Changes Over Time*
Using Darwin's theories of evolution and natural selection, identify what might be the relationship between the toxodon and the capybara.

15. *Nature of Science*
Using one of the traits of a pea plant studied by Mendel, explain how plants exhibiting dominant and recessive factors can be bred to develop plants that exhibit only recessive factors.

Thinking Critically

16. Some cockroaches have become resistant to pesticides. What might happen if this resistance to known pesticides continues and no new pesticides are developed?

17. The primary source of the giant panda's diet is bamboo. Giant pandas sometimes also eat fish and small rodents. What might happen to the giant panda if the bamboo is destroyed? How might the panda survive the change in its environment?

18. Mendel studied seven characteristics of pea plants. Quite accidentally, each of the seven was carried by a different chromosome. Imagine that Mendel had found that all tall plants had green peas and all short plants had yellow peas. What conclusions might Mendel have drawn about the relationship between plant height and pea color?

19. In some cattle, a cross between a red parent and a white parent will result in a calf that has both white and red hairs. What conclusions can you draw about the traits of white hair and red hair as a result of the offspring inheriting both traits?

20. Straight hair is a recessive trait in humans. If both parents have straight hair, could the offspring have curly hair? Explain your answer.

Discovery Through Reading

Hyndley, Kate. *The Voyage of the Beagle.* Wayland Limited, 1989. An account of the five years that English naturalist Charles Darwin spent traveling around the world.

CLASSIFICATION

*I*magine a world without organization. It would be a very confusing place. Fortunately, we classify and group just about everything. We group books in a library, shoes in a shoe store, tapes in a music store, and vegetables in a garden. And with over two million living things on Earth, scientists also have found it necessary to classify them into groups.

A dam was in the grocery store. He needed to buy milk, corn flakes, taco shells, lettuce, and oranges. He was in a hurry. His game was starting in 20 minutes. If he hurried, he could buy the groceries, take them home, and still get to the ball park on time.

When he entered the store, he became panic stricken. Nothing seemed to be organized. He frantically ran up and down the aisles looking for the items he wanted. He found milk near the bananas, and the lettuce was stacked near canned tomato soup. He looked everywhere, but he couldn't find the oranges. Just when he thought he would be late for sure, his alarm went off. Saved by the bell! Adam breathed a sigh of relief and mumbled, "It was only a dream." His heart, however, was beating rapidly as though he had actually been running up and down the aisles.

Adam's dream couldn't really happen. Grocery stores are well organized. Similar food items are grouped together. You normally would go to the dairy section for milk, the cereal section for corn flakes, and the produce section for lettuce and oranges. In fact, if you think about it, most stores organize their merchandise in some way. Think about your favorite store. How does it organize the merchandise it sells?

For Your Journal

- How does organizing merchandise help retailers? Shoppers?
- Why might scientists organize living organisms into groups?
- How would you organize living organisms? Why?

SECTION 1

The History of Classification

Objectives

State *the need for a common classification system.*

Summarize *how the classification system for living things has evolved over the years.*

Organize *objects into appropriate groups.*

In Adam's dream, none of the groceries he wanted were where he expected to find them. If Adam had actually been shopping in a supermarket, he would have found the corn flakes in the boxed cereal section. Milk would have been in a section with other dairy products. The lettuce and the oranges would have been found in the produce section. The aisles of the stores would have been labeled so that Adam could easily find the kinds of groceries that he was looking for.

In much the same way that grocers organize the items they sell in their stores, scientists have organized living things into groups. Grouping organisms helps to make studying them easier just as grouping groceries helps to make finding them easier.

Early Attempts to Classify

Figure 7–1. How do you think the supermarket would classify these foods?

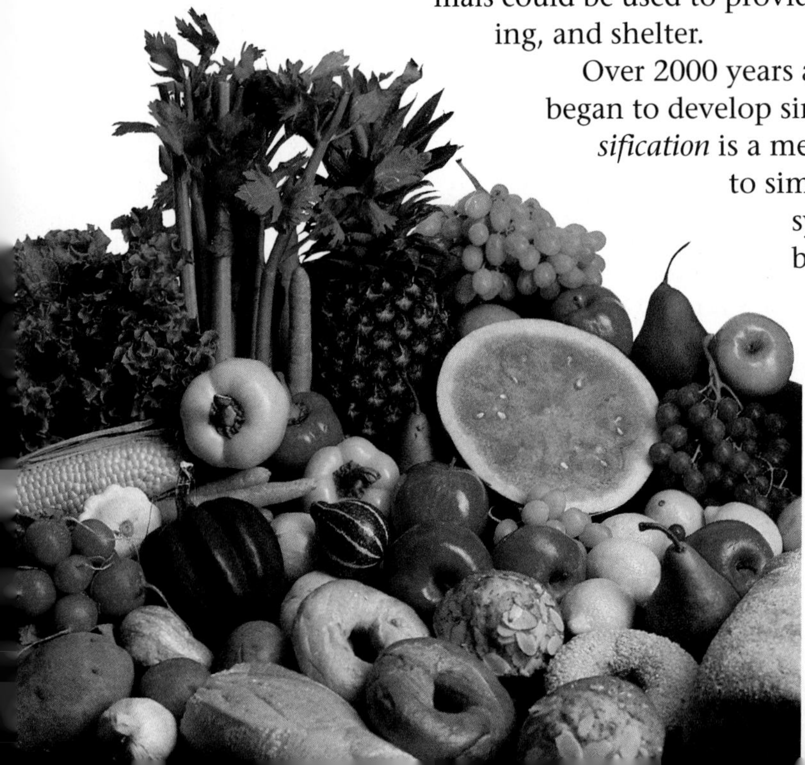

Evidence of early systems for organizing plants and animals into groups dates back about 7000 years. These systems were based on people's need to communicate about how plants and animals could be used to provide basic needs such as food, clothing, and shelter.

Over 2000 years ago, Greek and Chinese scholars began to develop simple systems of classification. *Classification* is a method of grouping things according to similarities and differences. The early systems of animal classification were based mainly on body coverings and body parts, such as legs, tails, and teeth. An ancient Chinese scholar noted that a system of classification must truly distinguish one animal from another.

Figure 7–2. Early Chinese scholars used body parts to classify animals such as the horse and the ox. What other characteristics could they have used?

The horse and the ox are different, but if someone says that a horse is not an ox because the ox has teeth and the horse a tail, that will not do. In fact, both have teeth and tails—one has to say that horses are not the same as oxen because the ox has horns and the horse does not. . . . Animals of four legs form a broader group than that of horses and oxen. Everything may be classified in broader and narrower groups. It is like counting fingers; each hand has five, but one can take one hand as one thing.

Quoted in Order: In Life *by Edmund Samuel*

Around 350 B.C., Aristotle, a Greek philosopher, was also developing a system for classifying animals. He identified over 500 animals and placed the different animals into 10 groups based on their body parts, life histories, activities, and character. He then organized the groups into two categories. Six of the groups he called "blooded," and four he called "bloodless." Into the blooded group he put humans, birds, and fish. He considered insects and mollusks bloodless. Why do you think he placed these animals in the groups he did?

Together with his student Theophrastus, Aristotle also classified plants. They divided plants into groups based on their size and appearance. Their classification systems continued to be used with some changes for about 2000 years.

Figure 7–3. Aristotle's and Theophrastus's classification systems are simple by today's standards but quite imaginative for the time.

DISCOVER BY *Observing*

Carefully examine two different plants. If possible, use a hand lens to observe their leaves, stems, and roots. You may want to examine two very different types of plants, for example, a house plant and a small tree. Make a chart identifying the similarities and differences in the structures of the plants. Based on your comparison, tell whether you would place the two plants in the same category. ✐

▼ ASK YOURSELF

What was the basis of most early classification?

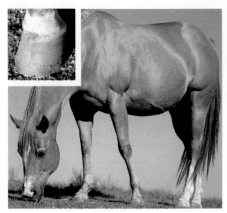

Figure 7–4. In Ray's hierarchy, each of these animals would be placed in separate subcategories of hoofed animals.

Improving the System

As more and more differences in organisms were recognized, scientists began to realize that they needed a better classification system. Eventually, some scientists began devoting full time to classifying living things. Soon classifying became a science. The science of classification has now become known as **taxonomy.**

A Place for Everything
During the late 1600s, English biologist John Ray set up a system that attempted to place living things in a hierarchy. In a hierarchy, items are placed in categories and subcategories based on their characteristics.

Think about the supermarket. Grocers identify different categories of food, such as meat, vegetables, bread, and so on. They further identify the different forms and types within each category. For example, bread products include rolls and buns as well as loaves of bread. Different kinds of bread—wheat, white, rye, raisin, and multiple grain—are also identified.

John Ray's hierarchy of organisms worked in the same way. He placed organisms into a hierarchy by dividing and subdividing categories. He examined organisms in detail. He also recognized that differences could exist among organisms of the same kind. For example, all house cats share common characteristics but vary in color, size, and shape. What other examples can you name of differences among animals of the same kind?

Ray's hierarchy put mammals into two groups. One group had toes and the other had hoofs. He then divided the hoofed mammals into those with undivided hoofs, two-toed hoofs, and three-toed hoofs. He further divided the animals with two-toed hoofs into three more groups. One group was cud chewers that had permanent horns. Another group consisted of animals that chewed their cud and shed their horns each year. And still another consisted of animals that did not chew their cud. Ray used structure and behavior to describe organisms.

What's in a Name?   In the late 1700s, a young Swedish scientist named Carolus Linnaeus (lih NAY us) picked up on the work done by John Ray. While in medical school, Linnaeus had a job taking care of a small botanical garden. He observed and collected plants and insects, writing detailed descriptions of each one. Later he was given a grand total of $50.00 to go to a region north of the Arctic Circle to collect and study plants. Through activities like these, Linnaeus began to develop his classification system for organisms. His system is the basis for today's system of classification. As a result, he is known as the father of taxonomy. You can read more about Linnaeus's life on page 201 at the end of this unit.

Figure 7–5. Carolus Linnaeus developed the classification system that is the basis of the modern system.

Linnaeus's classification system used just twelve Latin words to describe each kind of organism. The last two Latin words in each description were the specific name of an organism. Using two words to name an organism is called **binomial nomenclature** (by NOH mee uhl NOH muhn klay chuhr).

Why did Linnaeus use Latin instead of the Swedish language to classify organisms? At the time, Latin was the international language. By using Latin, he knew that the names he gave each living thing could be understood by scientists throughout the world, not just those who could read Swedish.

Look at the following table. It gives the word for *dog* in several different languages. You can imagine the confusion that would result if all animals were named without a single common language. Yet because of Linnaeus's nomenclature, scientists all over the world can recognize the word for *dog* by its Latin name *Canis familiaris*.

Table 7-1	**The Word *Dog* in Various Languages**				
English	*dog*	French	*chien*	Russian	*sabaka*
Spanish	*perro*	Japanese	*inu*	Polish	*pies*
Hebrew	*kelev*	German	*hund*	Italian	*cane*

Some languages have several different names for the same animal. For example, in English the mountain lion is also called a panther, a cougar, or a puma. But all scientists know it by its Latin name *Felis concolor*. Even today, some 300 years later, scientists still use Linnaeus's methods to name newly discovered organisms.

Figure 7–6. What is the common name you use for *Felis concolor*?

DISCOVER BY *Researching*

Make a list naming ten different animals. Find a picture of each animal. Then try to describe each animal using just twelve words as Linnaeus did. Don't use sentences. Instead use brief descriptions for each characteristic. Once you have described an animal, locate its Latin name in an encyclopedia or a science reference book. Share your descriptions with your classmates. Did any of your classmates describe the same animals? How are your descriptions alike? How are they different? ✎

Linnaeus grouped plants and animals according to likenesses and differences in their structures. He did his work about 150 years before the theory of evolution by natural selection was known. Today, scientists classify organisms according to how closely related they are to one another in their evolutionary development. The further two organisms are from a common evolutionary ancestor, the more unlike they will be.

► ASK YOURSELF

How was the work of John Ray similar to that of Carolus Linnaeus?

A Classification Key

By now you are probably saying to yourself, "So what if scientists have described and classified different kinds of organisms? How could I determine what something is if I have never seen it before?" That's a good question. Not even the best scientist can know even a small fraction of all the living things. In fact, any one scientist is lucky if he or she can know and identify a thousand of the over two million organisms. However, scientists have a way to help them identify living things. They have developed keys that can be used for identifying living things.

Unlocking Identity   The most common key used to identify organisms is made of paired descriptions. To identify something, you have to decide which description fits the thing you are trying to identify. That description will either give the name of the organism or send you to another number where there is another set of paired descriptions. When you go to those paired descriptions, you will select the one that best describes the thing you are trying to identify. That description will either give you the name or send you to yet another pair of descriptions. You continue in the same way until the

Figure 7–7. Scientists could use a classification key to identify this swamp buttercup as *Ranunculus septentrionalis.*

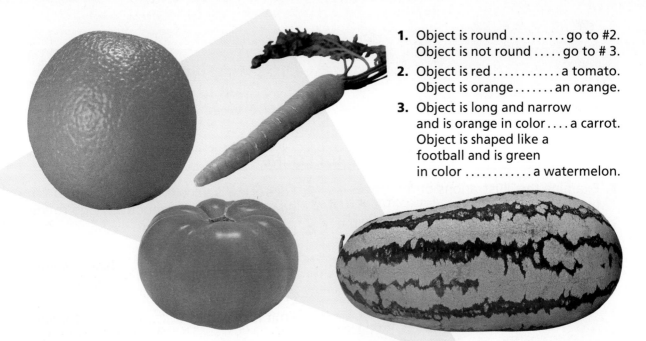

1. Object is round . . . . go to #2.
   Object is not round . . . . . go to # 3.
2. Object is red . . . . a tomato.
   Object is orange . . . . an orange.
3. Object is long and narrow
   and is orange in color . . . . a carrot.
   Object is shaped like a
   football and is green
   in color . . . . a watermelon.

description you choose from a pair is followed by the name of the organism.

Look at the illustration of the four objects. They can be identified using the simple key. To learn how the key works, pick out one of the objects. Then start at the top with the first paired descriptions and go to the next step until you have identified the object. Use the key to identify all the objects.

Scientists have written hundreds of different classification keys. There are keys for birds of the western states, mammals of the Rocky Mountain region, insects of North America, trees of the eastern United States, and so on. Scientists who specialize in certain organisms such as insects would use the key that describes the organisms they are studying.

 ASK YOURSELF

How does a classification key help you to narrow your choices and to identify an organism?

SECTION 1   *REVIEW AND APPLICATION*

Reading Critically
1. Why is Latin used to name organisms today?
2. What does the story of classification tell you about the system of classification?

Thinking Critically
3. Choose four items in your classroom. Create a system for classifying these items. What other items would you place in the same classifications?
4. Cite two examples of classification systems you use in everyday life, and describe the hierarchy within those systems.

SKILL Using a Classification Key

▶ **MATERIALS**
none

▼ **PROCEDURE**

1. Examine the illustrations of the leaves from different trees.
2. Use the classification key to identify the tree that produces each of the leaves shown in the illustration.

3. Edge of leaf is smooth.................go to #4.
   Edge of leaf is not smooth.......go to #5.
4. Leaf is made of several leaflets. Fruit is a round nut.......................black walnut, *Juglans nigra*.
   Leaf is heart-shaped. Fruit is a long pod.......northern catalpa, *Catalpa speciosa*.
5. Edge of leaf has deep, rounded indentations and sharp points at the end. An acorn is the fruit................................pin oak, *Quercus palustris*.
   Edge of leaf looks like the edge of a saw blade. Its seed is surrounded by a thin wing.....................American elm, *Ulmus americana*.
6. Needles arranged around the branch. Cones drooping, with scales that are open........................Douglas fir, *Pseudotsuga taxifolia*.
   Needles appearing mostly on one side of the branch. Cones upright, with scales that are tightly closed.............white fir, *Abies concolor*.
7. Needles grouped in threes..............................loblolly pine, *Pinus taeda*
   Needles grouped in fives....................western white pine, *Pinus monticola*.

CLASSIFICATION KEY

1. Leaf is needlelike.................. go to #2.
   Leaf is broad and flat............ go to #3.
2. Needlelike leaves are not in bundles.............................. go to #6.
   Needlelike leaves are grouped in bundles of three or five ...... go to #7.

▶ **APPLICATION**
Obtain a copy of a classification key for trees in your area. Use the key to identify trees around your school and home.

✳ ***Using What You Have Learned***
Were your descriptions helpful to you? Were you able to identify the leaves by using the key? If not, make changes to improve the key.

Modern Classification

Objectives

Define the word species.

Differentiate among the five kingdoms of living things.

Evaluate the worth of a system of classification.

Supermarkets receive many new products each week. Before displaying the new items, the grocers examine the products. They determine what each product is. Is it a new kind of pasta? They examine the packaging. Is the product in a box or a bag? They think about where the new product will fit on the shelves best. They most likely place the new product near products that are similar.

Changes Since Linnaeus

Grocers sort products based on what the products are and how similar the products are to other grocery items. Scientists classify organisms in much the same way. Classification, as we know it today, is similar to what Linnaeus was doing over 250 years ago. But there have been significant changes since Linnaeus's time.

So Many First of all, there are far more organisms known today than were known during Linnaeus's time. While they don't know how many different organisms there are, scientists estimate that there are over 2 000 000 kinds of organisms alive on Earth today. New discoveries add to this number.

Figure 7–8. These bacteria have been magnified by an electron microscope.

Look at This The second change is that there are better instruments, tools, and methods to use for identifying organisms. This is particularly true when you consider the microscope. Improvements in microscopes have allowed us to see far more detail. This means that characteristics otherwise unseen can be used to describe living things. With the improved tools and instruments, scientists are able to examine organisms in great detail.

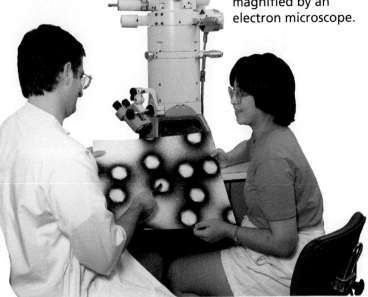

It's Relative   Scientific knowledge has grown tremendously since Linnaeus's time. Today, scientists have tools, instruments, and methods that enable them to examine the chemical makeup of organisms. They know that the more alike the chemical makeup of two organisms is, the more closely the two organisms are related.

The third significant change is that much of what scientists know now is based on evolution. This means that to identify organisms, scientists look at factors other than obvious physical and structural characteristics.

All of these factors have influenced our modern system of classification. What is more important to realize is that continued developments in science will likely lead to further changes in our modern classification system.

 ASK YOURSELF

What advances have contributed to the improvement of the classification system?

What Is a Species?

Grocers don't simply use the term *soup* to classify all soups. Instead, they label and arrange soup by variety. There are canned and dry soups. Tomato, noodle, vegetable, bean, split pea, and dozens of other soups are arranged on the shelves. Just as grocers don't call all soup *soup,* neither does a scientist classify all feathered creatures as just birds.

Figure 7–9. These animals are all birds, but they must be classified further to note differences among them.

Figure 7–10. All dogs are members of the same species even though they may look different.

As with varieties of soup, there is a term to use when speaking of different organisms. The term is *species*. But it is not enough to define species as different kinds of organisms. In fact, the concept of species is pretty complicated. In the study of life science, it is important to know what a species is. You might ask what makes one species different from another.

The obvious answer is that they do not look alike. But while that is often the case, it is not always true. As a matter of fact, some members of the same species look less alike than members of a different species look. For example, all dogs are members of the same species. But you know that all dogs do not look alike. Certainly, the poodle in the picture doesn't look like the dachshund. But, nonetheless, they are the same species. There are even major differences in the appearance of the male and female organisms within some species. For example, look at the male and female pheasants in the picture.

Figure 7–11. Males and females of the same species can look quite different.

Figure 7–12. The Asian and African elephants look alike, but they are members of different species.

Look at the two elephants. They look alike, yet they are members of different species. The one on the right is an African elephant, and the one on the left is an Asian elephant.

So then, what makes an organism a certain species? This remains a difficult question to answer. Generally, a **species** is a group of organisms that naturally mate with one another and produce fertile offspring. By "fertile," we mean that the offspring are able to reproduce. Take dogs, for example. While there are many types of dogs, they all can mate and produce fertile young. This is because they are members of the same species. Yet, the African and Asian elephants are not members of the same species and, therefore, do not mate with one another.

Figure 7–13. The lion and tiger do not mate in their natural environment. However, they can mate under artificial conditions to produce infertile offspring such as this liger.

Under captivity, members of different species are sometimes able to crossbreed and produce offspring. For example, a male lion and a female tiger can crossbreed and produce offspring that carry the characteristics of both species. A male donkey and a female horse can also mate under artificial conditions to produce a mule. But usually, the offspring of this crossbreeding are not fertile. Different species of plants have also been artificially crossbred. Unlike animals, however, plants resulting from such crossbreeding can be fertile.

Figure 7–14. Do you think this mule looks more like a donkey or a horse?

 ASK YOURSELF

Is it accurate to say that all members of a given species look exactly alike? Explain why or why not.

Beyond Species

Living things are not classified into species alone. In the same way grocers divide and subdivide products they sell, scientists have built a hierarchy for grouping living things. Take, for example, the produce section. In one part you have fruits, and in another part you have vegetables. The fruits could be divided into different types, such as apples, oranges, bananas, grapes, and so on. Then the apples could be further divided by types, such as Red Delicious, Golden Delicious, McIntosh, and Granny Smith.

Figure 7–15. Even apples are subdivided according to type.

DISCOVER BY *Writing*

Visit two different types of stores and analyze how they classify their products. Do they arrange products by style or type? Do they put all brand-name products together? Do they place products that would be used together near one another? Use your journal to record your findings. Construct a diagram to represent each system of classification. ✏

Figure 7–16. The upside-down pyramid shows the seven categories of classification. As you move up through the categories, you find more and more types of organisms as members of the categories.

The classification system for living things is made up of seven separate categories. These categories range from broad and general categories to narrow and specific categories. The broader the category, the more species you will find in it. Species, of course, is the narrowest category. The largest category is a **kingdom.** You might think of the categories as an upside-down pyramid with species at the bottom. The drawing of the upside-down pyramid lists the seven categories of living things. The list next to it traces the human being from kingdom through species.

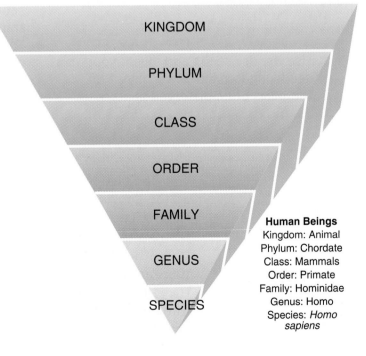

KINGDOM

PHYLUM

CLASS

ORDER

FAMILY

GENUS

SPECIES

Human Beings
Kingdom: Animal
Phylum: Chordate
Class: Mammals
Order: Primate
Family: Hominidae
Genus: Homo
Species: *Homo sapiens*

ACTIVITY

How many different ways can you classify seeds?

MATERIALS

eight types of seeds (half of the seeds should be soaked in water), hand lens, blank paper, glue

PROCEDURE

1. Begin the activity with eight seeds, one from each of the containers in the classroom. Carefully observe each seed. Look for similarities in structure, shape, color, texture, and size. Take the seeds that have been soaked in water and remove their coats. Use a hand lens to observe seed coats and internal structure.

2. Use the dry seeds and divide your seed kingdom into two "phyla" on the basis of an observed charac-

teristic. You do not need the same number of seeds in each phylum.

3. Design a chart on which to record your classification. Give your chart and columns or rows a title.

4. Glue the seeds next to their descriptions.

5. Obtain another set of the same eight seeds. Look for more specific characteristics that will help you further divide these two seed "phyla" into smaller categories. Write the description of the subdivision in your chart, and glue the seeds under the appropriate description. Obtain more seeds as needed. Continue writing new descriptions and gluing seeds until each seed is in its own group. Give a name or characteristic to each unique category.

APPLICATION

1. What characteristics did you use to separate the eight seeds into smaller and smaller categories?

2. Look at the completed seed charts of your classmates. Discuss the grouping characteristics that you think are the best.

The broadest category of classification is the kingdom. Of all the categories, it has by far the most organisms. For a long time, scientists placed all living things in just two kingdoms—plants and animals. But with the development of better and better instruments, tools, and methods, they have learned that there are greater differences between certain organisms than they first thought there were. Today, we classify all living things into five kingdoms. The chart shows the five kingdoms and examples of organisms found in each.

MONERANS	
PROTISTS	
FUNGI	
PLANTS	
ANIMALS	

Figure 7–17. Which of the five kingdoms do you think has the most species?

Monerans   The monerans became a separate kingdom when scientists, using powerful microscopes, discovered that some cells do not have a distinct nucleus. Bacteria, which often are incorrectly called "germs," are the most common members of this kingdom. They are one-celled organisms that live as separate cells or as many cells joined to form a colony. Scientists estimate that there are over 10 000 species of bacteria. Even so, the monerans make up the smallest kingdom.

Protists   Like the moneran kingdom, the protist kingdom is made up of simple one-celled organisms. But unlike monerans, the protists have distinct nuclei in their cells. Thanks to the microscope and other tools and methods, scientists have been able to do a better job of classifying these tiny organisms. The picture shows three common protists.

Figure 7–18. Bacteria are classified into three shapes—rod-shaped, ball-shaped, and spiral-shaped. What shape are these bacteria?

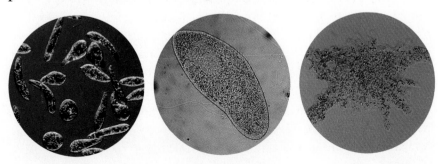

Figure 7–19. The *Euglena,* the paramecium, and the amoeba are common protists.

Figure 7–20.
Mushrooms, which are classified as fungi, help decompose dead plants.

Fungi   For a long time, scientists classified fungi with plants, calling them nongreen plants. In many ways, fungi seem like plants. For example, they do not move around but seem to be rooted. But unlike plants, fungi do not have chlorophyll and cannot make their own food. Instead, they absorb mostly dead, decaying matter from their surroundings. Mushrooms, molds, yeast, and mildew are fungi. Scientists have identified about 100 000 species of fungi.

DISCOVER BY Researching

With a partner, use encyclopedias and science reference books to identify the seven levels of classification for each of the following organisms: bacillus, moss, bread mold, rose, dog. Write the common name of each organism. On an inverted pyramid, write the kingdom, phylum, class, order, family, genus, and species for each organism.

Plants   With over 350 000 known species, the plant kingdom is the second largest kingdom. These species range from tiny green mosses to giant trees. Plants are made of many cells. Their cells contain chlorophyll, which they use to make their own food. Like fungi, they cannot move around from place to place.

Figure 7–21. Plants are both large and small.

Animals The animal kingdom is the largest kingdom with well over 1 000 000 known species. Animals have many cells. They are different from plants in that they cannot make their own food and most can move around from place to place. The animal kingdom is divided into two major groups—those with a backbone and those without. The pictures show examples of animals from these groups.

Figure 7–22. The bat and the polar bear are examples of animals with a backbone.

 ASK YOURSELF

List the five kingdoms. For each, tell one way in which it is different from the others.

Figure 7–23. The starfish, the mosquito, and the oyster are animals without a backbone.

SECTION 2 *REVIEW AND APPLICATION*

Reading Critically

1. What three factors contributed to changes in the classification system?

2. Explain the difference between species and kingdoms.

3. What contribution did Carolus Linnaeus make that continues to influence the science of taxonomy today?

Thinking Critically

4. Think about the living things you have seen since waking up this morning. From which kingdom did you see the greatest number of different species? Prepare a list to support your answer.

5. Explain why dogs that look so different can belong to the same species whereas doves and pigeons, which look so much alike, are members of different species.

6. Write your prediction of how the classification system might change in the next 500 years. Give reasons to support your prediction.

INVESTIGATION

Observing, Ordering, and Displaying a Collection of Organisms

▶ MATERIALS

- microscope ● slide ● coverslip ● pond water ● drawing paper
- pencil ● bread mold ● forceps ● plant parts ● worms and insects

▼ PROCEDURE

1. List characteristics of organisms in each of the five kingdoms. After you examine each of the specimens in this investigation, try to place it in the appropriate kingdom.

2. Use the microscope to view the pond water. Are there any organisms in the water? If so, draw the organisms. Label the drawing with the kingdom you think the organisms belong to.

3. Examine some bread mold under a microscope. What kingdom do you think it belongs to? Why?

4. Examine small parts of a house plant, such as leaves or stems. Sketch the parts of the plant you observe. In what kingdom does this organism belong?

5. Look for worms, ants, or other small organisms in areas near your home or school. Sketch these organisms and label your sketches with the kingdom to which the organisms belong.

6. Make a display of your drawings.

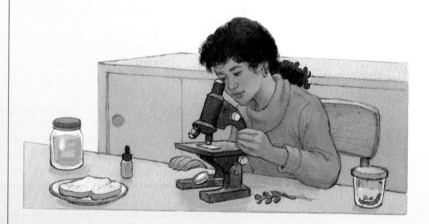

▶ ANALYSES AND CONCLUSIONS

1. Did you observe organisms from each kingdom? Why or why not?
2. Why is it necessary to display drawings rather than the actual organisms?
3. For which kingdom were organisms easiest to find? Why?

▶ APPLICATION

Suppose you have an organism that has chlorophyll and moves. How would you classify it? Why?

✳ Discover More

As you walk or ride to school, look for different types of organisms. Keep a list of your observations. What kingdom do most of the organisms you see belong to? Why do you think you see organisms from this kingdom more than others?

The Big Idea

Humans have been classifying living things for years. Because of the wide range of organisms, scientists classify organisms so that study of them can be simplified. As our knowledge of living things improved, our systems for classifying living things evolved into better systems.

New tools and instruments and new theories about living things not only changed the way we classify things, but also gave us many more organisms to classify. The classification system as we know it today is the result of the work of many scientists and scholars. The continued work of scientists today and in the future will likely bring even greater improvements in the system.

Think about what you have learned about classification in this chapter. Review what you have written in your journal, and revise or add to the entries to reflect your new understandings about scientific classification. Don't forget to include information about how the classification system has changed and what factors have influenced the changes.

Connecting Ideas

Copy the unfinished concept map into your journal. Complete the map by filling in the blanks.

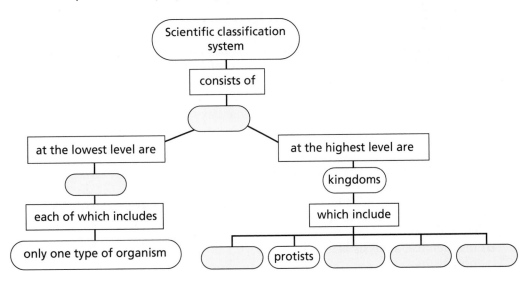

REVIEW

Understanding Vocabulary

Write a sentence that demonstrates your understanding of each of the terms below.

1. taxonomy (176) **2.** species (184) **3.** kingdom (185)
4. monerans **5.** protists **6.** fungi
7. binomial nomenclature (177)

Understanding Concepts

MULTIPLE CHOICE

8. Which of the following statements is *not* true of bacteria?
   a) Bacteria are monerans.
   b) Bacteria are single-celled organisms.
   c) Bacteria may live in colonies.
   d) Bacteria have a distinct nucleus.

9. One reason scientists no longer classify fungi as plants is because
   a) fungi do not have nuclei in their cells.
   b) fungi do not move around.
   c) fungi cannot produce their own food.
   d) fungi cannot grow.

10. Which of the following statements is true?
   a) Some members of different species are similar looking.
   b) Different species can never crossbreed.
   c) All members of the same species look alike.
   d) More than one kind of organism can be in the same species.

11. All of the following developed after Linnaeus classified organisms *except*
   a) binomial nomenclature.
   b) the theory of evolution.
   c) improved tools such as the microscope.
   d) biochemical analysis of organisms.

12. Organisms that are not in the same family *cannot* be in the same
   a) kingdom. **b)** class.
   c) order. **d)** genus.

SHORT ANSWER

13. Explain why poodles can mate with collies and produce offspring, but collies and poodles cannot mate with foxes to produce offspring.

14. The system for classifying living things will probably change in the future. Explain why.

Interpreting Graphics

15. What does the diagram below tell you about the things being classified? Give a detailed explanation.

16. Use the diagram below and add another level giving more specific descriptions.

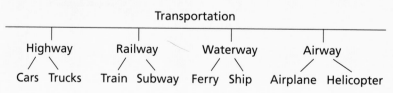

Transportation

Highway   Railway   Waterway   Airway

Cars Trucks   Train Subway   Ferry Ship   Airplane Helicopter

Reviewing Themes

17. *Systems and Structures*
Systems of classification have changed and evolved over the years. Why is it important for science to be open to the change and evolution of ideas?

18. *Changes Over Time*
The observable structure of living things has always played an important role in classification. Cite two reasons why scale and structure are not enough when classifying organisms.

Thinking Critically

19. Select a first name you hear often at school. Prove the need for a two-name system by making a list of all the people you know who have the same first name.

20. There are over 6000 kinds of ants. How does this fact help show the need for a classification system?

21. Devise a paired description key to identify the following organisms: pine tree, daisy, dog, elephant.

22. *Homo sapiens* is used to identify modern humans, and *Homo erectus* is used to identify an earlier form of humans. The English translation of these Latin terms is: *Homo*—"human," *sapiens*—"wise," and *erectus*—"upright." Give reasons why these two different names are used to identify different species of humans.

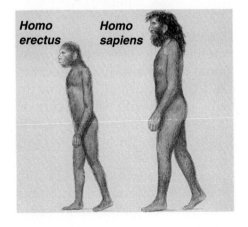

Homo erectus   Homo sapiens

23. Both monerans and protists are one-celled organisms. Why do you think scientists have placed them in separate kingdoms?

24. You learned about the theory of evolution in Chapter 6. During Linnaeus's time the theory of evolution was not an accepted theory. How might his system for classifying living things have been different had he known about and accepted the theory of evolution?

Discovery Through Reading

Stidworthy, John. *Mammals: The Large Plant-Eaters.* Facts on File, 1989. Articles focusing on individual species of mammals, their structure and size, and background information.

Science PARADE

Finding the Past

For nearly 17 centuries, the Roman city of Pompeii lay buried under 7 m of volcanic ash from an eruption of Mt. Vesuvius. The ruins of Pompeii, in southern Italy, were first discovered in the early 1700s by workers building an underground water pipeline.

Accidental Discovery

Pompeii, like many archaeological discoveries, was found by accident. Some important discoveries that have provided a great deal of information about the past have been made by people who were hiking or camping. Some finds have been made by workers building roads or digging foundations for houses or by farmers plowing fields.

Such accidental discoveries are not the only source of information for archaeologists and paleontologists. These scientists actively look for places where people

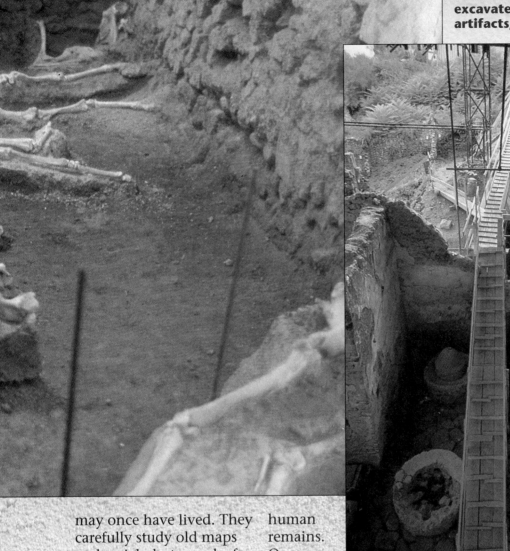

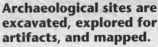
may once have lived. They carefully study old maps and aerial photographs for clues to the locations of likely places of habitation, abandoned villages, or burial sites.

Careful Work

When scientists find an area that looks as if it might have been inhabited, they explore it to see if they can find any evidence of human occupancy. They look for such things as building foundations, walls, pottery, and human remains. Once an actual site has been discovered, a report is made before any digging begins. Then a few test pits may be dug to find out what soil layers are present and how large an area the site covers. Wooden stakes and lengths of twine are used to divide the site into sections measuring one or two square meters. The site then looks a bit like a large piece of graph paper.

The actual digging process is slow. A hand trowel is used to remove the soil a little at a time so that any item that may be buried is not disturbed. Loose soil around delicate items is swept away by using a small brush. A screen might be used to sift through soil to find tiny pieces of bone or fragments of objects. Photographs of each

5-10 MILLION YEARS AGO
Aegyptopithecus

3-4 MILLION YEARS AGO
Australopithecus afarensis

2-3 MILLION YEARS AGO
Australopithecus africanus

item found are taken, and the position of each object in every layer of a square is carefully noted on a grid.

New Insights

These excavation and research skills have provided new insights into the origin of modern humans. Archaeologists and paleontologists have painstakingly applied these procedures at various locations around the world. These scientists have used their findings to trace the evolution of humans. Discoveries of bones, tools, and other artifacts have aided scientists in establishing a timeline of human evolution.

Humanlike fossils have been found that date back millions of years. A 2.5-million-year-old skull called *Proconsul*, found by Mary Leakey in Kenya, Africa, is believed to be the oldest early ancestor of humans. The remains of early humans, called *hominids*, were first discovered by scien-tist Raymond Dart at a limestone quarry in South Africa in 1924. His findings have been supported by the additional work of Mary and Louis Leakey and their son Richard.

In 1974 in Ethiopia, Africa, Donald Johanson found part of a female skeleton that he named "Lucy" or *Australopithecus afarensis* (aws trah loh PITH uh kuhs AF uh rehn sihs). Lucy walked on two legs and stood slightly over 1 m tall. She is the oldest evidence of a hominid found so far.

Pieces of pottery provide clues to how a people lived.

| 1.9 MILLION YEARS AGO | 35,000–150,000 YEARS AGO | PRESENT–100,000 YEARS AGO |
| Homo erectus | Homo sapiens (Neanderthal) | Homo sapiens (modern humans) |

 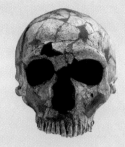 

Closest Ancestors

The closest ancestors to modern humans are the hominids named *Homo erectus*. *Homo erectus* means "upright human." Richard Leakey and his co-workers made the largest *Homo erectus* discovery in 1984. This was a skull and nearly complete skeleton of a 12-year-old male. The fossil remains of *Homo erectus* show that they had an ability to stand fully upright and also had a larger brain than other previous hominids had had. Remains of shelters, tools, and cooking fires have been found at the sites where *Homo erectus* was discovered.

The evolutionary history of humans is still under study. In 1987, a group of molecular biologists proposed that mitochondrial DNA could be used to trace the development of humans. Mitochondrial DNA is genetic material that is located outside the nucleus of a cell. This DNA is passed on only by females. It was theorized that by using a computer to trace backward, scientists could determine the original "mother" of all humans. In 1991, computers were used to analyze DNA similarities among people from around the world and to produce one hundred possible ways to find one common human ancestor. Scientists, however, are in disagreement over this research. Some think that DNA inside the cell's nucleus would provide different results. Still others contend that any number of possible outcomes could result depending on the DNA mix.

So the search for our past goes on. There are still gaps in time and many questions that remain to be answered. But thanks to the discoveries of archaeologists and paleontologists, we now have a much better picture of our distant past. ◆

The early hominid skeleton called "Lucy."

SHARK LADY

by Bet Hennefrund
from *Ranger Rick*

Down . . . down . . . down, into the cold, dark ocean depths, a huge whale shark glides like a giant toboggan. On the shark's back rides a small, dark-haired woman. She's Eugenie Clark, "the shark lady."

Eugenie holds tight to the shark's big back fin. Down the two go together . . . 100 . . . 130 . . . 150 feet. The water becomes dark as

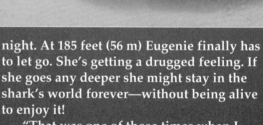

night. At 185 feet (56 m) Eugenie finally has to let go. She's getting a drugged feeling. If she goes any deeper she might stay in the shark's world forever—without being alive to enjoy it!

"That was one of those times when I wished I were a fish," Eugenie said later. "Then I could go wherever they go."

Eugenie has been hooked on fish since she was a child living in New York City. She spent hours with her nose pressed against the tanks in the New York City Aquarium. She wanted to be one of the fish even way back then.

When she was in college, Eugenie put on her first diving mask and flippers. Then she dove into the sea, where the fish live free. At the same time, she was studying fish. She learned enough about them to become *Doctor* Eugenie Clark, an expert fish biologist. Today she is a professor at the University of Mary-

Eugenie will never forget her first time under water with great white sharks. In her metal diving cage she didn't have to worry about being eaten — but the hungry sharks really tried their best to eat the cage!

land. She's written books and articles and has been on television.

How to Meet a Shark

To Eugenie, there's nothing more wonderful than swimming underwater for hours, nose-to-nose with angelfish and eels, octopuses and . . . sharks!

Eugenie says sharks aren't "mean," as many people believe. They attack only to eat or because they feel threatened. And if they sometimes mistake a person for a sea lion or an enemy, that's not the sharks' fault. Eugenie is very careful in the water and she has no fear of anything that lives there. And she doesn't see why *anyone* who is diving carefully should be afraid either.

"You just have to realize that you are a visitor in the fish's backyard," she says.

Her advice to anyone who happens to meet a shark: Be polite. Back slowly away

And in general: Stay clear of sharks' feeding grounds, especially if fish blood is in the water. Don't wear shiny things that might look like flashing fish movements to a shark. And don't thrash around if you're swimming where sharks may be.

Eugenie met her first big shark in the southern Pacific Ocean. She was swimming near a coral reef when a shark came straight toward her. It was so close she could have reached out and touched it. But she didn't. She just stayed perfectly still and stared. She couldn't understand why she wasn't afraid. As the shark gracefully turned and swam off, all Eugenie could think was, *How lucky I am to see that sleek, beautiful creature so close up!*

After that, Eugenie met lots of sharks face to face in many oceans. Once she spent ten days observing great white sharks from a metal diving cage. Her first moments underwater with the giant fish were unforgettable. One shark came straight at the cage with its mouth open so wide that Eugenie could see right down its throat. As the shark's jaws crunched down on the cage, its nose poked through the space between the bars. Eugenie pressed her body against the opposite side of the cage. But at the same time, another shark swam by and brushed against her back!

Even when there were five sharks around the cage at one time, Eugenie was too excited to be afraid. She just aimed her camera right into the sharks' faces or down their throats. She wasn't at all worried about those big, gaping jaws and sharp teeth.

Eugenie *has* been bitten though: One time she reached inside the body of a living, pregnant tiger shark she was examining in the lab. When she pulled out her hand, a baby shark was hanging onto her fingers with its tiny teeth!

Here a Shark, There a Shark

Eugenie got to know sharks best when she was the director of a marine lab in Florida. She and the other scientists caught and studied many different kinds of sharks. There were hammerheads, sometimes a white or sandtiger shark, bull and sand-

In an underwater cave, one of Eugenia's students studies a "sleeping" shark. Because they stay very still, the sharks look as if they're asleep. But they're really not sleeping at all.

bar sharks, nurse sharks, tiger sharks, dusky sharks, and lemon sharks.

Eugenie kept the live sharks in big swimming pens so that she could study how they behave. She watched them eat, swim, rest, and communicate with each other. Then she wrote about what she learned. Soon scientists came from all over to learn from her.

Sharks with "Smarts"

Sharks, everybody once thought, were not only mean, but stupid too. That was before Eugenie proved for the first time what sharks can learn. At her marine laboratory, two lemon sharks and a nurse shark were taught to press a target to get food. They even learned to tell white targets from red, and striped ones from plain-colored ones.

Eugenie discovered that sharks can re-member things too. After *not* working with the targets for two months, the sharks went right back to pressing targets for their food again—just as if they had been doing it every day.

The Mystery of "Sleeping" Sharks

Several times Eugenie has watched sharks "sleeping." She has found them in underwater caves off the coasts of Mexico and Japan, and in the Red Sea. The sharks aren't really sleeping, but they look as if they are. They

ing, closing, opening, closing. And their eyes follow every move divers make.

Sharks are supposed to *have* to keep moving so that water flows over their gills. That's how they get the oxygen they need to breathe. But these sharks proved they could breathe even when still.

To find these "sleeping" sharks seemed very strange to Eugenie, though. Why were they there? Did they come to the caves to be cleaned by *remoras*? Remoras are fish that often eat *parasites*, pesky creatures that cling to other fish and take food from them.

Eugenie studied the animals and tested the water. She found that the cave water was less salty than water in the open ocean. And she thinks it's likely that the water may act like a kind of drug on the sharks and their parasites. In less salty water, parasites loosen their grip. And that would make the cleaning easier.

The Shark's Best Friend

Eugenie believes that the more we know about sharks, the less we'll fear them. Her children grew up unafraid of sharks. They have even helped their mother in her work.

By writing her books and articles, teaching, making films, and giving talks, Eugenie spreads the word about sharks. She wants everyone to know what she knows: Sharks are magnificent animals that deserve respect,

Carolus Linnaeus (1707—1778)

In his book *Systemae Naturae,* Carl von Linné suggested a standardized system for naming plants and animals. Linné introduced the idea of "binomial nomenclature." He assigned to each organism a two-part name in Latin. In keeping with his system, he even changed his own name to Carolus Linnaeus.

Carolus Linnaeus was born in Roshult, Sweden, on May 23, 1707. He studied medicine and the natural sciences at several universities before receiving his medical degree.

Linnaeus based his classification on newly discovered facts about stamens and pistils. He grouped plants into 24 classes according to the number of stamens. He then broke these classes down according to the number of pistils. Counting the number of visible parts let scientists identify plants quickly and classify newly discovered plants. Linnaeus's binomial system quickly became the standard way to classify and name plants and animals. ◆

Rita Levi-Montalcini (1909—)

As the head of the cell research laboratory at Rome's Institute for Cell Biology, Rita Levi-Montalcini studies the growth and function of cell types in the body. Her work has been important for cancer research and may lead to advances in the treatment of Parkinson's disease. It may help explain the process by which one fertilized egg cell develops into an organism made of billions of cells.

Born in 1909 in Turin, Italy, Rita Levi-Montalcini grew up with an interest in neurology. In 1936, she began her career as a research biologist at the University of Turin Medical School. Since she was Jewish, Levi-Montalcini was forced by the Fascist regime to leave the university, but she continued her research in secret during the Nazi occupation.

Washington University in St. Louis, Missouri, offered her a research position in 1947. In St. Louis, Levi-Montalcini discovered the material that causes the growth of nerve cells. This material, found in mammals, birds, reptiles, and fish, is known as *nerve growth factor.* Her research led to a Nobel prize in physiology and medicine in 1986. She is the fifth woman to receive the prize in this category since 1901. ◆

Eduardo S. Cantu, Geneticist

Dr. Eduardo Cantu studies the role that genes and chromosomes play in determining a person's health. He is a geneticist at the Medical University of South Carolina.

What type of work do you do as a geneticist?

I am the Director of the Medical University of South Carolina Cytogenetic Section. We run a variety of tests that are helpful to physicians in diagnosing diseases, particularly genetic diseases. Most of what we do is chromosome work. Most of our chromosome work involves diagnosis of genetic disorders before birth.

How many genetic disorders are there?

There is a catalog of about 3000 to 3500 genetic disorders. Most of these are extremely rare. I would say 50 to 100 could be called common genetic disorders.

Where do you get the cells you examine?

The sample for a diagnosis before birth is taken by *amniocentesis* (am nee oh sehn TEE sihs). A physician inserts a hollow needle through the abdominal wall into the uterus of a pregnant woman. Some of the amniotic fluid is collected and then is examined under a microscope.

How did you prepare for a career as a geneticist?

I attended Pan American University in the lower Rio Grande Valley in Texas, where I was born and raised. I went to graduate school at the University of Michigan because I was interested in birds, and a famous bird expert was there. His course was wonderful, but I also took a genetics course, which was so fascinating that it replaced my interest in birds. I was happy that I could then go to the University of Hawaii to work with one of the pioneers in genetics. Later, I spent a year at Baylor College of Medicine in Houston before coming to South Carolina.

Is genetics a growing field that students should consider?

Yes, physicians are relying more and more on genetics. They realize that many diseases and disorders have a genetic component. It's a very promising, challenging, and exciting field to get into. You can do that by studying medicine and specializing in genetics, or by doing graduate work in biology and specializing in cell biology. I certainly recommend it. ◆

These Germs Work WONDERS

by Doug Stewart
from *Reader's Digest*

We know germs as hostile little things that spoil meat and spread disease. But to scientists, germs, or microbes, can be obedient servants, ready to cure illness, clean up oil slicks, and even make delicious chocolates. This microscopic labor force has come about thanks to biotechnology, which uses engineering techniques to study living organisms. Ten years ago this field didn't exist. Today it's a billion-dollar-a-year industry. Princeton University professor Freeman Dyson has predicted that biotechnology will change the world more than the Industrial Revolution did.

Dyson's prediction is based on the work of gene splicers like Stephen Lombardi, a microbiologist at the Army's Research, Development, and Engineering Center in Massachusetts. Lombardi can snip and rejoin strings of chemicals inside living cells. These chemical strings, or genes, map out how a cell will behave. They determine, for instance, whether a microbe will turn milk into yogurt or into cheese.

In the past 15 years, gene splicers have learned to re-

> *Ten years ago this field didn't exist. Today it's a billion-dollar-a-year industry.*

move microscopic pieces of genes from cells lining a human pancreas, for example. The scientists place the pieces of code inside ordinary bacteria cells. The microbes then have the ability of the pancreas cells to produce insulin. In the late 1970s, some 50 million animal pancreases a year were ground up for insulin, which is used to treat diabetes. Rumors spread that a pancreas shortage was on the way. Today a unit the size of a small refrigerator containing altered bacteria cells can pump out more insulin in a day than a hundred animals could provide.

Once a batch of altered microbes starts to multiply, it can serve as a chemical factory, continually producing the desired product without error. Amazingly, the altered microbes mass-produce

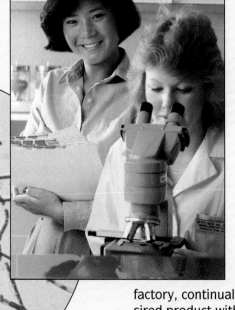

Microbes can convert waste material into inexpensive fuel.

ria into eating the sugars from saw-mill and paper-mill wastes. They eat this waste and produce ethanol, a chemical that can be used as a fuel. With oil prices climbing once again, ethanol could become a cheap alternative for running the family car. Its production could reduce our waste-disposal problems at the same time.

Douglas Dennis, a biologist at James Madison University, in Virginia, has long known about a naturally occurring microbe that produces plastic. Unfortunately, this talented creature thrives only on an expensive sugar diet. So Dennis has spliced the plastic-making part into a microbe that eats whey—a waste product of cheese-making. Plastics made in factories, by contrast, have oil as their raw material. There is even a bonus for Dennis's plastic—every bit of it decomposes.

Microbes' appetites for scraps is one reason manufacturers find them so attractive. Even candy can be made from throwaway ingredients. Since cocoa beans aren't grown in the United States, chocolate makers must pay to import them from the tropics. But scientists have discovered a strange little fungus, related to bread mold, that converts animal fats and vegetable oils into something almost identical to cocoa butter. Unfortunately, the fungus is hard to grow in bulk. So scientists have taken a gene from the fungus and transferred it to a microbe that can mass-produce the "cocoa" all day long.

Microbes' appetites for scraps is one reason manufacturers find them so attractive.

perfect copies of themselves. A single microbe can become two, then four, then eight, then a billion in less than a day. All that's needed is enough food, which is easily provided.

The first generation of gene-spliced products—nearly all of them medicines—has been a trickle. The next generation will be a flood. The biotechnology industry is hoping eventually to make every-day products like anti-freeze, animal feed, and detergent by the tank load. Earle Harbison, president of a huge chemical company, expects biotechnology to be as important in the 21st century as chemistry and physics have been in the 20th. "To compare the old biology to the new," Harbison says, "is like comparing a mule to a tractor."

Microbes may supply us with clean-burning fuel for our cars. At the University of Florida, microbiologist Lonnie O. Ingram has engineered a new microbe. With a bit of gene-shuffling, Ingram "brainwashed" bacte-

Some microbes are more prized for what they eat than for what they make. After the

Exxon Valdez oil spill in Alaska, the Environmental Protection Agency used some naturally occurring bacteria with a craving for oil to help in the cleanup. Early studies suggest that the bacteria may have done as good a job in cleaning some of the lightly soiled beaches as high-pressure hoses and detergents would have.

Often, what nature's hard-working microbes can do well, a custom-made microbe

Often, what nature's hard-working microbes can do well, a custom-made microbe can do better.

can do better. Bacteria with altered genes may someday clean up an oil spill before it has a chance to hit the beach. Ananda Chakrabarty, a microbiologist at the University of Illinois at Chicago, identified bacteria that had a taste for industrial wastes. Chakrabarty implanted genes from them into bacteria suitable for spraying. In powdered form, these new microbes could be sprinkled onto oil slicks that are just starting to spread.

Just 20 years ago the thought of microbes making medicines, plastics, or chocolates would have seemed like science fiction. No one is predicting that microbes will ever be smart enough to do your homework. But in the next century, thanks to the work of biotechnology, the tiny creatures you carefully wash off your hands may be responsible for a new industrial revolution. ◆

Bacteria were used in the Exxon Valdez cleanup.

SIMPLE LIVING THINGS

Y ou're a little naive about it [AIDS] and you think it could never happen to you. You only thought it could happen to, you know, other people and so on . . . It can happen to anybody, even me—Magic Johnson.

Earvin "Magic" Johnson, Jr.

Science PARADE

VIRUSES AND MONERANS

*I*n the 1300s, about one-fourth of the European population died from the bubonic plague. No one knew at the time that the disease was caused by a moneran, a microscopic bacterium. Today, people have a better understanding of microscopic organisms, such as viruses and bacteria, and their impact on people.

During the Middle Ages, the black rat, so numerous throughout Europe, carried the plague into both the cities and the countryside. Plague kills a rat even faster than it kills a human. But the plague bacillus doesn't pass directly from rat to rat, or from rat to human. For this job, we need the rat flea....

For hundreds of years, the bubonic plague mystified people. Victims did not understand what it was they had, or how they came down with the disease. All that most people knew was that they had the same symptoms. The cause of the disease was unknown until it was finally discovered by doctors less than a hundred years ago.

The disease produces very high fever and chills and makes victims feel weak and delirious so they must stay in bed. Many parts of the body may ache, and the patient will be in great pain as the lymph glands become affected.

The lungs also may become infected with plague. The disease is then called pneumonic plague and can be transmitted by the breath. It is believed that this is how the disease is spread so rapidly among humans. A person who is infected with pneumonic plague dies quickly. For other plague victims, hemorrhages turn black, and in the later stages of the disease, black marks show upon various parts of the body. This is why the plague often is called "The Black Death."

from *The Black Plague*
by Walter Oleksy

For Your Journal

- Explain why diseases caused by viruses are difficult to treat.
- Why do you think diseases caused by microorganisms are more likely to occur in some countries than in others?
- How can people reduce their likelihood of contracting diseases caused by microorganisms?

Viruses

Headaches, chills, fevers, aches, and pains—you think you have the flu. If you've ever had the flu, you've probably had viruses inside you similar to that pictured here. When compared to other viruses, influenza viruses are not particularly large, nor are they particularly small. To get some idea of what size an influenza virus really is, let's compare one to a penny. You would need about 950 000 of these viruses laid end to end just to reach across one side of a penny!

The Microscopic World of Viruses

The size of viruses is measured in nanometers, or billionths of a meter. Suppose it were possible for you to lay 950 000 influenza viruses end to end across the surface of a penny at the rate of one virus each second. It would take you 950 000 seconds—about 11 straight days and nights—to complete the task!

If you think that 950 000 isn't a very large number, and that viruses really aren't very small, consider a different question: How many influenza viruses laid end to end could you fit along the length of a new pencil? This time the answer is a larger number—it would take about 9 500 000 viruses! Placing one virus per second, it would take you about 110 days, or almost four months, working day and night, to complete such a task!

Let's try one more question: How many influenza viruses laid end to end could you fit along Earth's equator? This time the answer truly is a large number—it would take about two quadrillion viruses. Quadrillion is a number containing 15 zeros. It would take you about one million lifetimes to finish laying two quadrillion viruses end to end!

The influenza virus is an average-sized virus. Other viruses are much smaller, like the virus that causes hoof-and-mouth disease in cattle. Still other viruses are much larger, like the human smallpox virus. Regardless of their size, viruses can pose a threat to you and your health.

Did you know that other animals and even plants can also become the victims of a virus attack? However, the viruses that attack plants do not attack people, and the viruses that attack people do not attack plants. You will never see a tree suffering from chicken pox!

 ASK YOURSELF

What size are viruses?

Figure 8–1. The white fly carries a virus that attacks crop plants such as lettuce.

Characteristics of Viruses

A-a-a-choo! Your head aches, you have the sniffles and a cough, and your throat tickles. You have a cold, and you feel miserable. Did you know the common cold is caused by a virus? When it invades your body, the virus multiplies and soon you show all the symptoms of the common cold. A **virus** is a very small particle that can reproduce only inside a living cell. When a virus reproduces inside the living cells of your body, you don't feel too well.

If you could look at viruses using a very powerful microscope, you would see many different shapes. Some viruses are spherical, some are tube shaped, some look like miniature spaceships, and others look as though they are made from sticks and spools! Viruses do not look like any other organism. Besides unusual shapes, they have other unusual characteristics.

Dead or Alive?   Viruses are neither living things nor nonliving things, yet they have some characteristics of both. Viruses are like nonliving things in several ways. They cannot use energy, and they are not made of cells. Viruses cannot live on their own. They can reproduce only when they have infected a cell that is living.

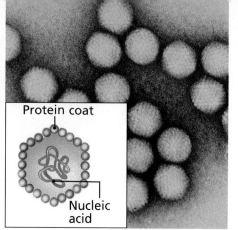

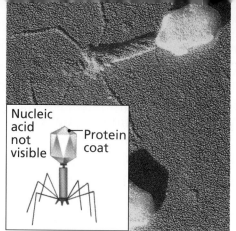

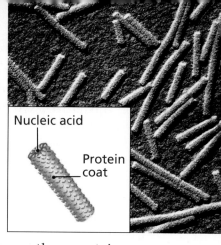

Protein coat

Nucleic acid

Nucleic acid not visible — Protein coat

Nucleic acid

Protein coat

Figure 8–2. Although these viruses have different shapes, each virus is made of a core of nucleic acid surrounded by a protein coat.

Viruses are similar to living things because they contain proteins and nucleic acids. The cells of your body also contain proteins and nucleic acids. *Nucleic acid* is the material that carries genetic, or hereditary, instructions. The nucleic acid is found in the center of the virus, and protein forms a coat around it. This protein coat is responsible for the many different shapes of the viruses shown in the illustrations.

Public Enemies   Viruses are considered hostile because they invade and then attack the cells of living organisms. An organism that is invaded by a virus is called a **host.** Plants, animals, monerans, protists, and fungi all serve as hosts to different kinds of viruses. Once a virus invades a host cell, the virus can begin to produce new viruses. This occurs when the attacking virus releases its nucleic acid into the host cell. The host cell cannot tell the difference between its own nucleic-acid instructions and those of the virus. So the host cell begins following the virus's instructions to make new viruses.

Figure 8–3. Viruses reproduce in a host cell until the cell can no longer hold them and bursts.

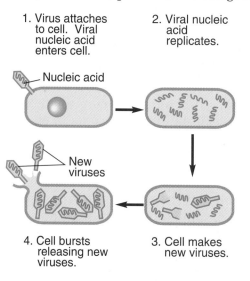

1. Virus attaches to cell. Viral nucleic acid enters cell.

2. Viral nucleic acid replicates.

Nucleic acid

New viruses

4. Cell bursts releasing new viruses.

3. Cell makes new viruses.

If you overfill a balloon with air, it bursts. The same thing happens to cells invaded by viruses. Eventually, the host cell fills with viruses and bursts, the new viruses are released, and they then attack other cells.

ASK YOURSELF

Is an organism ever a willing host to a virus? Why or why not?

Viruses and Disease

Viruses are present everywhere in the world around you. The fever blister that seems to develop just in time for school pictures is caused by a virus. Even a wart is the result of a virus invading your body.

Different viruses can invade the body in different ways. Many are airborne and are breathed in by a person. Once in the body, the viruses can begin to reproduce and to cause disease. Sometimes the effects of an invading virus create nothing more than a nuisance to the host. If you have a cold, your body is a host for invading viruses, but these viruses usually cause discomfort only for a short period of time. However, some invading viruses can cause much more serious illness. Did you have the chicken pox when you were younger? Remember how the pox itched? The chicken pox are caused by a virus. Other once-common childhood diseases, such as measles and mumps, are also caused by viruses.

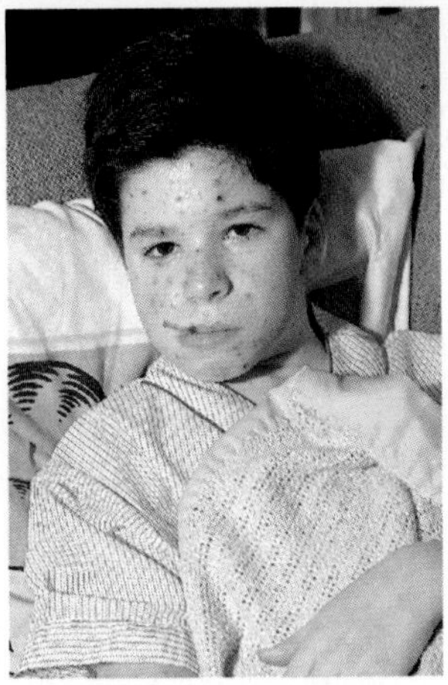

Viruses also cause polio, a disease that can result in paralysis. Polio once swept through countries and communities in epidemics. An **epidemic** is the rapid spread of a disease through a large area. Older family members or friends may remember polio epidemics that swept the United States. Even a President

of the United States was a victim of this disease. Franklin D. Roosevelt, the thirty-second President, spent most of his adult life in a wheelchair and wore leg braces because he had had polio. But he battled the disease and became a great success and an inspiration to others. Polio epidemics no longer occur because most people are vaccinated against the disease. In the next activity, you can find out more about how epidemics have been stopped.

Figure 8–4. Chicken pox is an infectious disease that is caused by a virus.

Figure 8–5. Franklin D. Roosevelt, who served as President from 1933 to 1945, battled polio courageously.

DISCOVER BY Researching

The stories that describe successful battles against epidemics can be exciting and inspiring. Use reference materials to discover more about our successes in halting the spread of epidemics, such as smallpox or polio epidemics, and share what you have learned with your classmates. 🖉

A Deadly Viral Disease   In recent years, acquired immune deficiency syndrome (AIDS) has received a great deal of attention. AIDS is caused by a virus called HIV (human immunodeficiency virus). HIV is often referred to as the "AIDS virus." AIDS is a serious infection that destroys the body's immune system. A healthy **immune system** provides the body with the ability to fight infection. HIV weakens the immune system so much that disease-causing organisms become life threatening.

People who have AIDS die from diseases that noninfected people can usually resist. A simple infection, for example, may prove deadly to an AIDS patient whose body can no longer stop the infection from spreading.

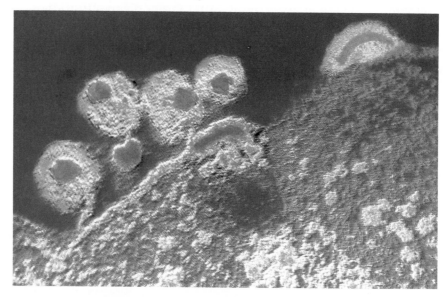

Figure 8–6. These AIDS viruses, shown as red and yellow spheres, weaken the immune system so much that disease-causing organisms become life-threatening.

AIDS was first recognized in the early 1980s. The virus is transmitted through semen (the fluid carrying sperm) and blood. Scientists know that AIDS is not transmitted by casual contact, such as shaking hands with an infected person or being in the same room with him or her. Sexual contact, contaminated blood products, and sharing of contaminated needles, however, can and do transmit HIV from person to person.

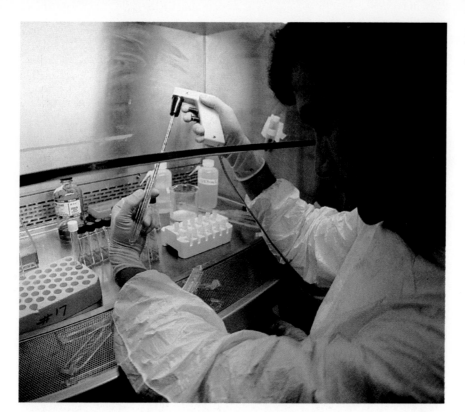

Figure 8–7. Research to find a cure for AIDS is ongoing.

At this time, a cure for AIDS has not been discovered. Vaccines are not yet available to prevent people from getting this disease. However, great effort and millions of dollars are being spent to find a way to prevent this fatal disease. For more information about AIDS, see page 248 at the end of the unit.

Stopping Viral Infections

Any illness caused by a virus is difficult to treat or cure. Because viruses are not living, they are very difficult to destroy. Since treatment of viral diseases is difficult, great energy is spent trying to prevent them. A **vaccine** is a substance given to people to keep them from getting a disease. You have probably had several vaccines against diseases such as measles, mumps, and polio. In the following activity, you can learn about vaccines and their makeup.

Figure 8–8. Today, vaccines are available to prevent diseases such as mumps and measles.

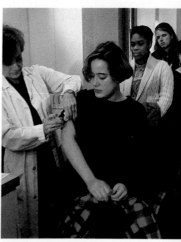

DISCOVER BY *Writing*

Have you ever wondered what a vaccine was made of or how it worked? When you receive a vaccination against polio, for example, you actually receive a dose of weakened viruses. Use reference materials to find out why people seldom contract a disease after receiving a vaccination against that disease. Summarize the information in your journal.

Table 8-1	Viral Diseases for Which Vaccines Exist
Hepatitis A and B Influenza Polio Rubella (German measles)	Measles Mumps Rabies

In 1796, Edward Jenner observed that people did not get smallpox, a deadly disease, if they first had cowpox, a minor disease. Jenner conducted an experiment in which he used material from a cowpox sore as a vaccine against smallpox. His experiment was successful, and since Jenner's time, billions of people worldwide have been vaccinated against smallpox. As a result, there have been no cases of smallpox anywhere in the world since 1981.

Figure 8-9. Since the discovery of a smallpox vaccine by Edward Jenner, vaccines against many other viruses have been developed.

 ASK YOURSELF

How are viruses like and unlike living organisms?

SECTION 1   *REVIEW AND APPLICATION*

Reading Critically

1. How are viruses like living things? How are they different?

2. List several diseases that are caused by viruses.

3. Why is AIDS such a serious disease?

Thinking Critically

4. What would happen if one generation of chicken pox viruses never found a host?

5. At one time, some people believed they could get warts by touching a toad. What is now known about viruses that indicates that this is probably not true?

6. Describe several ways in which an epidemic would pose serious health concerns.

SKILL

Identifying Viral Variables

▼ PROCEDURE

1. Study the table of various health problems and symptoms caused by viruses.
2. Examine each case study carefully, then infer from the symptoms which health problem might be involved.

Case Study 1: Patient felt nauseated, had a fever, felt weak, and ached all over. Several other family members felt the same way.

Case Study 2: Patient had a fever and painful facial swelling. Patient also had a headache and muscle aches.

Case Study 3: Patient had a headache and fever. Irritating spots, which turned into blisters and scabs, appeared on the patient's cheeks, arms, and legs.

Case Study 4: Patient suddenly experienced a fever, a splitting headache, and nausea. A student at the patient's school was hospitalized earlier that day with the same symptoms.

SOME HEALTH PROBLEMS CAUSED BY VIRUSES	
HEALTH PROBLEM	**SYMPTOMS**
Common cold	Runny nose; chills; sore throat
Influenza	Headaches; contagious; similar to a very bad cold; fever; general weakness
Meningitis	Inflammation of membranes covering the brain and spinal cord; sudden fever; painful headache; vomiting; contagious; potentially fatal if untreated
Chicken pox	Crops of spots appearing on trunk, then on face and limbs; rash blisters and scabs; fever; headache
Mumps	Painful swelling of glands in cheeks; fever; headache; muscle aches

▶ APPLICATION

How can a knowledge of common health problems and their symptoms be helpful to a parent? Why is it important that people *not* make diagnoses of diseases without the help of medical professionals? Explain why a traveler to foreign countries is sometimes required to receive one or more immunization shots before departure.

※ Using What You Have Learned

Just like people, pets sometimes do not feel well. When you don't feel well, a family member or doctor will usually ask you what's wrong or where it hurts. Pets don't have the ability to speak. How do you think the practices of a veterinarian would be similar to or different from the practices of your doctor?

Monerans

Compare and contrast *viruses and bacteria.*

Identify *five diseases caused by bacteria.*

Summarize *the effects of bacteria on humans.*

Although bacteria are microorganisms, they are enormous in size when compared to viruses. The size of viruses is measured in billionths of a meter; the size of bacteria is measured in millionths of a meter. While they are much larger than viruses, bacteria are still too small to be seen without a microscope.

Recall that the conditions required for the reproduction of viruses are limited—they need to be inside living cells. Bacteria are much different—some bacteria can exist rather comfortably in a cup of steaming hot coffee!

Characteristics of Monerans

You just learned your best friend has strep throat. Now your throat is scratchy. You have a fever, and your tonsils are swollen. You may have strep throat, too. If you do, you have an infectious disease caused by bacteria. *Bacteria* are a common example of monerans. **Monerans** are organisms that do not have a nucleus but do have a cell wall. Bacteria are commonly found in three shapes: spheres, rods, and spirals. While some bacteria live singly, many are found in groups, pairs, chains, or clusters. The bacteria that cause strep throat are spheres and appear in chains.

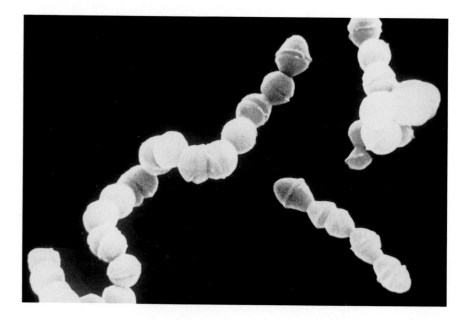

Figure 8–10. Strep throat is caused by bacteria that are spread from person to person through moisture sprayed by the nose or mouth.

Figure 8–11. Paint pots, or hot springs that contain boiling water and mud, are often brightly colored by bacteria living in the boiling water.

Hot or Cold   Bacteria have the ability to live in a variety of conditions. The illustration shows bacteria living successfully in boiling water. Bacteria have also been found in very cold environments and even buried 5 m deep in the soil! Because monerans have been found in some very unusual places, it's easy to see that they must adapt successfully to less-than-perfect environmental conditions.

Home Sweet Home   Monerans can be found almost everywhere. While some monerans live freely, others live in plants or animals. Still others live in products made from plants and animals. Monerans are found in many foods, such as cheese and yogurt. Many monerans are helpful to plants and animals. Others, however, are harmful.

 ASK YOURSELF

Under what conditions can bacteria live?

Bacteria—Common Monerans

Bacteria reproduce quickly by splitting into two cells. Under proper conditions, reproduction can occur every 20 minutes. If conditions were perfect, one bacterium could produce so many others in 24 hours that the entire group would weigh 2 million kilograms, approximately the weight of a train locomotive!

DISCOVER BY Calculating

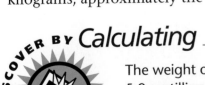

The weight of the earth is estimated to be about 5.9 sextillion (10^{21}) tonnes. If one bacterium could produce a group of bacteria weighing 2 million kilograms in 24 hours, how long would it take that group to reproduce so that it weighed as much as the entire earth?

Fortunately, perfect conditions for bacteria reproduction are rare because the environment in which bacteria live does not have enough food to support so many organisms. The amount of waste products produced by the bacteria also would interfere with their continued growth and reproduction.

In Harm's Way   Like viruses, some disease-causing bacteria invade other organisms. Since bacteria reproduce so quickly, the illnesses they cause can spread throughout the body rapidly. Bacteria cause disease in two ways. Some bacteria destroy cells. When cells in your body are destroyed, you feel weak and often have a fever. Other bacteria give off poisons, or toxins, that can damage your body. In the next activity you will identify some of the disease-causing bacteria and viruses.

ACTIVITY

What microorganisms cause various diseases?

MATERIALS

unlined paper, colored pencils

PROCEDURE

1. Find the name of the first disease listed in this activity (boils). Then note the type of organism that causes that disease (staphylococcus).
2. Use a reference book to find out what the microorganism looks like. Draw the microorganism and label it with its name and the disease it causes.
3. Repeat the procedure for the remaining diseases listed.

APPLICATION

1. Which diseases listed in this activity are caused by monerans?
2. For which diseases listed have vaccines been developed?
3. Why is it important for you to know about diseases, their causes, and their treatments?

SOME DISEASE-CAUSING MICROORGANISMS

Disease	Microorganism
Boils	Staphylococcus
Chicken pox	Virus
Pneumonia	Bacillus or virus
Rabies	Virus
Scarlet fever	Streptococcus
Tuberculosis	Bacillus (singly or in a V shape)
Whooping cough	Bacillus (singly or in pairs)
Yellow fever	Virus

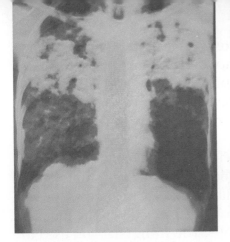

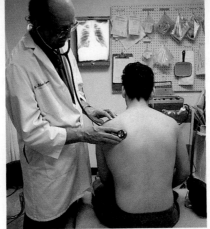

Figure 8–12. Improved preventive and treatment methods have reduced the number of people who die from tuberculosis. However, the disease remains a major concern, especially in developing countries.

Table 8-2	**Some Diseases Caused By Bacteria**	
Diphtheria	Strep throat	Tuberculosis
Leprosy	Tetanus	Whooping cough

Bacteria cause many illnesses, such as those listed in the activity. Fortunately, most bacterial diseases respond to treatment. If you've ever been sick with a bacterial infection and received medication for it, you probably received an antibiotic. **Antibiotics** are chemical substances used to kill or slow the growth of bacteria. The word *antibiotic* comes from two Greek words that mean "against life." Antibiotics interfere with the cell processes of bacteria and, as a result, kill them. Unfortunately, antibiotics may also kill some healthy cells. Occasionally, antibiotics may cause unusual side effects, such as rashes, diarrhea, or sensitivity to sunlight. These side effects, however, are tolerated by many patients because the side effects represent less of a threat to long-term health than the bacterial infection does.

With a Little Help
Even though it appears that many harmful bacteria inhabit our world, there are actually many more helpful bacteria than harmful bacteria. You may find this hard to believe after learning about some of the illnesses caused

Figure 8–13. These photomicrographs show (left to right) the bacteria that cause pneumonia, tuberculosis, boils, and cholera.

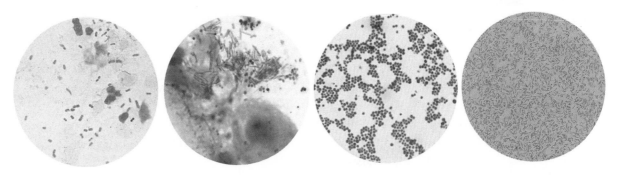

by bacteria. However, many bacteria have useful functions. For example, bacteria are necessary for decay to take place. Imagine what life would be like if all of the dead plants and animals in the world around you never decayed! Garbage and waste would pile up and fill every city, state, and country with material that could not be used and would never go away. Bacteria decompose these materials and even help recycle valuable compounds contained in garbage and waste back into the environment.

Bacteria are also used in industry and in the production of food. In industry, for example, bacteria break down the fibers of plants used to make linen and rope. The string that you used to fly a kite was probably created with the help of bacteria. If you have ever cooked with vinegar or eaten cheese or sauerkraut, you have used products that were produced with the help of bacteria. Bacteria are even used to create antibiotics that kill other bacteria!

ASK YOURSELF

Name several ways that bacteria positively and negatively influence your life.

Cyanobacteria

Have you ever wanted to swim in a pond or lake but could not because green "scum" was covering the surface of the water? The "scum" is really a large group of organisms that may be cyanobacteria.

 Doing

 If you live near a pond that contains a large amount of "scum," ask your teacher for permission to bring a small amount to class for observation under a microscope. Try to identify which organisms are cyanobacteria. Draw what you see under the microscope. ✐

You may have seen some organisms in your sample that were not cyan (blue-green) in color and wondered what these organisms were. Cyanobacteria are not always blue-green, or even green. They can also be red, violet, or yellow, though they are always called cyanobacteria. Sometimes they are even called blue-green algae, but they are really monerans, not protists as true algae are.

Figure 8–14. Yogurt and cheese are produced through the action of bacteria on milk and cream. This photograph shows cheese being made.

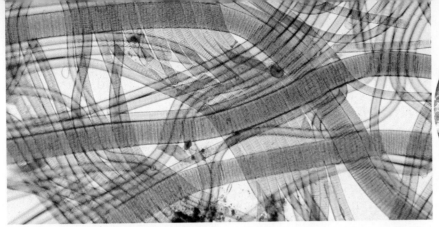

Figure 8–15. These moneran cyanobacteria are often mistaken for algae.

If you think it's unusual that cyanobacteria are not always cyan in color, think of a violet. Violets may have flowers of a violet color, but the flowers can also be colored pink, white, or blue. Can you think of other instances in which the name given to something is somewhat misleading?

The various colors of the cyanobacteria represent different species. The colors come from the pigments in the cells. **Pigments** are chemicals that give color to the tissue of living organisms. *Chlorophyll,* the pigment that makes plants green, is the best-known pigment. Chlorophyll enables cyanobacteria to make their own food.

Cyanobacteria reproduce by simple cell division. They are so plentiful that they are considered a nuisance by many pool owners and by gardeners who try to scrape them off damp flowerpots. However, cyanobacteria also serve a useful purpose. They help restore nitrogen to the soil. Nitrogen is necessary for healthy plant growth and is often added to soil in fertilizers. The presence of cyanobacteria allows plants to grow with very little added fertilizer.

 ASK YOURSELF

How can cyanobacteria improve the soil?

SECTION 2 *REVIEW AND APPLICATION*

Reading Critically
1. Compare and contrast the characteristics of viruses with those of bacteria.
2. How are cyanobacteria different from other kinds of bacteria?
3. How are bacteria useful to humans?

Thinking Critically
4. Why are antibiotics effective against bacteria but not against viruses?
5. When certain chemicals are added to a swimming pool, the bacteria growing there die. Explain why it would not be good to add these chemicals to a pond or stream to kill the cyanobacteria growing there.

INVESTIGATION

Testing Disinfectants

▶ **MATERIALS**

● safety goggles ● laboratory apron ● nutrient agar-filled Petri dish ● wax pencil ● cotton swab ● filter-paper disks (4) ● forceps ● disinfectant solutions ● hole punch

▼ **PROCEDURE**

1. CAUTION: Put on safety goggles and a laboratory apron and leave them on throughout this investigation. Turn your Petri dish upside down and use a wax pencil to label the dish as shown.
2. Moisten a cotton swab with water and rub it lightly across a surface in the classroom, such as a desk, doorknob, or window.
3. Lift the top of the Petri dish just enough to insert the swab. Gently, but quickly, move the swab over half the agar surface. Make sure the swab touches all parts of that surface. Then turn the Petri dish one-half turn and repeat the procedure. Replace the cover.
4. Use your hole punch to make four 5-mm disks of filter paper. Use your forceps to dip one in a disinfectant solution. Place the disk on section 1 of your Petri dish. Dip two more disks in different disinfectant solutions and place them on sections 2 and 3 of your Petri dish. Place the last disk without disinfectant on the fourth section.
5. Tape your Petri dish shut and incubate it upside down for 48 hours in a warm, dark place.
6. CAUTION: Do not remove the cover of the Petri dish while observing the bacterial colonies. Wash your hands thoroughly after this investigation. After incubation, compare the diameters of the clear areas around each disk.
7. Dispose of the Petri dish as directed by your teacher.

ANALYSES AND CONCLUSIONS

1. Which section(s) of the dish contained the variable(s)? The control(s)?
2. Which disinfectant appears to be the most effective? How can you tell?
3. The agar and the Petri dish were sterilized before this investigation. Why do you think this was necessary?
4. What is the purpose of a control in an investigation?

APPLICATION

Why do you think it is important to wash your hands thoroughly at the end of this investigation? List several ways in which this procedure applies to your everyday life.

✴ Discover More

People sometimes use homemade disinfectant solutions instead of solutions that are available commercially. **CAUTION: Do not mix any disinfectant with any other disinfectant. To make a disinfectant solution, just add water to a disinfectant.** Disinfectant solutions typically include water mixed with bleach, water mixed with ammonia, or water mixed with vinegar. With the approval and guidance of your teacher, create one of these homemade disinfectant solutions and repeat the steps of this investigation. Then compare and describe the results of each investigation.

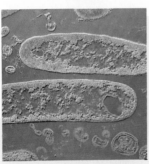

The Big Idea

People can see most organisms without the use of microscopes. However, they cannot see viruses and monerans. Both are tiny organisms that can be seen only by using a microscope. Although they are both microscopic organisms, viruses are much smaller than bacteria. Despite their size, viruses and bacteria have a huge impact on the environment.

Many viruses impact the environment in negative ways by causing diseases and other sicknesses in plants and animals. Monerans, such as bacteria, may benefit or harm organisms.

For Your Journal

Review the journal entries that you wrote at the beginning of the chapter. Revise your entries to reflect what you have learned. Include information that explains why bacterial and viral diseases are more likely to break out in some areas than in others. Make recommendations for reducing the risk of contracting viral or bacterial diseases.

Connecting Ideas

Copy this unfinished concept map into your journal. Complete the concept map by writing the correct term in each blank.

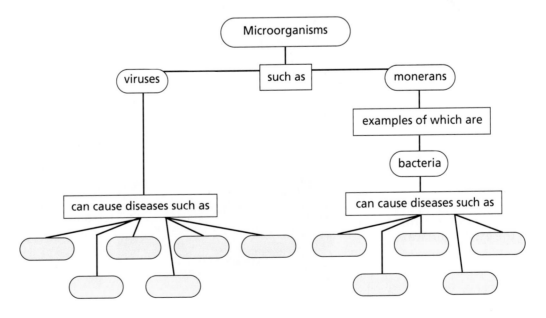

Microorganisms

such as

viruses — monerans

examples of which are

bacteria

can cause diseases such as

can cause diseases such as

Understanding Vocabulary

For each set of terms, explain the similarities and differences in their meanings.

1. virus (211), host (212)
2. immune system (214), antibiotics (221)
3. monerans (218), bacteria (218)
4. vaccine (215), epidemic (213)

Understanding Concepts

MULTIPLE CHOICE

5. Viruses are different from monerans because they
   a) are microorganisms.
   b) may cause diseases in humans.
   c) can be cured by antibiotics.
   d) can reproduce only in living cells.

6. All of the following are true about bacteria *except* that they
   a) have a distinct nucleus.
   b) can survive in very hot climates.
   c) can be treated with antibiotics.
   d) come in three different shapes.

7. Viral infections are difficult to treat because
   a) viruses are difficult to detect, even with microscopes.
   b) viruses have many different shapes.
   c) viruses do not appear to be living things.
   d) symptoms of the infections are difficult to diagnose.

8. Choose the location in which bacteria would be least likely to be found.
   a) in the atmosphere
   b) under water
   c) at the core of the earth
   d) on the earth's surface

9. Why is it not likely that people today will contract polio?
   a) The virus that causes polio no longer exists.
   b) Vaccines against polio have been developed.
   c) The polio virus exists today, but in a weakened form.
   d) Antibiotics have slowed the spread of the disease.

SHORT ANSWER

10. What is an epidemic?
11. Explain why some scientists say that viruses are not living things.
12. Compare and contrast antibiotics and vaccinations.

Interpreting Graphics

13. Look at the illustrations. In which environments could some bacteria reproduce? In which could some viruses reproduce?

Reviewing Themes

14. Systems and Structures
Compare the size and complexity of viruses and monerans to their role in the world around us.

15. Environmental Interactions
Explain how monerans influence you and your environment in positive and negative ways.

Thinking Critically

16. Why do bacteria respond to antibiotics, while viruses do not?

17. Describe some of the problems humans would face if there were no bacteria.

18. Polio is now controlled. Why, then, is a polio vaccine still required in many states?

19. Scientists and researchers have created vaccines for various diseases such as polio and smallpox. List several reasons why you think scientists and researchers have not been able to create vaccines for other diseases such as AIDS.

20. This photograph shows AIDS viruses attacking a healthy cell. The AIDS viruses are shown in blue. Why is AIDS so dangerous? Could AIDS be considered an epidemic? Why?

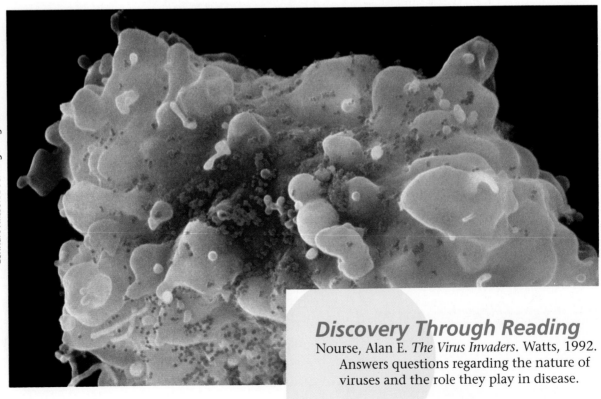

Lennart Nilsson/Boehringer Ingelheim International Gmbh

Discovery Through Reading
Nourse, Alan E. *The Virus Invaders*. Watts, 1992. Answers questions regarding the nature of viruses and the role they play in disease.

PROTISTS AND FUNGI

In the United States, people think of mosquitoes mostly as a nuisance. In other parts of the world, however, mosquitoes can carry and transmit microscopic organisms that cause disease and sometimes death. Malaria, a disease transmitted by mosquitoes, is the cause of death for more than one million people each year.

People are sometimes afraid of things they cannot see. But when we think of things we cannot see, we usually think of large things. Seldom do we think of microscopic things.

The microscopic world of organisms contains a great variety of living things—all of which are too small for you to see. Some of these microscopic organisms are helpful, and others are harmful. Yet they are everywhere in the world around you.

The Anopheles mosquito shown in the photograph carries and transmits the disease malaria. Characterized by symptoms such as chills, fever, and weakness, malaria is a parasitic disease caused by microscopic organisms, which the mosquito injects into a person's bloodstream. Once in the bloodstream, the organisms invade the liver, causing the liver cells to produce more organisms. Ultimately, the liver cells burst open, releasing the organisms to invade healthy red blood cells, and the process is repeated over again. Within a short time, these organisms or the poisons they produce are present in enormous amounts. The organism that causes malaria is just one of many microscopic organisms in the world around you. Even though some of these organisms are harmful, some are very helpful, and you would not be able to live without them.

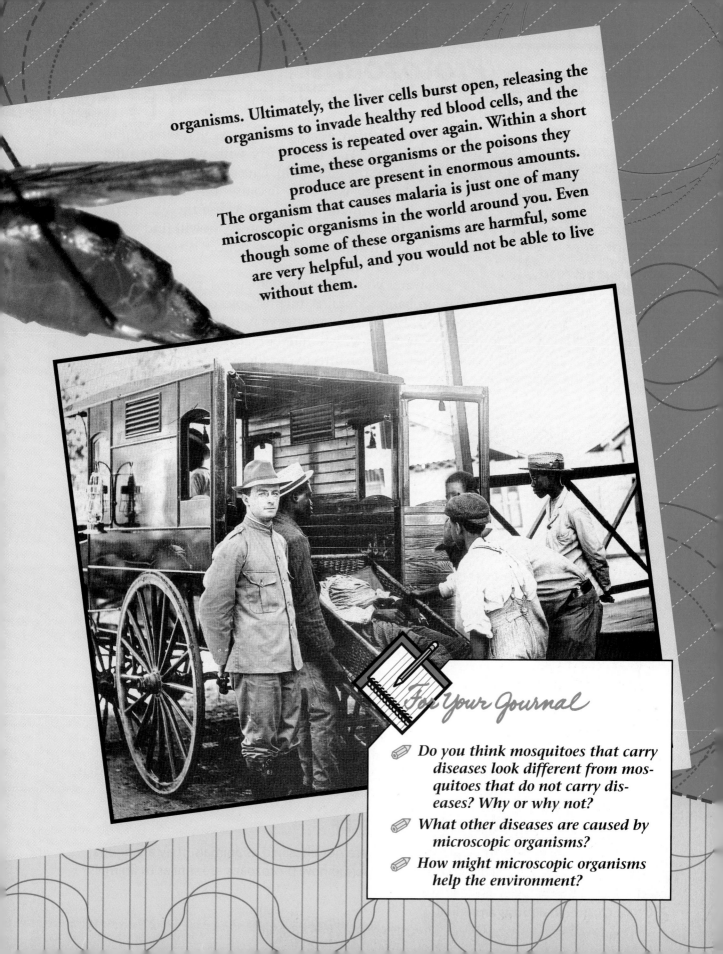

For Your Journal

🖊 *Do you think mosquitoes that carry diseases look different from mosquitoes that do not carry diseases? Why or why not?*

🖊 *What other diseases are caused by microscopic organisms?*

🖊 *How might microscopic organisms help the environment?*

Protozoans

Objectives

Compare *protozoans and animals.*

Identify *several features that make protozoans different from each other.*

Sketch *a protozoan and* **label** *its major features.*

Imagine that you are going on a two-week hiking trip in a wilderness area. You must carry with you everything you will need for the entire trip. One of your most important requirements is water. Because carrying enough water would be difficult, you hope to use the water you will find in the wilderness. You know you will find water in rivers and streams, or you can collect it by catching rain or perhaps by melting snow.

The water you find in the wilderness may look clean. But it probably contains microscopic organisms called *protozoans*. Drinking the water may make you ill. This is just one way in which microorganisms—living things you cannot see—can affect your life.

Figure 9–1. Although this stream looks clear, it may contain many protozoans.

Characteristics of Protozoans

Have you ever heard of African sleeping sickness or malaria? These diseases and many others are caused by protozoans (proht uh ZOH uhnz). **Protozoans** are microscopic organisms that are members of the protist kingdom. Like other protists, protozoans are single-celled organisms, but they usually do not have cell walls. Most protozoans are found in water.

Just as you have special structures in your body that help you move and eat, protozoans have special structures that help them get food and move from one place to another. Protozoans are able to take in oxygen and give off carbon dioxide directly through their cell membranes.

Protozoans are similar to animals in other ways. Just as animals can react quickly to changes in their environment, so too can many protozoans. Protozoans usually move toward food and better environmental conditions as animals often do. However, some protozoans cannot move on their own.

 ASK YOURSELF

Although protozoans are not classified as animals, they are animal-like in some ways. Describe how protozoans are similar to animals.

Types of Protozoans

Protozoans are classified, or grouped, according to their ability to move and the way in which they move. The four basic groups of protozoans are shown in the following table.

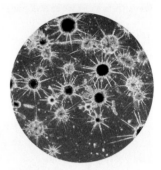

Figure 9–2. These radiolarian protozoans are known for their complex and beautifully sculpted skeletons.

Table 9-1	Types of Protozoans	
Protozoan	**Locomotive Structure**	**Example**
Sarcodines	Pseudopods	Amoeba
Ciliates	Cilia	Paramecium
Flagellates	Flagella	Trypanosome
Sporozoans	(none)	Plasmodium

Sarcodines *Sarcodines* are protozoans that move from one place to another by using **pseudopods** (SOO duh pahdz), or "false feet." Pseudopods are projections made of cytoplasm. Perhaps you have heard of or seen a picture of an amoeba (uh MEE buh). An example of a common sarcodine is the amoeba in the illustration. Though you might think that an amoeba looks something like uncolored gelatin, it is actually a small mass of cytoplasm surrounded by a cell membrane. Amoebas have no specific shape because they change shape almost constantly. Think of what your life would be like if you constantly changed shape! As the cytoplasm in an amoeba moves, new pseudopods are formed and the amoeba "creeps" along from place to place.

An amoeba also uses its pseudopods to capture food. When an amoeba finds a food particle, the pseudopods surround and close around the food particle, forming a food vacuole. Digestion takes place within the food vacuole. The digested food moves across the membrane of the vacuole and into the cytoplasm, where the nutrients are used. Food vacuoles can then move to the cell membrane and discharge undigested food or waste material from the cell.

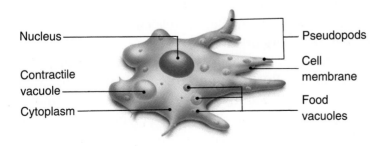

Nucleus
Contractile vacuole
Cytoplasm
Pseudopods
Cell membrane
Food vacuoles

Figure 9–3. The movement of an amoeba is caused by the outward "flow" of cytoplasm to pseudopods.

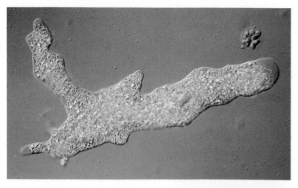
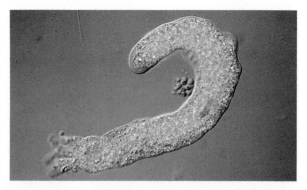

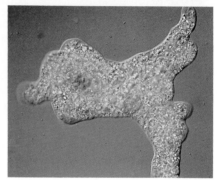

Figure 9–4. When an amoeba comes in contact with a food particle (top left), the pseudopod surrounds the food particle (top right) and slowly engulfs the particle (bottom left), forming a food vacuole (bottom right).

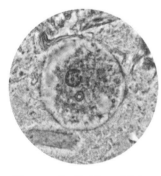

Figure 9–5. The thick protective covering, or cyst, surrounding this amoeba forms when the pond in which the amoeba lives dries up during a summer drought or freezes in the winter. When conditions are favorable, the amoeba will become active again.

Amoebas have other vacuoles called *contractile* (kuhn TRAK tuhl) *vacuoles.* These vacuoles collect extra water from the cytoplasm and release it through the cell membrane.

Most types of amoebas are harmless to you, but some cause disease. For example, one kind of amoeba causes amoebic dysentery (DIHS uhn tehr ee). Amoebic dysentery is a severe type of diarrhea. People may get the dysentery-causing amoeba by eating contaminated food or drinking contaminated water. Usually, such contamination occurs in areas with poor sanitation facilities. Another kind of amoeba lives in warm, still water. If swimmers inhale the amoeba, it causes a fatal inflammation of their brain and spinal cord.

Ciliates   *Ciliates* (SIHL ee ihts) are the most complex protozoans. A ciliate has hundreds of short, hairlike structures called **cilia** that often cover the entire cell. These cilia are used to move the organisms through water in much the same way that your arms move you through water when you swim. In most ciliates, the beating cilia create a current of water that brings food to the organism.

The paramecium (par uh MEE shee uhm) [plural, *paramecia*], shown in the illustration on the next page, is an example of a common ciliate. Paramecia and other common ciliates are often found in pond water. You can see paramecia in action in the next activity.

Discover by Doing

Your teacher will prepare a culture of paramecia to which stained yeast cells have been added. Make a wet mount of the paramecia. After locating the red-stained food vacuoles, describe in your journal the process by which yeast cells are taken into the paramecia. Using what you have learned, predict how waste products are removed. ✎

Most ciliates are free-living and do not cause disease in humans. Ciliates feed on dead plants and other small protozoans in the water, helping to continue the food chain. As cilia move ciliates through water, food enters the organisms. Food vacuoles, which form around the food particles, digest the food particles. Digested food is then absorbed by the cytoplasm through the membrane of a food vacuole, and waste material and undigested food are given off. As in the amoeba, contractile vacuoles remove extra water from the cell.

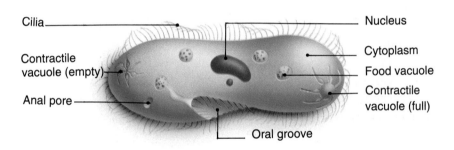

Cilia

Contractile vacuole (empty)

Anal pore

Nucleus

Cytoplasm

Food vacuole

Contractile vacuole (full)

Oral groove

Figure 9–6. The cilia of a paramecium propel the organism quickly through the water. Cilia also help sweep food particles into the mouthlike oral groove.

Flagellates   *Flagellates* (FLAJ uh layts) move by using long, whiplike structures called **flagella** (fluh JEHL uh). Flagellates are found in water and soil and even inside animals.

Have you ever wondered how termites eat wood? Within their digestive systems, they have a particular species of flagellate. These flagellates have enzymes that can digest wood. The digested wood is then used by the termites as food. Termites could not digest the wood without the flagellates.

Some flagellates are parasites. *Parasites* live in or on a host organism and cause harm to it. Some parasites cause diseases in humans or other organisms. The flagellate that causes African sleeping sickness is carried from one person to another by the tsetse (TSET see) fly. This flagellate gives off a toxin, or poison, that causes weakness in humans and can even cause death if the disease is not treated.

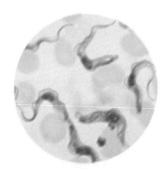

Figure 9–7. The flagellate that causes African sleeping sickness is shown above in human blood.

Sporozoans All *sporozoans* (spawr uh ZOH uns) are parasites. They cannot move from one place to another on their own. Sporozoans may live in the muscles, the kidneys, or other organs of humans or other animals. They may also live in the blood. The best-known sporozoan lives part of its life in the red blood cells of humans, causing the disease malaria.

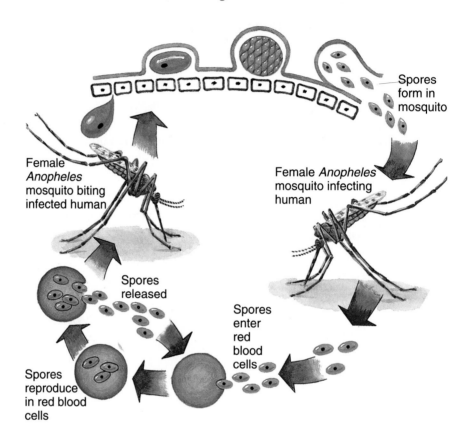

Spores form in mosquito

Female *Anopheles* mosquito biting infected human

Female *Anopheles* mosquito infecting human

Spores released

Spores enter red blood cells

Spores reproduce in red blood cells

Figure 9–8. The *Anopheles* mosquito is responsible for transmitting the sporozoan that causes malaria.

ASK YOURSELF

Some protozoans are harmful to humans. What steps might be taken to help reduce our exposure to these protozoans?

SECTION 1 *REVIEW AND APPLICATION*

Reading Critically
1. Name three structures that enable protozoans to move.
2. What is the function of food vacuoles? Of contractile vacuoles?
3. What is a parasite?

Thinking Critically
4. Which protozoan is the least animal-like? Explain.
5. How is the relationship between flagellates and termites different from the relationship between a sporozoan and a red blood cell?

INVESTIGATION

Observing Protozoans

▶ MATERIALS

- medicine dropper ● microscope slide ● stock cultures or pond water
- coverslip ● compound light microscope

▼ PROCEDURE

1. Copy the table shown and complete it for each water sample you examine.
2. Using a medicine dropper, place one drop of water, provided by your teacher, on a microscope slide.
3. Cover the drop of water carefully with a coverslip.
4. Using low power on your microscope, observe the drop of water. Draw and label what you see.
5. Switch to high power. Again draw and label what you see.

6. Try to identify the types of organisms you find. Use the illustrations below to help you.

7. Repeat steps 2-6 with one or more additional water samples.

TABLE 1: OBSERVATION OF PROTOZOANS				
Protist Name	Sketch	Description (Size/Shape/Color)	Type of Movement	Type of Feeding

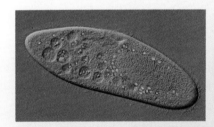

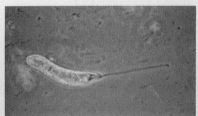

▶ ANALYSES AND CONCLUSIONS

1. What types of protozoans do you see? Flagellates? Ciliates? Amoebas? Use the photographs to help you identify the protozoans.
2. Compare your drawings with those of your classmates. How are the organisms in your samples similar to and different from the organisms in their drawings?
3. How do the organisms from the different water samples compare?

▶ APPLICATION

Why is it important for a scientist or a technician working in a water purification plant to be able to identify protozoans? Explain your reasoning.

Discover More

Ask your teacher to supply you with a different liquid and repeat the investigation. What microorganisms were present in the new liquid? Compare the results of your investigations and explain the similarities and differences.

Algae

Objectives

Identify *the general characteristics of algae.*

Name *and* ***describe*** *kinds of algae.*

Summarize *several ways in which algae are important to humans.*

Do you know where you might find algae while hiking in the woods? Find some still water and look for the green stuff floating on the surface—that's algae. Believe it or not, you might be able to find algae more easily at home than in the woods! Here's a riddle: What might toothpaste, ice cream, mayonnaise, salad dressing, and instant pudding mix have in common? They may contain extracts of algae that are used to make them thick and "creamy." The next time you brush your teeth, read the ingredient label on the toothpaste while you brush. It's possible that your toothpaste contains algae!

ᴅɪꜱᴄᴏᴠᴇʀ ʙʏ *Researching*

Carrageenin, mannitol, and agar are some of the extracts of algae. Read the labels of several food and cosmetic containers. Identify and list products in your home that contain algae extracts. ✎

Figure 9–9. These algae are sometimes called *sea lettuce* because they look like something you might find in a salad bowl.

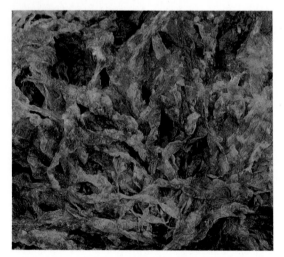

Characteristics of Algae

Algae are plantlike organisms that contain chlorophyll and perform photosynthesis. They are characterized by several important features. Algae cells contain chloroplasts and have cell walls. Chloroplasts are the organelles in which food, in the form of sugars, is made in the process of photosynthesis. The sugars are then converted into starches, fats, and other compounds. These foods provide energy not only for the algae but also for the many types of organisms that eat algae.

 ASK YOURSELF

What might happen to other ocean organisms if all of the algae in the oceans suddenly disappeared?

Types of Algae

Not all scientists agree on which organisms are protists. Generally, though, all algae are classified as protists. Algae are grouped according to the different pigments they contain—green, red, or brown.

Euglenas   One type of freshwater alga is the *Euglena* (yu GLEE nuh). *Euglenas* have characteristics of both animals and plants. They are animal-like because they are able to move from place to place. They are plantlike because they have chloroplasts that contain chlorophyll, giving them the ability to make their own food. When a *Euglena* is in an area that does not have enough light for photosynthesis to take place, it can absorb food from the water. Look for *Euglenas* when you do the next activity.

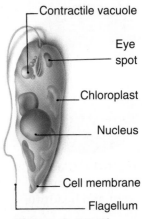

Figure 9–10. The presence of chloroplasts in *Euglena* makes the organism unique. It can move like an animal and make its own food like a plant.

DISCOVER BY *Observing*

Collect specimens of algae from ponds, ditches, or the ocean if possible. Your teacher can help you look at your specimens under a microscope. In your journal, draw and describe what you see. ✎

Dinoflagellates   *Dinoflagellates* (dy noh FLAJ uh layts) are found in the ocean. A microscopic view of these organisms would reveal that each dinoflagellate moves using two flagella. The two flagella sometimes work together to spin the organism like a top!

One type of dinoflagellate is responsible for a situation known as *red tide*. Red tide occurs when dinoflagellates reproduce very rapidly. The toxins, or poisons, that they produce spread through the water, sometimes causing widespread fish kills. After an episode of red tide, beaches are often littered with dead fish. Humans can become ill if they eat shellfish that have absorbed the toxins produced by these dinoflagellates. You can read more about red tide in the article on page 254 at the end of this unit.

Some dinoflagellates have a very interesting characteristic. When the water in which they are living is disturbed by a ship or even a fish, these "fire algae" glow like tiny fireflies! Dinoflagellates contribute in a very helpful way to the balance within ecosystems. They serve as a major food source for animals in the ocean.

Figure 9–11. Although they are extremely small, dinoflagellates are a primary source of food for many fishes.

Figure 9–12. Diatoms can be found in many different and beautiful shapes.

Diatoms

Diatoms (DY uh tahmz) are golden brown algae and are the most common of all the single-celled organisms in the oceans. Diatoms are another important food source for animals that live in the water.

If you looked at a diatom with a microscope, you would see an organism that is very beautiful. The cell walls, or shells, of diatoms are made of a chemical similar to glass. The shells of diatoms are made up of two parts. The parts are of unequal size, with the larger part fitting over the smaller like the lid of a box.

When diatoms die, their empty shells collect on the ocean floor to form *diatomaceous* (dy uh tuh MAY shuhs) *earth*. Diatomaceous earth is a useful resource; it is collected, processed, and used in making insulation, toothpaste, and silver polish.

Green, Red, and Brown Algae

Algae live in many places, not just in water. You may have seen green algae in forests, growing on the trunks of trees. Green algae are also found in oceans, lakes, ponds, and swimming pools.

Although all algae are classified as protists, some are multicellular and grow to be quite large. Kelp, a type of brown algae that lives in the oceans, can grow to 50 m or more in length. Red algae also live in the oceans, attached to rocks on the ocean bottom. Red and brown algae are often used as thickening agents in ice cream and puddings.

Figure 9–13. Red algae (right), brown algae (center), and green algae (left) all contain chlorophyll. However, the red and brown algae also contain other pigments.

 ASK YOURSELF

What useful products contain algae?

SECTION 2 *REVIEW AND APPLICATION*

Reading Critically
1. How are algae like plants?
2. How are *Euglenas* like protozoans? Like algae?
3. Describe a diatom.

Thinking Critically
4. How can algae in an aquarium be helpful?
5. Why are large deposits of diatomaceous earth found inland despite the fact that diatoms live in the ocean?

SKILL Organizing Data

Sometimes it is easier to remember data if it is organized in some logical way. Scientists often have huge amounts of data that they need to use for their research. To organize the data you have read in a chapter, you can use some of the skills scientists use. Then, in addition to learning a skill, you will be studying at the same time.

▼ PROCEDURE

1. Go over the material you need to organize, and decide on the method you wish to use. You may wish to make an outline, a table, or a graph.
2. If you are going to use an outline, decide on the major ideas. Label the main ideas with Roman numerals.
3. Beneath each main idea, list topics related to the idea. Label these topics with capital letters.
4. Beneath each topic, list two concepts. Label each concept with an Arabic numeral.

5. If you are going to use a table, chart, or graph, sketch the way in which you are going to present the data. Decide on a title, headings, and the number of columns or sections.
6. Under the proper headings, list the material to be presented.
7. If measurements are needed, be sure to include the correct units.
8. Go through the material step by step to include all the important facts.
9. You may need to modify your organizational design as you actually add information.

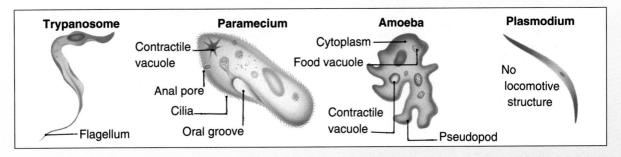

▶ APPLICATION

1. Organize the information about protozoans in a table. In your table, include shape, movement, and location of the organisms.
2. Organize the information about algae in an outline. In your data, include type of pigment, appearance, and uses.

✳ Using What You Have Learned

1. Which method of organization did you find most useful? Why?
2. Which method of organization made it easiest to locate information later?
3. Look at the diagrams and labels. Use any method described in this activity to organize the data in the diagrams. List the advantages and disadvantages of the method you chose.

Fungi

Objectives

Describe how fungi are different from other organisms.

Sketch the body structure of a typical fungus.

Summarize some of the ways in which fungi are important to humans.

What would you say if you went to the school cafeteria for lunch one day and you were given a plate of fungus? You'd probably say, "No, thank you!" The idea of eating fungus does not interest most people. However, some types of fungi can be very tasty. Have you ever eaten mushrooms? Mushrooms are a type of fungus. Have you ever eaten a slice of bread? Bread dough rises with the help of a fungus. Have you ever eaten cheese? Certain types of fungi help produce cheese. So the next time you are offered a plate of food that was produced by fungi, you might want to say, "Yes, thank you!"

Characteristics of Fungi

Fungi (FUHN jy) [singular, *fungus*] are organisms that decompose organic, or once-living, material. Sometimes called nature's re- cyclers, fungi cannot produce their own food the way plants can, and they do not have mouths to take in food as you and other animals do. Fungi must absorb their food through their cell walls.

Fungi may be *saprophytes* (SAP ruh fyts) or parasites. Saprophytes obtain food from dead organisms or from the waste products of living organisms. Parasites obtain food from a living host, usually causing injury or disease in the host.

The bodies of fungi are made of a network of threadlike structures called *hyphae* (HY fee). Fungi absorb food through their hyphae. Many hyphae together form a group of in- terlocking threads called the *mycelium* (my SEE lee uhm).

Most fungi reproduce by forming *spores*. The structures of a fungus that form spores are usually located above the surface on which the fungus grows. Because of this, the re- productive spores can be easily blown to different places by the wind. In the next activity, you can find out what some fungus spores look like.

DISCOVER BY Doing

You will need at least one cap from a gilled mushroom. (*Gilled* means the cap has ridges under it.) Place the cap gilled side down on a piece of paper. Put a damp cotton ball on top of the cap. Place a glass or plastic cup over the cap. Several hours later, remove the cap. The spores will have made a distinctive pattern on the paper. This pattern can be used to identify the mushroom. ✎

▼ ASK YOURSELF

How do fungi obtain food?

Types of Fungi

The common groups of fungi are the sporangium (spaw RAN jee uhm) fungi, the club fungi, the sac fungi, and the imperfect fungi. Lichens and slime molds are grouped separately even though they are similar to fungi in many ways.

Sporangium Fungi Have you ever seen a neglected piece of bread become "fuzzy"? This fuzzy substance is a fungus called *bread mold*. Molds grow best in warm, moist, dark places. However, if you have ever seen spoiled food from a refrigerator, you know that molds can also grow in cold temperatures.

Sporangium fungi are named after the spore case, or *sporangium* [plural, *sporangia*], in which the spores are produced. As you can see in the illustration, the sporangia are held in the air by hyphae. When a spore case breaks open, the spores are released. These spores may remain inactive for months. When the conditions are right, the spores will produce new hyphae.

Figure 9–14. Magnified bread mold resembles tiny spheres sitting on top of thin stems. The sphere-shaped structures are the sporangia. The "stems" are the hyphae.

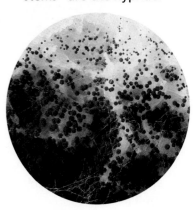

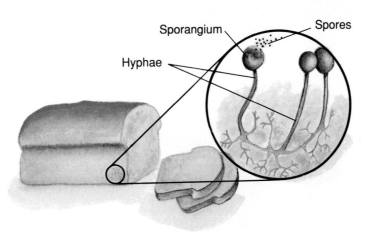

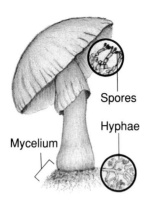

Figure 9–15. Mushrooms (left and right) and puffballs (center) are club fungi. Mushrooms produce spores under the cap.

Club Fungi

The mushroom, the puffball, and the bracket fungus are all examples of club fungi. They are called *club fungi* because the structures in which spores are produced look like clubs, as shown in the illustration.

Sometimes fungi destroy agricultural crops. The rusts and smuts, shown in the photographs, are parasitic club fungi that destroy grain. Each year rusts and smuts cost the agricultural industry thousands of dollars in lost grains.

Figure 9–16. Corn smut (left) and wheat rust (right) destroy many crops. Scientists are working to develop types of corn and wheat that are resistant to these fungi.

Sac Fungi

Sac fungi produce spores in sacs. Yeasts, one type of sac fungi, are important in the process of fermentation.

During the fermentation of bread dough, yeasts release carbon dioxide causing bread dough to rise.

Some sac fungi are parasites. Dutch elm disease and chestnut blight are tree diseases caused by sac fungi. In Dutch elm disease, the fungus is spread by bark beetles.

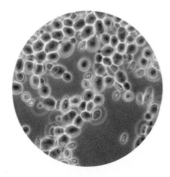

Figure 9–17. The morel (right) is an edible sac fungus. Yeasts (above) are used to make many food products.

MATERIALS
medicine dropper, slides (2), forceps, samples of fungi, coverslips (2), compound light microscope

PROCEDURE
1. Refer to page 582 of the Reference Section to learn how to make a wet mount. Using a medicine dropper, place one drop of water in the middle of a slide. With forceps, place a thin piece of mushroom or other fungus on your slide.
2. Put a coverslip on the slide. Look at your slide under the microscope. Make a sketch of what you see. Try to identify the hyphae and spores in your sample.
3. Using an additional sample of fungus, make another wet mount.
4. Repeat step 2.

APPLICATION
1. What structures were you able to identify in your samples?
2. How do the two samples differ? How are they similar?
3. Why do fungi successfully grow and reproduce in so many different places?

Imperfect Fungi Imperfect fungi are called *imperfect* because their complete reproductive cycles are unknown. The fungi that cause ringworm and athlete's foot belong to this group.

Although many imperfect fungi cause disease, one type of imperfect fungus can help cure disease. Have you ever had an illness for which you took an antibiotic? Maybe you were given penicillin. The imperfect fungus *Penicillium* is the source of the antibiotic penicillin. The antibiotic properties of *Penicillium* were discovered by British scientist Alexander Fleming in 1928. Fleming noticed that a mold was stopping the growth of bacteria in laboratory cultures. After many tests, Fleming isolated penicillin, and its use since then has saved hundreds of thousands of lives.

Figure 9–18. *Penicillium* mold is shown under a microscope (left) and on fruit (above).

Figure 9–19. Lichens are often the first organisms to grow on bare ground.

DISCOVER BY Writing

Choose the fungus that you find most interesting. Then in your journal draw and label the parts of that fungus. Describe the importance of the fungus.

▼ ASK YOURSELF

Are fungi helpful or harmful to the environment? Explain your answer.

Lichens and Slime Molds

A **lichen** is an organism that is part fungus and part alga. This combination produces an organism very different from either one. The alga in the lichen makes the food for the organism. How is this accomplished? The fungus gives the lichen support and helps it retain water. Lichens can be found in places where neither fungi nor algae can be found alone. Lichens can often be found attached to bare rocks.

Slime molds are very unusual organisms. They may look and reproduce like fungi, but they move like amoebas. Slime molds are often brightly colored and are found in damp soil and on rotting leaves or logs. The body of a slime mold behaves like an amoeba. It creeps along the ground or other surfaces and engulfs, or captures, food particles. When food is scarce, the slime mold becomes a spore-forming body, like a fungus. The spore-forming body releases spores that grow into new amoebalike organisms.

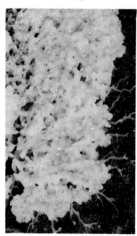

Figure 9–20. In one stage of its life cycle, a slime mold is an amoebalike growth.

▼ ASK YOURSELF

If you had to identify a slime mold as either a fungus or a protist, which term would you choose? Why?

SECTION 3   REVIEW AND APPLICATION

Reading Critically

1. Describe the body of a common fungus.
2. Relate several reasons why lichens and slime molds are difficult to classify.

Thinking Critically

3. What advantages do fungi have as saprophytes?
4. How are lichens able to grow on the surfaces of bare rocks?

HIGHLIGHTS

The Big Idea

Protozoans, algae, and fungi are very small organisms that make a large contribution to the world around us.

Protozoans comprise a diverse group of organisms that interact with the environment in helpful and harmful ways. While some protozoans cause disease, others feed on dead material, helping to replenish the supply of nutrients in the water environment.

Algae represent a primary food source for various aquatic organisms. Photosynthesis by algae also contributes tremendously to the earth's supply of oxygen. Some scientists estimate that algae release more oxygen into the atmosphere than all other land plants combined.

Without fungi, the speed with which organic materials decay would be decreased. Fungi also serve as a food source but are perhaps most noteworthy for their contribution to the world of germ-fighting antibiotics.

For Your Journal

Think about what you wrote in your journal at the beginning of the chapter. What have you learned about microscopic organisms? Revise your journal entry to reflect this new information. Did you include your ideas on the importance of protists and fungi to the environment?

Connecting Ideas

Copy this unfinished concept map into your journal. Complete the concept map by writing the correct term in each blank.

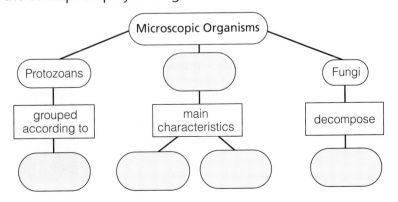

REVIEW

Understanding Vocabulary

For each set of terms, explain the similarities and differences in their meanings.

1. protozoans (230), algae (236)
2. algae (236), fungi (240)
3. fungi (240), protozoans (230)
4. flagella (233), pseudopods (231)
5. ciliates (232), sporozoans (234)
6. lichens (244), slime molds (244)

Understanding Concepts

MULTIPLE CHOICE

7. Which of the environmental conditions below is most likely to affect a protozoan?
   a) availability of food particles
   b) presence of light
   c) absence of water
   d) lack of oxygen

8. Choose the relationship that best describes the parasite/host relationship.
   a) A parasite seldom harms its host.
   b) A host seldom harms its parasite.
   c) A parasite frequently benefits its host.
   d) A host frequently benefits its parasite.

9. What advantage might a flagellate organism have when compared to an amoeba?
   a) efficiency of digestion
   b) efficiency of reproduction
   c) ease of respiration
   d) ease of movement

10. Which organism moves in a way that looks similar to a fish's movements?
    a) flagellate
    b) sarcodine
    c) sporozoan
    d) slime mold

11. Choose the way in which protists, algae, and fungi least benefit humans.
    a) They are a source of food.
    b) They replenish the supply of oxygen.
    c) They are a source of disease.
    d) They serve as a food source for other organisms we eat.

SHORT ANSWER

12. In what ways are amoebas similar to plants? In what ways are amoebas similar to animals?

13. Is it accurate to refer to fungi as decomposers? Explain your reasoning.

Interpreting Graphics

14. Why is it not a good idea to drink water from a mountain stream like the one shown in the picture?

15. Under what conditions would it be acceptable to drink water from a mountain stream?

Reviewing Themes

16. Systems and Structures
Choose a specific type of alga and explain how that alga interacts positively and negatively with its environment.

17. Energy
Explain how forest fungi replenish the supply of energy to the forest community.

Thinking Critically

18. Why is amoebic dysentery a common problem in less developed countries?

19. Why is diatomaceous earth a good silver polish?

20. Lichens are often the first organisms to grow on cooled lava after a volcanic eruption. What characteristics of lichens enable this growth to occur? What organisms might grow and what processes might occur after the lichens are established on the lava?

21. You may have heard of ways to avoid getting athlete's foot. Based on what you know about how fungi grow, explain why these preventive measures are effective.

22. Why is it so difficult to control the spread of fungi such as mildew or bread mold?

Discovery Through Reading
Morgan, Nina. *Louis Pasteur*. Bookwright Press, 1992. This biography details Pasteur's life, dreams, family, and accomplishments in chemistry and microbiology.

Science

PARADE

AIDS Research

Scientists have called it the modern plague. By the early 1990s, it had infected more than nine million people worldwide. This disease is AIDS. AIDS destroys the body's ability to fight infection. More than a decade after discovering it, scientists are still seeking ways to treat and prevent this deadly disease. Medical science still has not solved the mystery of AIDS.

Mystery Killer

The events surrounding the discovery of AIDS unfolded like a detective story. Physicians and scientists observed the first signs of the disease in the late 1970s. They noticed patients dying from a normally rare skin cancer and a usually mild form of pneumonia. Scientists suspected that something was weakening the patients' natural defenses against these infections. At the Centers for Disease Control in Atlanta, Georgia, physicians named the new disease AIDS and began looking for a cause.

A group of researchers was already investigating a family of viruses that attacked white blood cells, or lymphocytes. Lymphocytes control the body's immunity. Any virus that destroys lymphocytes leaves the body defenseless against pneumonia, cancer, and many other diseases. Researchers suspected that a virus of this type caused AIDS.

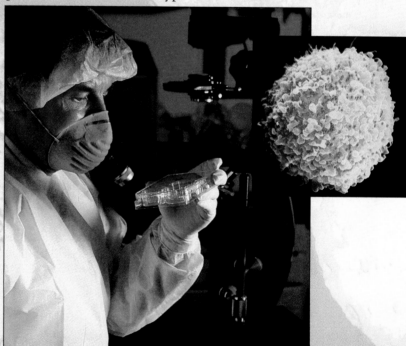

Left: Dawn Marcal, infected as a teenager in high school, was a speaker for AIDS Education. She died in August, 1992.
Center: Magic Johnson, former basketball star, revealed that he is HIV positive in November of 1991.
Below: AIDS research

In 1984—less than three years after the first AIDS cases were identified—researchers successfully isolated the AIDS virus in the laboratory. They named it *Human Immuno-deficiency Virus* (HIV). Scientists soon learned how the virus works. When HIV first enters the body, it stimulates the growth of antibodies, chemical substances that fight infection. However, the antibodies cannot stop the virus. Soon the virus attacks the body's lymphocytes, and the body cannot fight off infections and disease.

Possible Vaccines

The search for an AIDS vaccine illustrates the difficulties of AIDS research. One problem is making the vaccine itself. Scientists have tried deactivating the HIV virus and then injecting it into the body. They hope the body will respond by producing antibodies to fight HIV infection. In another approach, researchers alter the genetic makeup of living HIV. The altered virus does not cause AIDS but can reproduce in the patient's body, stimulating the immune system.

No Easy Task

Scientists may have to develop several AIDS vaccines. Recent findings indicate that a vaccine against intravenous infection may not guard against sexual transmission. A different vaccine may protect a fetus in the womb from infection by an HIV-positive mother. Doctors would like to vaccinate HIV-positive patients against the onset of AIDS. Others believe that an effective vaccine can only prevent an HIV carrier from passing the virus to others.

The nature of HIV complicates the search for a vaccine. First, the AIDS virus is highly variable. It can alter itself genetically within the person's body. In fact, the AIDS virus may have evolved genetically just in the years since its discovery.

Second, the human virus infects only humans, which complicates testing. Researchers can infect monkeys with a simian variety of HIV. However, a vaccine that is then found effective on monkeys may not work on humans.

Finally, reliable human testing requires that scientists give a vaccine to a test group and withhold it from a control group. If the vaccine fails, the test group might contract HIV. If it succeeds, AIDS patients might demand the vaccine before a reliable test is completed.

AIDS Drugs

The same problems apply to AIDS drug research. The antiviral drug AZT has become the standard treatment for HIV infection and AIDS. AZT apparently blocks the virus from taking over the DNA of white blood cells. Researchers are still testing AZT to determine its effectiveness. While the drug may slow AIDS onset, it does not necessarily extend the patient's life. Researchers are also seeking drugs to treat AIDS-related infections and diseases. ◆

Through the Microscope

by Arthur S. Gregor
from *Man's Conquest of Nature from Ancient Times to the Atomic Age*

"Gentlemen of the Royal Society, Honored Sirs. My name is Anton van Leeuwenhoek and I am the sheriff's deputy and janitor of the town hall here in the city of Delft in Holland, where I was born. I write to you because I have a hobby in which you might be interested. Much of my time I spend cleaning the town hall, but every moment I have to spare I arrange lenses in such a way that the tiniest things in the world are enlarged. In fact, I build microscopes, 247 of them so far.

Through my microscopes I look at hairs, at seeds, at the sting of bees and at the brains of flies. Gentlemen, I have seen things no man has ever seen before!"

*A*t first [Anton van Leeuwenhoek] could not believe his eyes. He called out to his family to come and see the marvel. Under the lenses of his microscope, imprisoned in the water, were tiny creatures, turning, twisting, and leaping about.

Some swam with the help of tiny hairs. Some rolled over and over. Some actually changed their shape as they went lumbering from one side to the other. The drop of water was alive with thousands of little animals.

Leeuwenhoek had stumbled upon the world of microbes!

Here was a new world of which men had never dreamed, a world so small that a million of these creatures could fit into a thimble. Here was the cause—though Leeuwenhoek did not know it—of many of the diseases that have plagued mankind since the beginning of time.

He was too good a scientist to rest content with his discovery. Question after question rushed to his mind.

Where did these amazing little creatures come from? Could they have come out of the sky? How could he find out?

Well, he'd experiment as he always did when he came upon a new problem.

He cleaned and scoured a dish until it shone. Then

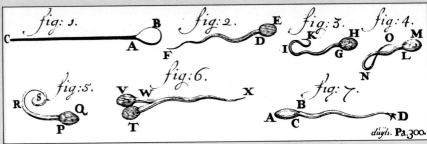

he caught a drop of rainwater directly in it. Now again he looked. But this time not a trace of the creatures!

He repeated the experiment over and over. Still no trace of them. So they did not come out of the sky. Well, then where did they come from?

He put the dish aside for several days and then looked again.

Ah, there they were, full of life, frolicking amidst bits of dirt and dust.

Now Leeuwenhoek turned detective and set to work to track down his creatures. He examined water from housetops, from mud puddles, from barrels, from wells, from the canals of Delft. And wherever there was dust and dirt he was sure to find his creatures.

Then he had a curious thought. "Why go out to find them. Can I have them come to me?"

He allowed a few grains of pepper to soak in water. And what he saw several days later sent him scribbling like mad to the Royal Society. The pepper water was alive with the creatures, millions of them, more than there were people in the whole of Holland.

Leeuwenhoek's simple microscope

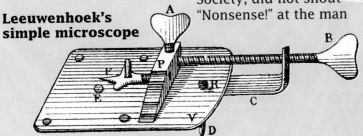

Anton van Leeuwenhoek had discovered one of the most important of all laboratory methods: a way of growing microbes!

"Nonsense!" cried some of the members of the Royal Society when Leeuwenhoek's letter came in 1680. "How much of this foolishness do we have to take?"

But Robert Hooke, a great experimenter himself and member of the Royal Society, did not shout "Nonsense!" at the man

Leeuwenhoek's drawings

who had seen the circulation of the blood with his own eyes.

Hooke built a new and powerful microscope and brewed pepper water himself. And what *he* saw sent him dashing back to the Royal Society.

"Gentlemen, our Dutch friend has not lied. Come and see for yourself."

The scientists and the professors crowded around Hooke's microscope. It was just as Leeuwenhoek had said; the water was teeming with microbes.

It was the greatest day in Anton van Leeuwenhoek's life when the reply from the Royal Society arrived. For along with it was a magnificent diploma in a beautiful silver case.

The sheriff's deputy and janitor of the town hall of the city of Delft was now an honored member of the Royal Society! On his death in 1723 he repaid the Society by leaving it his twenty-six most precious microscopes. ◆

ROBERT KOCH
(1843—1910)

In 1890, Robert Koch (KAWK) thought he had a cure for tuberculosis in a bacterial extract called tuberculin. He was wrong, but he continued his research. Later, tuberculin was found useful for diagnosing tuberculosis.

Robert Koch was born in Clausthal, Germany, on December 11, 1843. He attended Göttingen University, where he studied medicine and the natural sciences. Koch became a physican but then devoted himself to research. He identified bacterial causes of diseases, such as anthrax, a deadly disease that occurs in sheep. Koch also discovered that bacteria were responsible for the major cause of death in his century, tuberculosis. A research institute for the study of tuberculosis was set up by the Prussian government and headed by Koch.

Koch developed methods that showed scientists how to search for and identify disease-causing bacteria. He also made contributions to the study of cholera, sleeping sickness, leprosy, cattle plague, and bubonic plague. For his work on tuberculosis and other diseases, the founder of bacteriology received many honors, including the Nobel Prize in Medicine in 1905. ◆

LYNN MARGULIS
(1938—)

In 1966, Lynn Margulis theorized that all multicellular organisms are the result of the joining together of monerans. Her ideas were not widely accepted then, but today, many scientists agree with her.

Lynn Alexander Margulis was born in Chicago on March 5, 1938. After receiving her bachelor's degree from the University of Chicago in 1957, she did graduate work at the University of Wisconsin and the University of California at Berkeley. Since receiving her Ph.D. in genetics in 1965, Margulis has taught at Boston University.

Margulis has written many articles on the classification and evolution of living organisms. She believes that many cellular organelles were originally bacterialike cells that over millions of years began to work together.

Not all scientists agree with Margulis's theories. But many biologists recognize that our understanding of the organization of cells has been greatly expanded through her work. In 1975, Margulis received the Dimond Award from the Botanical Society of America for her work on living bacteria, and in 1981, she received the Public Service Award from NASA. ◆

HELENE GAYLE,
Public Health Physician

Dr. Helene Gayle provides information to people about how to prevent disease. She is a public health physician. Gayle began her career in disease control at the Centers for Disease Control in Atlanta, Georgia. The Centers for Disease Control, or CDC, is a part of the United States Public Health Service. Its main purpose is to prevent early death among the citizens of the United States. The CDC does this by finding ways to prevent and control diseases.

This work involves learning the ways a disease is spread. Gayle kept records of how many people had a disease, where the people lived, and what their lifestyles were like. Then, studying this information, Gayle looked for patterns to help explain how the disease was being spread and how to prevent it from spreading further.

Gayle is currently Chief of the AIDS Division for the United States Agency for International Development, a foreign assistance agency of the United States government. The AIDS Division is part of the agency program designed to improve health care in developing countries in Africa, Asia, Latin America, and the Caribbean. Through the agency,

Gayle provides health information and medical resources to assist in HIV and AIDS prevention.

Gayle believes that education is the best way to prevent the spread of AIDS. She wants young people to realize that the choices they make now may affect their entire lives.

It is Gayle's hope that her work will make a difference. The more people who know about AIDS and how it is spread, the better they can protect themselves. Gayle wants people everywhere to know that AIDS can be prevented. ◆

AIDSCOM
Education is not enough . . .

Algal Blooms

Algae can be found in all bodies of water. When conditions are favorable, algae reproduce rapidly and form colored patches called *blooms* on the surface of the water. Algal blooms can cause problems. They can damage aquatic plants and animals, certain industries that depend on marine life, and supplies of drinking water.

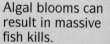

Algal blooms can result in massive fish kills.

Ecosystem Overload

Algae multiply rapidly and form blooms when nutrients enter fresh water. The nutrients may come from treated sewage or from the runoff of fertilized farm soil. When large numbers of algae die and decompose, they cause an oxygen shortage in the water. This condition is called *eutrophication*.

Algal blooms also limit the amount of sunlight that can reach the bottom of a lake or river. As a result, plants that need sunlight die, and the water ecosystem is damaged still further.

Deadly Consequences

A species of algae called *dinoflagellates* is responsible for toxic ocean blooms called *red tides*. These algae give the water a red color, and their toxin can contaminate marine animals such as mussels and clams. Eating this contaminated seafood can be fatal. Ocean spray that contains red-tide dinoflagellates can even cause respiratory problems in humans.

Search for Solutions

It is possible to control algal blooms in fresh water, but doing so is expensive. The water can be cleaned up by removing the algae and the bottom sediment. If the dumping of sewage into fresh water cannot be stopped, silica added to the sewage can help. Silica encourages the growth of diatoms, a type of algae that does not bloom.

The diatoms slow the growth of other algae and increase the food supply of fish.

So far no clear-cut control has been found effective against red tides in the oceans.

Now researchers are developing models for predicting red tides. Prediction rather than control may turn out to be the only good solution.

Scientists are also studying the possible benefits of red tides on the ocean ecosystem. There have been reports of larger-than-usual catches of shrimp in years following a major red tide. Scientists also hypothesize that dinoflagellate blooms are responsible for oil deposits in the North Sea. ◆

Bloom-causing dinoflagellates (top) can be offset by adding diatoms (center and bottom) or draining water.

UNIT 4 PLANTS

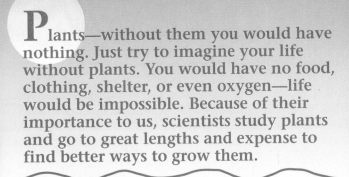

Plants—without them you would have nothing. Just try to imagine your life without plants. You would have no food, clothing, shelter, or even oxygen—life would be impossible. Because of their importance to us, scientists study plants and go to great lengths and expense to find better ways to grow them.

Science PARADE

CHAPTER 10

An Introduction to Plants

Don't look now, but plants are everywhere around you. You know that the geraniums decorating the windowsill are plants. But what about the wood in your pencil, the cotton in your shirt, or the apple in your lunch? Wood, cotton, and apples all come from plants. And that's just three of the many ways people use plants.

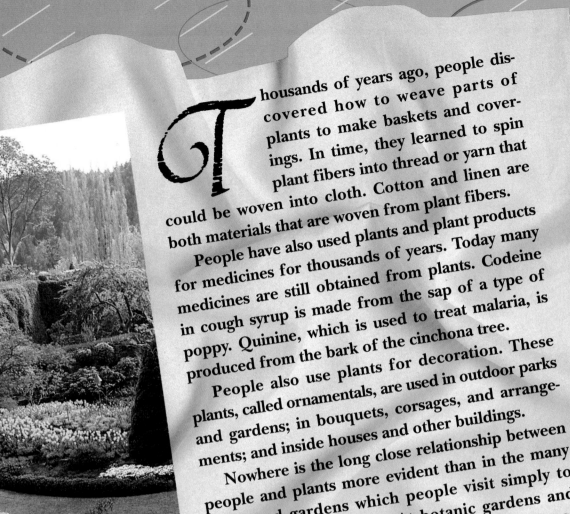

Thousands of years ago, people discovered how to weave parts of plants to make baskets and coverings. In time, they learned to spin plant fibers into thread or yarn that could be woven into cloth. Cotton and linen are both materials that are woven from plant fibers.

People have also used plants and plant products for medicines for thousands of years. Today many medicines are still obtained from plants. Codeine in cough syrup is made from the sap of a type of poppy. Quinine, which is used to treat malaria, is produced from the bark of the cinchona tree.

People also use plants for decoration. These plants, called ornamentals, are used in outdoor parks and gardens; in bouquets, corsages, and arrangements; and inside houses and other buildings.

Nowhere is the long close relationship between people and plants more evident than in the many parks and gardens which people visit simply to enjoy looking at plants. At botanic gardens and arboretums, plants are displayed for people to see and enjoy. But these institutions have other important purposes. They collect, grow, and study plants to increase our knowledge about plants, to improve the varieties of plants, to conserve rare plants, and perhaps to discover new ways people can use plants in the future.

For Your Journal

- What do you visualize when you think about plants?
- How do you use plants in your daily activities?
- How do plants grow, make food, and reproduce?

Types of Plants

Objectives

Describe *the structures of vascular plants.*

Identify *some of the groups of vascular plants.*

Distinguish *between vascular and nonvascular plants.*

One warm spring day your class visits a botanic garden, where hundreds of varieties of plants are grown, studied, and displayed. Your guide through the garden is a botany student named Beth. She explains that you will visit greenhouses that recreate the kinds of environments that different plants need. Then Beth holds up a petunia plant. She points to the different parts of the plant as she talks about the functions of the parts.

Vascular Plants

Like the petunia, most of the plants you see every day are vascular plants, or plants with a system of vessels that carry materials throughout the plant. Inside the roots, stems, and leaves of these plants are two types of *vascular* tissues that transport nutrients and water. **Xylem** (ZY luhm) transports water and dissolved minerals from the roots to the leaves. **Phloem** (FLOH ehm) transports the food that is made in the leaves to all parts of the plant. Why is it necessary for plants to have both xylem and phloem?

Figure 10–1. Many varieties of plants may be found in a botanic garden.

Anchors Away Beth gently removes the petunia from the soil. The many roots of the plant are thin and branching—soil particles cling to the roots.

Roots anchor many plants into the ground, so the plants are not blown away by wind or washed away by rain. Roots also absorb water and minerals from the soil. Thick roots, such as those of carrots and radishes, store food.

Plants have one of two types of root systems. Plants with taproots have one main root that grows almost straight down. *Taproots* can be long, like carrots, or rounded, like radishes. Plants with *fibrous roots* have many thin

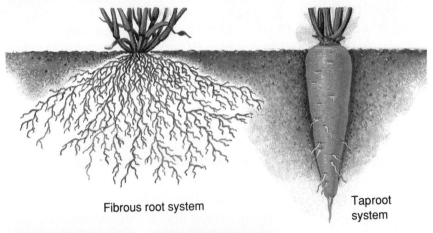

Fibrous root system

Taproot system

Figure 10–2. Fibrous roots spread out over a large area. Taproots are thick, single roots.

roots in a dense network and no one main root. Fibrous roots hold soil together and help keep it from eroding. Both types of roots have very fine, threadlike structures called *root hairs*. Root hairs absorb water and minerals from the soil by osmosis and diffusion. You can observe root hairs in the following activity.

DISCOVER BY *Observing*

Obtain some seeds that have germinated, or sprouted. Observe the roots with your unaided eye, with a hand lens, and under low power of a microscope. Describe your observations. How does magnification affect your observations? ✎

Roots grow in both length and width. In plants, new cells are produced only in tissue called *meristem*. Cells produced in the root meristem begin to grow in the region of elongation. After the cells get longer, they develop into the other tissues of the root, including vascular tissue. The root cap protects the growing root tip as it pushes through the soil. Locate these root parts in the diagram.

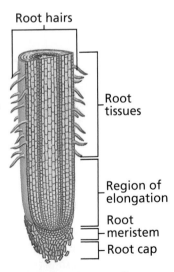

Root hairs

Root tissues

Region of elongation

Root meristem

Root cap

Figure 10–3. Cell division occurs only in the meristem of the root tip. If a root tip is cut or broken, it cannot grow any longer.

An Upright Job

An Upright Job   Beth asks you to observe the petunia's stem. It is green and flexible. The functions of stems are to support the leaves and to connect the leaves to the roots. In some plants, such as cacti, stems produce and store nutrients.

Plant stems are sometimes divided into two types: *herbaceous* (hur BAY shuhs) and *woody*. Herbaceous stems are green and flexible. They depend on water pressure in their cells to hold them upright. Woody stems are harder and stiffer than herbaceous stems and contain wood cells that give them support.

Most plants with herbaceous stems grow, mature, reproduce, and die in one growing season and are called *annuals*. Many kinds of vegetables, including corn, peas, and radishes, are annuals. Many ornamentals, such as petunias, pansies, and marigolds, are also annuals.

Plants with woody stems and some plants with herbaceous stems live for more than one growing season. These plants are called *perennials* (puh REHN ee uhlz). Many grasses, shrubs, and trees such as oak and pine are perennials.

During each growing season, the stem, or trunk, of a tree gets thicker because new wood is deposited between the outer covering of the trunk, called *bark*, and the internal tissue. The new growth for each year forms a growth ring. The rings vary in thickness from year to year depending on the weather and other growing conditions. A rainy year may cause a tree to grow faster, producing a thick growth ring. Dry weather may cause a tree to grow more slowly, producing a thinner growth ring.

Stems contain both xylem and phloem. In a herbaceous plant, xylem and phloem are arranged in bundles, as shown in the following illustration. The long, stringy fibers you see when you bite a stalk of celery are the vascular bundles. In a woody stem, the xylem and phloem are arranged in a ring. The phloem is on the outside of the ring, and the xylem is on the inside.

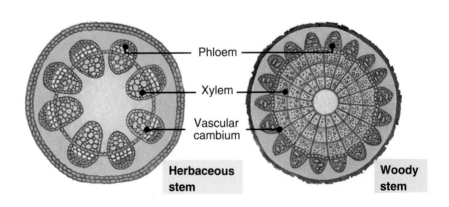

Figure 10–4. In a herbaceous stem (left), xylem and phloem are grouped together in bundles. In a woody stem (right), the xylem is toward the inside and the phloem is toward the outside.

Phloem

Xylem

Vascular cambium

Herbaceous stem

Woody stem

Between the xylem and the phloem is another tissue, called the *vascular cambium*. **Vascular cambium** is the growth tissue that produces new xylem and phloem cells in the stems of woody plants. As the xylem accumulates, the woody plant grows in diameter.

A stem, with its attached leaves and branches, grows from a region of dividing cells at its tip called the *apical meristem*. The apical meristem produces leaves that grow around it, forming the apical bud. Other buds form that become branches or flowers. In the mature region of a stem, the cells form into the vascular tissues and other tissues of the stem. You can observe the functioning of vascular tissue in the following activity.

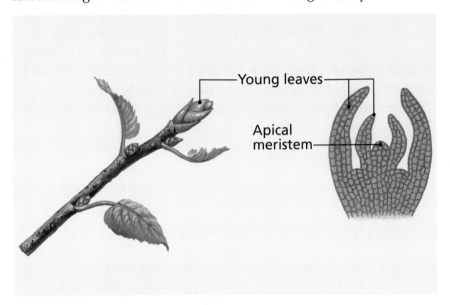

Figure 10–5. Just as roots grow at their tips, stems also grow at their tips.

Young leaves

Apical meristem

DISCOVER BY *Doing*

You will need a fresh white carnation, food coloring, scissors, and a glass of water. Cut the stem of the carnation to about 10 cm in length. Add several drops of food coloring to the water in the glass. Place the carnation stem in the colored water and let it stand for 24 hours. What happens to the carnation? Explain. 🖉

Food Factories   Leaves are the major food-making structures of a plant. They are the parts of the plant in which most photosynthesis takes place. **Photosynthesis** is the process by which green plants use chemicals from the environment and energy from the sun to make their own food. You will learn more about photosynthesis in the next section.

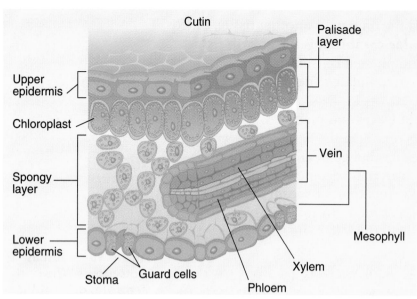

Cutin

Palisade layer

Upper epidermis

Chloroplast

Spongy layer

Lower epidermis

Stoma

Guard cells

Phloem

Xylem

Mesophyll

Vein

Figure 10–6. A leaf is many cell layers thick.

The illustration shows a cross section of a leaf. The outer layer of the leaf, or the *epidermis* (ehp uh DUR mihs), is a layer of thin, flat cells—like the cells of your skin. In some plants, the epidermis is covered with wax. The wax protects the inner parts of the leaf and reduces water loss. In the center of the leaf are the layers of cells in which photosynthesis takes place. Veins connect to the roots and stems. Openings in the epidermis, called *stomata*, allow air and water to pass into and out of the leaf. Each stoma's size is controlled by guard cells.

 ASK YOURSELF

Trace the path of water from the root hairs to the leaf stomata.

Complex Vascular Plants

At the botanic garden, Beth walks over to a small pine tree. She points out that both the petunia and the pine tree are complex vascular plants. This means that they produce seeds. Many of the plants with which you are familiar are complex vascular plants. Some produce flowers and fruits, while others do not. But they all produce seeds. Name some plants that produce seeds.

There are two groups of complex vascular plants: the *gymnosperms* (JIHM nuh spurmz) and the *angiosperms* (AN jee uh spurmz). Gymnosperms lack a protective covering around their seeds, and they do not produce flowers. Plants that have a protective covering around their seeds and produce flowers are angiosperms. Which kind is the pine tree? Which kind is the petunia?

Gymnosperms   There are over 500 species of gymnosperms, including cycads, ginkgoes, and conifers. Most gymnosperms keep their leaves throughout the year. As a result, they are often called *evergreens*. Gymnosperms produce seeds in cones.

Angiosperms Most of the plants that live on Earth are angiosperms. Angiosperms produce seeds from flowers, and all produce some kind of fruit. Crop plants, hardwood trees, shrubs, grasses, and desert plants are all angiosperms. Angiosperms are divided into two groups: *dicots* (DY kahtz) and *monocots* (MAHN uh kahtz). Some of the differences between dicots and monocots can be seen in the diagram below.

Figure 10–7. What angiosperms can you identify in this picture?

The vascular tissue bundles are arranged differently in dicots and monocots. In dicots, they are arranged in a circle. In monocots, vascular bundles are found throughout the stem. The leaves of dicots are broad and have veins branching throughout them. Monocots have long leaves with veins that run from one end of the leaf to the other. Dicots include oaks, maples, elms, roses, beans, cabbages, tomatoes, and petunias. Monocots include grasses, palms, daffodils, irises, and lilies.

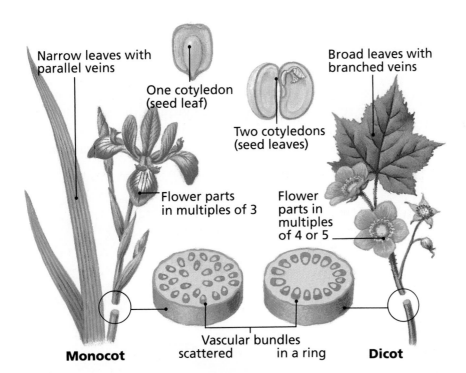

Narrow leaves with parallel veins

One cotyledon (seed leaf)

Two cotyledons (seed leaves)

Broad leaves with branched veins

Flower parts in multiples of 3

Flower parts in multiples of 4 or 5

Vascular bundles scattered

in a ring

Monocot

Dicot

Figure 10–8. Monocots and dicots have characteristics that make them easily identifiable.

ASK YOURSELF

What are the differences between gymnosperms and angiosperms?

Figure 10–9. Plants such as horsetails (inset) and ferns have poorly developed vascular tissues and do not produce seeds.

Simple Vascular Plants

Beth leads your class into a greenhouse where the air is moist and warm. Here some of the plants have leaves shaped like fiddleheads; others have thin, feathery leaves. Beth says that all of these plants are ferns. Plants such as ferns, horsetails, and club mosses are simple vascular plants. Simple vascular plants are seedless. Because many have simple, poorly developed xylem and phloem, these plants need a lot of moisture in their environment to survive and to reproduce.

Ferns   If you have ever walked through a warm, damp forest, you have probably seen ferns. They are the most common seedless plants. Ferns have stems, roots, and leaves, but they do not make seeds. There are about 12 000 different kinds of ferns; they live in areas where the soil is moist and the climate is temperate. The life cycle of ferns includes both sexual and asexual phases.

During the asexual, or sporophyte, generation, small brown or orange structures can be seen on the bottom of the fern leaf. These groups of spore-bearing cases hold many spores. When the spores are released, they fall to the ground or are carried away by wind or water. When conditions for growth are right, a spore can grow into the sexual, or gametophyte, generation, which produces gametes—sperm and ova.

Figure 10–10. The life cycle of a fern

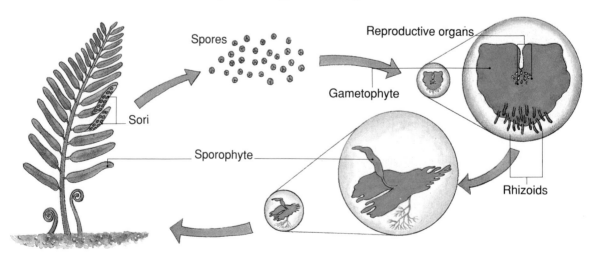

Rootlike structures grow into the ground to anchor the tiny plant and supply it with minerals and water. On the bottom of the gametophyte, male and female reproductive organs form. When sperm cells are released, they must swim through water—dew or rain—to reach the female part of the plant. One sperm combines with an ovum to start a new sporophyte generation. In the following activity, you can observe the spores of ferns.

DISCOVER BY Doing

You will need a fern leaf, a hand lens, a toothpick, a slide, and a microscope. Using the hand lens, examine the underside of a fern leaf. Locate the spore cases and make a drawing of them.

Scrape several spore cases onto a slide and make a wet mount of them. Examine them under the microscope and make a drawing of what you see. Try to germinate some spores in moist soil. ✎

Other Seedless Plants

Horsetails and club mosses are also seedless plants. Like ferns, these plants undergo sexual and asexual phases during their life cycles. Horsetails are found in areas where the weather is usually warm and moist. Their shoots look like the tails of horses. Only one genus of horsetails grows on Earth today. All of the others are extinct. Ancient horsetails, which lived over 300 million years ago, were as large as trees and lived in swampy areas.

The other seedless vascular plants are the club mosses. There are about 800 different species. The first club mosses lived about 350 million years ago. Club mosses are also found in moist areas.

ASK YOURSELF

How do ferns reproduce?

Figure 10–11. Horsetails (left) and club mosses (right) are among the most primitive vascular plants.

267

Figure 10–12.
Nonvascular plants such as mosses (left), hornworts (center), and liverworts (right) must grow in moist areas in order to survive.

Nonvascular Plants

In the next greenhouse, the air is very moist, the ground is very wet, and the plants are very small and low-growing. Beth points to what looks like a clump of green velvet and explains that it is a kind of moss. The moss is an example of a nonvascular plant. Nonvascular plants are simple plants that do not produce seeds and do not have vascular tissue. Some nonvascular plants have structures that look like but are not true roots, stems, or leaves.

Mosses, liverworts, and hornworts are simple land plants that need a moist environment. These plants have underground structures that anchor them in the soil and absorb water and minerals, but these structures are not roots. Other structures look like stems and leaves. However, the structures have no vascular tissue. Because food and water must pass from cell to cell, these plants don't grow very large.

 ASK YOURSELF

Why do nonvascular plants need a moist environment?

SECTION 1 *REVIEW AND APPLICATION*

Reading Critically
1. What is the function of vascular tissue in plants?
2. What is the difference between simple and complex vascular plants?

Thinking Critically
3. How would you distinguish between a moss and a fern?
4. In what ways might a sexual and an asexual phase of reproduction be an advantage to mosses and similar plants?

SKILL

Making A Bar Graph

▶ **MATERIALS**
graph paper

▼ **PROCEDURE**

1. Often it is necessary for scientists to collect and compile data. They may use the data to make a bar graph. Bar graphs can display a lot of information about the subject or process being studied in a very concise, efficient way.

2. To make a bar graph to show information, you must first decide on a title. Label the graph with the title and draw the vertical and horizontal axes so that they cover most of the paper.

3. Then you must decide how your axes should be labeled. Once you decide, make each square on the graph paper equal to a particular unit of measure. That unit could be 1 cm or 100 m. It does not matter, as long as each square stands for the same amount.

4. When you make a bar graph, the vertical axis usually indicates how many of something are the same. For instance, in this Skill activity, you might fill in three squares vertically if three of the plant leaves measured 10 mm in length.

▶ **APPLICATION**

1. **CAUTION: Do not attempt to collect leaves from any plants that are considered hazardous or endangered.** Choose a plant to study from an area near your home. Collect five leaves from this type of plant.

2. Decide on the unit that is best for the plant leaves you have chosen and measure the length of each leaf. Be sure that you are using an SI unit. Record this information in a table on the chalkboard. Copy the information for the leaves of the entire class.

3. Organize the leaves into categories according to length. For example, all leaves that are at least 6 cm long but less than 7 cm might be grouped in one category.

4. Make a bar graph of the class data. Compare your graph with those of your classmates. Did everyone in the class graph the information in the same way? Why or why not?

5. What general patterns do you see in the length of plant leaves for your class plant?

6. How does the bar graph help you see patterns and relationships in the data you collected?

7. After examining all the class graphs, which method used do you think was the best for representing the information?

✳ ***Using What You Have Learned***
Select another characteristic of plants and create another bar graph. You may wish to consider height of plants, number of leaves on a stem, or number of petals on a flower.

Photosynthesis

Standing in one of the largest greenhouses, you can see hundreds and hundreds of green plants. Overhead the sun streams through the glass panels. It is very quiet, but Beth says that actually there is a lot of activity going on. All the plants are busy, making, storing, and eating food! What does she mean?

Beth explains that unlike animals, plants make their own food. For a long time, people believed that plants ate soil just as animals ate food. In the 1600s, a man from Belgium, J. B. van Helmont, conducted an experiment to test the idea. First, some soil was dried and weighed. Then van Helmont planted a willow tree in a large tub filled with the soil. For five years, nothing was added to the soil but water. After five years, van Helmont removed the tree from the tub. Again he dried and weighed the soil. Although the tree had grown much larger, the soil weighed almost the same as it had five years earlier! What do you think van Helmont concluded? He concluded that plants don't eat soil but that they may eat water since that was all he had added to the tub.

Figure 10–13. The botanic garden in Oklahoma City

How Plants Get Food

Plants absorb sunlight. To understand how this works, Beth asks you to look at one green plant in the greenhouse. The sun shines on the plant. Energy from the sunlight is absorbed by chlorophyll in the cells of each leaf. **Chlorophyll** is the pigment that gives plants their green color. Chlorophyll traps solar energy and converts it into chemical energy. You can try the next activity to test the importance of sunlight for plants.

DISCOVER BY Doing

You will need two bean plants that are the same size. Put one in a dark place and the other in light. Wait for several days. Then bring out the bean plant that was in the dark place. Compare the two plants. What color is each plant? What other differences do you notice? ✏

Plants need sunlight to stay green. Only green plants can make food. Therefore, plants can't make food without sunlight. The process that enables a plant to make its own food is called *photosynthesis.*

Figure 10–14. Whether green plants grow on land or under water, chlorophyll converts sunlight into chemical energy.

Besides making the food that plants use themselves, photosynthesis is important to all animals, including humans. Since most animals eat plants, or animals that have eaten plants, photosynthesis provides animals with food as well. In addition, during photosynthesis, plants remove carbon dioxide from the air and give off oxygen. Animals could not live without the oxygen that plants produce.

Photosynthesis takes place in organelles called *chloroplasts.* Chloroplasts contain chlorophyll. Most chloroplasts are located in leaves. In the next activity, you can see where most of the chloroplasts are located in a leaf.

ACTIVITY

What does a leaf look like when observed under a microscope?

MATERIALS

prepared slide of leaf cross section, compound light microscope

PROCEDURE

1. Obtain a slide of a leaf cross section from your teacher.

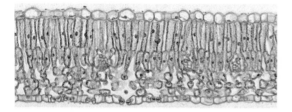

2. Observe the cross section under low and high power. Draw what you see.
3. Label the following plant tissues on your sketch: epidermis, vein, chloroplasts, stomata, and guard cells. Where are most of the chloroplasts located?

APPLICATION

1. What is the function of each part of the leaf that you labeled?
2. How is each of the tissues of the leaf especially well suited to the job it does?

Yellow, red, and orange pigments, in addition to chlorophyll, are also found in plants. However, in spring and summer, the leaves appear green because of the large amount of chlorophyll present. In the fall, the plant stops making chlorophyll and begins to use the existing chlorophyll as food, so the other colors become visible.

Other plant parts contain colored pigments, too. Carrots, for instance, have a great deal of orange pigment. Tomatoes and cherries have red pigment. What other examples of pigmented plant parts can you think of?

Figure 10–15. In the fall, plants stop making chlorophyll, so other colors show in the leaves.

 ASK YOURSELF

Why is photosynthesis important to plants and animals?

How Photosynthesis Occurs

The materials necessary for photosynthesis are water (H_2O) and carbon dioxide (CO_2). The most common products of photosynthesis are glucose ($C_6H_{12}O_6$) and oxygen (O_2). From glucose and other sugars, compounds such as starches and proteins are made. The chemical equation for photosynthesis is:

$$6CO_2 + 6H_2O \xrightarrow[\text{chlorophyll}]{\text{light}} C_6H_{12}O_6 + 6O_2$$

$6CO_2 + 6H_2O$ carbon water dioxide   light chlorophyll   **$C_6H_{12}O_6 + 6O_2$** glucose oxygen

Photosynthesis is a complicated process that occurs in two stages. During the first stage, light is absorbed by chlorophyll, and energy-rich chemicals are produced. These energy-rich chemicals are needed for the second stage of photosynthesis. Oxygen is released during the first stage, and glucose and water are produced during the second stage. In the following activity, you can learn more about photosynthesis.

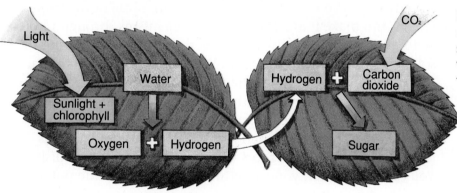

Figure 10–16. The reactions of photosynthesis provide plants with nutrients necessary for survival.

DISCOVER BY Doing

You will need a tall thin jar (like an olive jar), a small bowl, some water, a sprig of *Elodea,* and a desk lamp. Put some water in the bowl. Fill the jar with water and drop the *Elodea* into it with the cut end at the top of the jar. Put your hand over the top of the jar and turn it over. Place it in the bowl. Try to keep as much of the water inside the jar as possible. Put the lamp close to the jar and turn the lamp on. Watch the *Elodea* for 20 minutes. What happens? ✎

▼ ASK YOURSELF

How is light used during photosynthesis?

Figure 10–17. The energy stored by plants is transferred to humans and other animals when plants or animals are eaten.

Energy Transfer

Green plants are very important to humans as well as other animals; the survival of animals depends on green plants and other organisms that carry out photosynthesis. Most energy, except for geothermal and nuclear energy, comes either directly or indirectly from the sun. Green plants capture some of the sun's energy and make it available to other organisms through photosynthesis. When animals eat plants, the animals are getting this solar energy in the form of food. When, in turn, those animals are eaten by other animals, the energy is passed on through the food chain. Without plants, the energy from the sun could not be transferred from one organism to another.

You have probably eaten many different types of plants and plant parts. You eat roots when you eat carrots or radishes. You eat leaves when you eat lettuce. If you have white potatoes or asparagus with your dinner, you are eating stems. If you have cauliflower or broccoli, you are eating flower buds. If you have string beans or cucumbers, you are eating fruits. Peas and corn kernels are seeds.

 ASK YOURSELF

How does the sun supply energy to humans?

SECTION 2 *REVIEW AND APPLICATION*

Reading Critically

1. What is the source of energy for photosynthesis?
2. Name the materials needed for photosynthesis to occur.

Thinking Critically

3. In what way did photosynthesis make animal life on Earth possible?
4. What eventually happens to a tree's colorful fall leaves? Explain your answer.

Reproduction in Seed Plants

At the botanic garden, Beth brings your class back to the pine tree and the petunia plant that she showed you when you first arrived. Do you remember what the pine tree and the petunia have in common? They are both complex vascular plants; they have roots, stems, and leaves, and they produce seeds.

The development of seeds is an adaptation that increased the ability of vascular plants to survive unfavorable environments. Seeds protect plant embryos from harsh conditions; seeds can remain dormant for years and still produce healthy plants.

Two groups of seed plants have evolved. Gymnosperms generally produce their seeds in cones and keep their leaves all year. Angiosperms produce flowers, protect their seeds in fruits, and usually lose their leaves in the fall. The pine tree is a gymnosperm, and the petunia is an angiosperm. How are their seeds different?

Gymnosperms

Have you ever looked closely at a pine tree? If you have, you may have noticed that a pine tree has two different kinds of cones. It has male cones and female cones. The male cones, which contain pollen, are the smaller of the two.

Pollen grains produce sperm cells that fertilize the eggs within the larger female cones. Pollen is carried from the male cones to the female cones by wind. After fertilization, seeds mature and are released from the female cones. When conditions are right, the seeds grow into new gymnosperm plants.

Figure 10–18. The pine tree (above) is a gymnosperm. The petunias (left) are angiosperms.

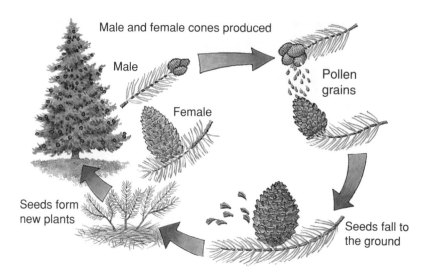

Figure 10–19. The life cycle of a typical gymnosperm

Male and female cones produced

Male

Female

Pollen grains

Seeds fall to the ground

Seeds form new plants

Conifers

The word *conifer* means "cone-bearer." Pines, spruces, and firs have stiff seed cones and needlelike leaves. Forests of conifers are found mainly in northern climates and other regions with sandy soil and moderate rainfall.

Conifers survive in harsh conditions because their needles have little exposed surface area. Needles retain moisture through hot, dry summers and cold winters. Instead of losing their leaves all at once, conifers shed and replace needles throughout the year.

Other Gymnosperms

Although they look nothing like the familiar conifers, cycads and ginkgoes are also gymnosperms. Cycads resemble palms but are unrelated. Their reproduction is similar to that of conifers, but instead of separate male and female cones on one tree, cycad trees are either male or female—that is, some cycad trees bear only male cones while others bear only female cones.

Figure 10–20. Conifer needles lose little moisture.

Figure 10–21. Cycads (left) are palmlike trees from the tropics, while ginkgoes (right) are native to China.

The ginkgo is the last living species of what was once a family of related trees. Ginkgoes have flat, fan-shaped leaves. Male ginkgoes produce pollen in small, conelike structures that hang down from the branches. After fertilization, female trees produce berrylike seeds. Ginkgoes are sometimes called *living fossils* because there are none left living in the wild.

ASK YOURSELF

How does the life cycle of a gymnosperm compare with that of a fern?

Angiosperms

Angiosperms are flowering plants, like the petunia. All angiosperms have flowers at some time during their life cycles. Flowers are important because they contain the plant's reproductive organs. These organs produce the pollen and egg cells that will eventually develop into seeds. The seeds grow inside protective fruits.

Figure 10–22. The African violet is an angiosperm commonly found as a houseplant.

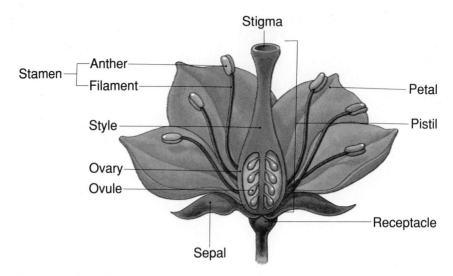

Figure 10–23. The parts of a typical flower are shown here. However, not all flowers contain all of these parts.

Stigma

Stamen — Anther
Filament

Style

Ovary
Ovule

Sepal

Petal

Pistil

Receptacle

Parts of a Flower

The parts of a typical flower are shown in the illustration. The male part of the flower is the **stamen.** The stemlike part of a stamen is the filament. At the top of a stamen is the anther, which contains the pollen grains. Most flowers have more than one stamen.

The female part of the flower is called the **pistil.** The slender part of the pistil is the style. At the tip of the style is the stigma. The stigma produces a sticky material that traps pollen grains. The base of the pistil is the ovary. The ovary contains ovules. Each ovule contains the female gamete, or ovum.

The base of the flower is the receptacle. Leaflike sepals grow from the receptacle and enclose the flower before it opens. If you look underneath the petals of a flower, you may be able to see the sepals. Petals grow above the sepals and protect the stamens and pistils. Petals usually have colors, shapes, or odors intended to attract certain insects or other animals.

The number of structures varies in different types of flowers. You may recall that monocots have flower parts in multiples of threes, and dicots have flower parts in multiples of four or five. In the following activity, you can observe the different parts of a flower.

DISCOVER BY Doing

You will need a simple flower, such as a tulip or a gladiolus, and a forceps or a toothpick. First use the diagram in your book to help you identify the flower's parts. Then carefully remove each part, tape it to a sheet of paper, and label it. How is your display different from the diagram in your book? ✎

Figure 10–24. One function of a flower is to attract insects. When an insect lands on a flower, pollen sticks to the insect's body. When the insect moves to another flower, the pollen is transferred to that flower, where pollination may occur.

Pollination and Fertilization

The transfer of pollen from the anther to the stigma is called *pollination*. Pollen is moved from anther to stigma by wind, water, or insects or other small animals, depending on the kind of flower. When pollen is carried from the anther of one flower to the stigma of another flower, cross-pollination takes place. When a stigma receives pollen from the anther of the same flower, self-pollination occurs.

Immediately after pollination, a pollen tube grows from a cell in the pollen grain, through the stigma, to an opening in an ovule. Sperm cells move down the pollen tube and enter the ovule. Fertilization occurs when one sperm combines with each ovum to form a zygote. The zygote, or fertilized ovum, eventually develops into an embryo.

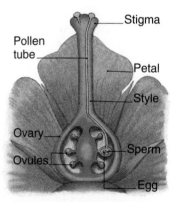

Figure 10–25. When pollen falls on the stigma, a pollen grain forms a pollen tube. The sperm inside the pollen grain moves down the pollen tube to fertilize the ovum.

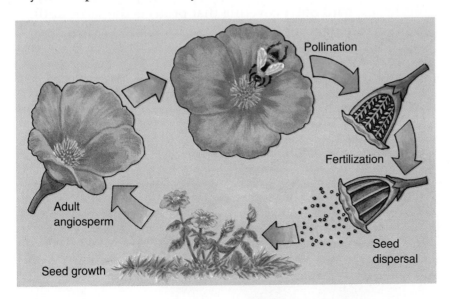

Figure 10–26. The life cycle of a typical angiosperm

Fruit and Seed Development A seed consists of three parts: the embryo, the stored food, and the seed coat. At the same time that the ovules are developing into seeds, the ovary is developing into a fruit. As the fruit swells and ripens, it supports and protects the developing seeds. The mature fruit will be either fleshy, like apples and tomatoes, or dry, like walnuts and pecans. Many fleshy fruits store sugars and, therefore, taste sweet. Dry fruits do not have sweet flesh; instead, starch, oils, and fats are stored in the seeds. In the following activity, you can make a list of the seeds, fruits, and other plant parts that people eat.

DISCOVER BY *Researching*

In your journal, make a table that has four columns with these headings: "Plant Name," "Plant Part Eaten," "Fruit or Vegetable," "Scientific Name." Under "Plant Name," list 20 plants that people eat. In the second column, list the part of the plant that is eaten. In the third column, indicate if the part eaten is commonly called a *fruit* or a *vegetable.* In the last column, write the scientific name of the part eaten.

Both dry and fleshy fruits play an important role in helping seeds disperse. The seeds in a fleshy fruit complete their development at about the same time that the fruit ripens and is ready to be eaten. Animals carry the fruits away, eat them, and leave the seeds in new locations. When the seeds are released from the fruit, they contain very little moisture and will not begin to grow until environmental conditions are favorable. Most dry fruits have a hard outer covering that protects the seeds until conditions are right for growth.

Figure 10–27. Apples and tomatoes are fleshy fruits. Walnuts and pecans are dry fruits.

Fruits can be classified as simple, aggregate, or multiple. A simple fruit is formed from a single ovary. Beans, peaches, tomatoes, and oranges are all simple fruits. The number of seeds in the fruit is the same as the number of ovules in the ovary.

Aggregate fruits form from several ovaries. All the ripe ovaries join to form a single fruit. Examples of aggregate fruits are blackberries, raspberries, and strawberries. Each ovary contains only one seed.

A multiple fruit has many single fruits growing close together. Pineapples and figs are examples of multiple fruits.

Figure 10–28. Apples are simple fruits, raspberries are aggregate fruits, and pineapples are multiple fruits.

Seed Germination   After a period of inactivity, mature seeds germinate, or sprout. Seeds germinate when environmental conditions are good and the seeds have plenty of water. The seed coat breaks open, and the young seedling emerges. The stems and roots begin to grow, leaves and branches begin to form, and the plant tissues begin to develop. You will learn more about seed germination and plant growth in the next chapter.

 ASK YOURSELF

How does a sperm cell reach an ovum in an angiosperm?

SECTION 3   *REVIEW AND APPLICATION*

Reading Critically
1. How does reproduction occur in gymnosperms?
2. Describe the male and female parts of a flower.

Thinking Critically
3. How is pollination different from fertilization?
4. Why are angiosperms more successful at growing in a variety of environments than are gymnosperms?

INVESTIGATION

Predicting the Effect of Colors of Light on Plant Growth

▶ MATERIALS
- four healthy plants (same kind, grown in the same conditions, each with several leaves) • large sheets of colored cellophane (yellow, red, blue)

▼ PROCEDURE

1. Number the plants 1 through 4. Measure and record the height of each plant. Also record your observations about each plant's appearance. Use a table similar to the one shown below.

2. Use Plant 1 as the control. Place it in direct sunlight.
3. Design a way to test the effect of yellow, red, and blue light on plant growth. Remember to keep all other conditions the same.
4. Place Plants 2, 3, and 4 in the same sunlight as Plant

1. Remember to water the plants as necessary.
5. Allow the plants to grow for about a week.
6. Measure and record the heights of the plants on the table. Also record your observations about the plants' appearance.

OBSERVATIONS OF PLANT GROWTH		
Date		
Plant 1 Control		
Plant 2 Blue Filter		
Plant 3 Red Filter		
Plant 4 Yellow Filter		

▶ ANALYSES AND CONCLUSIONS
1. Which plant grew the most? Which plant looks the healthiest?
2. Which plant grew the least? How does it look?
3. The cellophane filtered out the matching color of light. What can you conclude about the effects of certain colors of light on plants?

▶ APPLICATION
Fluorescent lights can be used to grow plants in places where there is no direct sunlight. How is that possible?

✳ Discover More
Try the experiment again using additional colors of cellophane to filter out different combinations of colors.

Design and carry out an experiment to find out what color of light speeds up the germination time of lettuce seeds.

The Big Idea

Plants come in many varieties. They range in size and appearance from the tallest redwood to a tiny green hornwort. They have adapted to live in practically every environment on Earth. Certain structural features can be used to divide plants into categories. Plants are classified as vascular or nonvascular, simple or complex vascular plants, or gymnosperms or angiosperms. Even within those divisions, plants can still vary significantly in form and appearance. Yet all plants, no matter what their structure, color, size, or shape, share two important characteristics: they contain chlorophyll and they can make their own food.

For Your Journal

Look back at the answers you wrote in your journal before you began this chapter. Would you change your answers now? Revise your journal entry to reflect what you have learned about plants. Include information about the many kinds of plants and their differences and similarities.

Connecting Ideas

Copy this unfinished concept map into your journal. Complete the concept map by writing the correct term in each blank.

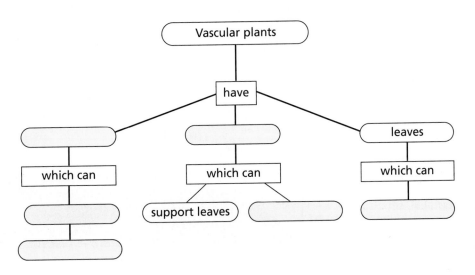

REVIEW

Understanding Vocabulary

Explain how the term or terms in the first column play a role in the area of a plant's life listed in the second column.

1. xylem (260), phloem (260) ———— transport
2. vascular cambium (263) ———— growth
3. chlorophyll (270) ———— photosynthesis (263)
4. stamen (278), pistil (278) ———— reproduction

Understanding Concepts

MULTIPLE CHOICE

5. Complex vascular plants' distinguishing characteristic is their ability to
   a) grow roots.
   b) produce leaves.
   c) grow stems.
   d) produce seeds.

6. Which of the following is a vascular plant?
   a) moss
   b) fern
   c) hornwort
   d) liverwort

7. During photosynthesis, plants
   a) take in oxygen and give off carbon dioxide.
   b) take in glucose and give off carbon dioxide.
   c) take in carbon dioxide and give off oxygen.
   d) take in oxygen and give off water.

8. The parts of a flower's pistil include
   a) the stigma, style, and ovary.
   b) the stamen, filament, and anther.
   c) the receptacle, sepals, and petals.
   d) the anther, stigma, and pollen.

9. Which of the following is _not_ generally a characteristic of an angiosperm?
   a) produces flowers
   b) its reproductive organs are in its flowers
   c) produces cones
   d) protects its seed in a fruit

SHORT ANSWER

10. How is a fruit formed, and from what flower parts might it develop?

11. Describe pollination and fertilization in the angiosperm.

12. What is meant by the statement that plants produce food?

Interpreting Graphics

13. Look at the illustration below. Which plant would you choose to plant in a sloping field? Explain your answer.

Reviewing Themes

14. *Systems and Structures*
Compare and contrast the reproductive methods for gymnosperms and angiosperms. Use this information to explain why there are more types of angiosperms than there are gymnosperms.

15. *Diversity*
Dividing plants into categories, such as vascular and nonvascular, complex and simple, angiosperms and gymnosperms, and so on, is helpful in learning about the wide variety of plants. Now review what all plants have in common.

Thinking Critically

16. Why are fruits considered to be a means of seed dispersal? How effective are they in this role? Can you think of other types of seed dispersal that might be more effective or efficient? Explain.

17. Pesticides are sometimes used on crops to kill harmful insects. But pesticides may also kill helpful insects. What impact could this have on angiosperms?

18. Why is a moist environment important for the survival of simple vascular plants such as ferns?

19. Why can vascular plants grow much larger than nonvascular plants? Draw sketches of the plants to support your answer.

20. The photographs show three different kinds of plants. Determine whether each plant is woody or herbaceous, vascular or nonvascular, and a monocot or a dicot. Explain your choices.

Discovery Through Reading

Facklam, Howard, and Margery Facklam. *Plants: Extinction or Survival?* Enslow, 1990. The importance of plants as a source of food, medicine, and other products is integrated with research contributions by scientists.

PLANT GROWTH AND ADAPTATIONS

From a stately oak to a colorful wildflower, all plants must reproduce, grow, and mature to survive. Plants depend on the environment for their livelihood. They need light and water to make food. They need to spread their seeds. They need to protect themselves from other organisms in their environment and often from the environment itself. Generation after generation of plants have survived because of the remarkable ways they have adapted.

In Florida the shore zone is usually dominated by red mangrove, chiefly young plants growing on ground that is periodically under water. . . . In meeting the ecologic challenge of life in the difficult, storm-wracked unstable zone where tidal salt water laps the edge of the land, the red mangrove has evolved into one of the most bizarre of all vegetables. Because the ecologic demands of the various kinds of seaside environments are very different, this species varies markedly in form. Some red mangroves are tall and straight-trunked; some are low-crowned domes standing on thin stilts; some spread out horizontally, and repeatedly reroot as they go, like vegetable centipedes.

Two of the fundamental adaptations that fit the red mangrove for its perilous life at the edge of the sea are salt tolerance and an ability to cling to unstable ground. A third adaptive achievement is a reproductive device that allows the

species to disperse widely and colonize new habitat. Red mangroves have little yellow, waxy flowers; the seeds that these produce germinate before they leave the tree, generating a cigar-shaped seedling 6 to 12 inches long. When these seedlings fall, some may lodge and take root beneath the parent tree and become part of a growing forest there, but most of them drift away with tides and currents and may travel for hundreds or even thousands of miles before they strand or die. When a seedling strands in shallow water it quickly grows roots, and these pull the little stem erect. A few leaves appear, and in a short while a new little mangrove stands in the shallows. . . . A single, well-grown, tidal-zone red mangrove standing high on its thin legs is a strange-looking plant. A forest of such trees is one of the most offbeat kinds of vegetation to be found anywhere.

from *The Everglades: The American Wilderness* by Archie Carr and The Editors of Time-Life Books

For Your Journal

- How does a tiny seed become a giant tree?
- How are plants adapted to growing in a wide variety of climates?
- How do seeds disperse?

Plant Growth

Objectives

Describe the internal structure of seeds.

Tell how plants respond to their environment.

Compare and contrast sexual and asexual reproduction in plants.

Some plants grow so quickly they almost appear to do so right before your eyes. For example, the red mangrove grows about 2.5 cm per hour. Red mangroves live in brackish waters along the coasts of southern Florida, Mexico, and the islands of the West Indies. Other plants hardly seem to grow at all. Bald cypress trees, which live in swampy areas of the Southeast from Delaware to Texas, may grow less than 2.5 cm in an entire year.

Starting Out

Plant growth begins with a seed. You already know something about seeds: they contain an embryo plant and they store food for the embryo's growth. Seeds also provide food for humans. Corn, peas, beans, and peanuts are all seeds. You can review the structure of a seed in the following activity.

DISCOVER BY Doing

You will need several raw peanuts. Use your fingers to open the peanuts. Spread the parts out on a paper towel, and look at them with a hand lens. Make a diagram of a seed and label the parts. ✎

Figure 11–1. A seed provides food for an embryo plant's growth.

When a seed sprouts, or *germinates,* the first thing it does is take in water. This causes the seed to swell as the *embryo,* or developing plant, inside the seed starts to grow. The root tip is usually the first part of the embryo to emerge from the seed. Since seeds store little water, it is important for the water-absorbing part of the plant—the root—to begin growing first. As the root continues to grow downward, the embryo's stem begins to grow up through the soil. The rapidly growing embryo, now called a *seedling,* faces many difficulties. If it fails to reach the surface and the light it needs for photosynthesis, it will die. The poem *Seeds* describes what a seedling can do.

Figure 11–2. The root is the first thing to sprout from the seed.

The young stem and leaves are the next things to emerge in a seedling. These first leaves are called seed leaves, or **cotyledons** (kawt un LEE duhns). In the next activity, you can discover the importance of cotyledons.

~~DISCOVER BY~~ Doing

You will need four small paper cups filled with potting soil and four soaked bean seeds. With a pencil, make a hole about 1 cm deep in each cup of soil. Place a soaked bean seed in each hole and cover it with soil. Place all the pots in sunlight and keep them watered. After the plants have pushed their way through the soil, remove the cotyledons from two of the seedlings. During the next few days, compare the growth of the seedlings.

Cotyledons are different from the leaves that will form later. In beans, peas, and peanuts, the cotyledons are thick and store a lot of food. The cotyledons of squash and radishes are thinner—more like other leaves. They store less food.

As the seedling emerges, its first true leaves begin to grow. In some seeds, these leaves are already present inside the embryo. The leaves turn green in the sunlight and grow even larger. The seedling continues to use the food from the cotyledons until the true leaves can provide enough food for the seedling. Eventually, the cotyledons shrivel and die.

Seeds
by Walter de la Mare

The seeds sowed—
For weeks unseen—
Have pushed up
pygmy
Shoots of green;
So frail you'd think
The tiniest stone
Would never let
A glimpse be shown.
But no; a pebble
Near them lies,
At least a cherry-stone
In size,
Which that mere
sprout
Has heaved away,
To bask in sunshine,
See the Day.

▼ ASK YOURSELF

Why is it so important for a new seedling to reach sunlight quickly?

Figure 11–3. The seedling (left) is no longer dependent on food stored in the seed. The sapling (right) grows rapidly.

Responding to the Environment

An embryo inside a seed grows by stretching, or elongating, its cells. However, for further growth to occur, new cells must be added by cell division. A seedling continues to grow in length at its stem tips and root tips. As the plant grows into a *sapling,* new cells elongate and mature. The processes of cell division and elongation continue as long as the plant is growing. For herbaceous plants, this is the main type of growth.

Woody plants, such as trees, shrubs, and many vines, have a lot of supporting tissue—xylem—in their stems and roots. In addition to growth at the tips of these organs, woody plants also have a layer of dividing cells under their bark that allows for growth in diameter.

If you have ever observed a plant growing on a window sill, you may have noticed that many of its leaves are turned toward the light. Why do the leaves of a plant grow toward light? Why do roots grow down and stems grow up? Why does a Venus' flytrap close when touched by a fly? All of these are examples of plants responding to the environment.

Figure 11–4. The Venus' flytrap responds to touch, quickly entrapping unsuspecting insects.

When a Venus' flytrap closes around an insect, the plant is responding to the stimulus of touch. A *stimulus* is anything in the environment that causes a response from an organism. When a plant grows toward a window, it is responding to the stimulus of light. What is a plant responding to when its stems grow up and its roots grow down? In the next activity, you can observe firsthand a plant's response to a stimulus.

DISCOVER BY Doing

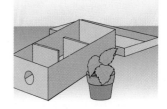

You will need a shoe box with a lid, a pair of scissors, two pieces of cardboard half as wide as the box, and a potted bean seedling. Cut a hole in one end of the shoe box. Then tape the cardboard strips to opposite sides of the box as shown. Put your plant in the end of the shoe box away from the hole. Put the lid on the box, and place it near a sunny window. Observe the growth of the plant for the next few days. Describe the growth of the plant. ✐

The plant in the activity bends because it needs light. It is responding to the light available. Plants' responses to the environment are called **tropisms** (TROH pihz uhmz). *Geotropism* is a plant's response to gravity. This tropism keeps the roots of a plant growing downward and the stems growing upward. *Phototropism*—response to light—is very important. A seedling must reach sunlight before its stored food runs out. It needs a stimulus so that it can grow in the right direction. That stimulus is the light itself.

Chemicals made in the tips of stems cause plants to respond to light. These chemicals make cells divide and lengthen, causing the stems to grow. Cells on the shady side lengthen more quickly, causing plants to bend toward the light. Even more of these chemicals are produced in the dark. Without these chemicals, a seedling would die underground.

Light is one of the most important stimuli for plants. When light and chemicals work together, interesting things sometimes occur. For example, you know that the number of daylight hours changes from season to season. The length of darkness affects the production of chemicals that control plant flowering.

Figure 11–5. When a plant bends toward a window, it is responding to light.

Figure 11–6. Violets and poinsettias are short-day plants.

Some plants, called *short-day plants,* will not flower unless they have a long period of uninterrupted darkness. In fact, they should really be called "long-night" plants. Short-day plants, such as poinsettias, flower in the early spring, late fall, or winter when the days are short. Other kinds of plants, such as spinach and clover, need eight to twelve hours of darkness each 24-hour period in order to flower. These are called *long-day plants.* Long-day plants flower during the longer days of early summer. Still other plants, such as corn, are *day-neutral plants.* These plants will flower in varying amounts of sunlight.

ASK YOURSELF

Why is it important for a plant's roots to always grow down and its stems to always grow up?

Reproduction

Remember that flowers are the reproductive organs of angiosperms and cones are the reproductive organs of gymnosperms. Both angiosperms and gymnosperms produce seeds by sexual reproduction. Some plants can reproduce by asexual reproduction. Cuttings of roots, stems, or leaves can grow into new plants. This type of growth is called **vegetative propagation.**

Vegetative propagation is often used to grow plants faster and more reliably than with seeds. Many seedless varieties of flowering plants have been developed in this way. These plants, such as oranges and grapes, produce seedless fruits. They can be reproduced only by vegetative propagation.

A different kind of vegetative propagation takes place in the white potato plant. The potato is an underground stem. The "eyes" of the potato are tiny buds on the stem. Farmers cut up whole potatoes and plant the pieces that have eyes. A new plant quickly sprouts from each of these buds. Growing potatoes from seeds takes a much longer time.

Figure 11–7. Potatoes can be grown from seeds or from "eyes."

Bulbs are another example of vegetative propagation. A bulb is a short, underground stem. Planted in the fall, a plant, such as a lily, grown from a bulb will flower the following spring. A lily grown from a seed will not bloom until the second year. By the end of the growing season, bulbs will have divided into several pieces, which can be separated and planted— each piece producing a new plant the following year. Tulips, daffodils, onions, and garlic can also be grown from bulbs.

The strawberry plant and some grasses can reproduce by runners. A runner is a stem that "runs," or grows, along the ground. When one of its buds touches the ground, it forms roots and produces a new plant. Even pieces of stems and leaves can sometimes grow into new plants. The plant in the picture produces "babies," or tiny plants, at the edges of its leaves.

Plants that are grown through vegetative propagation have the same characteristics as the plant from which they were taken. For example, if you grow an African violet from a leaf cutting, that African violet will have the same kind of leaves, flowers, and other characteristics as the plant from which you cut the leaf. The new plant was once part of the old plant, so it has exactly the same traits as the old plant. You can find out more about vegetative propagation by doing the next activity.

Figure 11–8. Bulb-grown plants flower much sooner than those grown from seeds.

Figure 11–9. This kalanchoe is a "mother" plant with her "babies."

DISCOVER BY Doing

Place carrot tops in a shallow dish and water them regularly. Observe the growth of the roots. Make cuttings from a begonia and put them in water. Observe the development of new roots.

Get a spider plant, cut off a "baby," and plant it in a pot. Observe how it grows. Record your observations in your journal. Find out how many ways you can grow plants by vegetative propagation. ✐

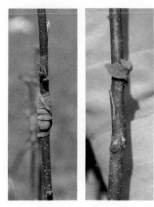

Figure 11–10. Grafting uses the root system of one plant and the branches of another.

Were you able to grow new plants from plant parts? If so, you were demonstrating regeneration. **Regeneration** is the ability to grow or replace missing parts. Many plants have this ability; some animals do, too. Other methods of asexual reproduction have been developed by scientists. These methods help farmers grow new plants without having to start them from seeds. *Grafting* is one of these methods.

Grafting enables farmers to make use of the best characteristics of two plants. For example, some types of orange trees are very hardy, but their fruit is extremely sour. Other types of orange trees are not as hardy, but their fruit is sweet and juicy. If the branches of the tree with sweet fruit are grafted to the roots of the sour orange tree, the hardy root system will support branches producing good fruit. It is also possible to graft branches from several different kinds of trees onto one root system, thereby producing a tree that bears several varieties of fruit.

Figure 11–11. This tree has five different varieties of apples. Grafting allowed one tree to represent an entire orchard.

 ASK YOURSELF

Describe several means of vegetative propagation.

SECTION 1 *REVIEW AND APPLICATION*

Reading Critically

1. How do plant embryos get energy for growth?
2. What are plant tropisms?
3. What is a cotyledon and why is it important to the growth of a new plant?

Thinking Critically

4. Wind pollinates many kinds of trees. How might wet weather conditions during flowering hinder the formation of seeds?
5. Any plant that reproduces asexually has exactly the same characteristics as the plant it grew from. What advantages and disadvantages could this have?

SKILL Measuring Plant Growth

▶ **MATERIALS**

● a bean plant with at least an 8-cm stem ● millimeter ruler ● fine-point permanent marker ● germinating bean seeds with roots 1 cm long ● straight pins ● paper towel ● jar with lid ● piece of cardboard

▼ **PROCEDURE**

1. Place the ruler next to the stem of the bean plant. Starting at the tip of the stem, make an ink mark every 5 mm along the stem. Be sure that you mark only the stem and not part of the youngest leaf.

2. Wait one or two days. Then complete step 3. While you are waiting, complete step 4.

3. Measure the spaces you marked the day or so before. Record your results.

4. Cut the piece of cardboard so that it will fit standing up inside the jar. It should be able to stand without falling over. Set this cardboard aside to use later.

5. Lay a germinated seed on the table. Starting at the tip of the root, make an ink mark every 2 mm. Work quickly so the root does not dry out.

6. Stick a pin through the center of the seed. Then pin the seed to the cardboard so that the root points downward. Place the cardboard in the jar.

7. Add enough water to the bottom of the jar until the root tip is 2.5 cm above the water.

8. Screw on the jar lid and keep the jar in a shady place. Wait one day.

9. The next day remove the seed from the jar. Measure the spaces you marked on the roots the day before. Record your results.

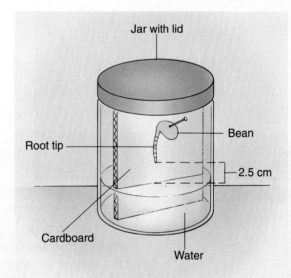

Jar with lid

Root tip

Bean

2.5 cm

Cardboard

Water

▶ **APPLICATION**

Did any of the spaces between marks on the stem get larger? Where does growth appear to occur on the stem? Did any of the spaces on the root increase in size? Where does growth appear to occur on the roots?

✳ *Using What You Have Learned*

What processes were you able to observe in this activity? What statement can you make about which parts of plants grow in length?

Plant Adaptations

Objectives

Describe *several plant adaptations.*

Summarize *how different plants survive harsh weather conditions.*

Observe *various types of seed dispersal.*

Plants are found growing nearly everywhere on the earth. Not every kind of plant, however, is able to grow in every environment. Some plants grow only in wet environments. Other plants, such as those pictured, grow in the desert, where water is scarce. Desert plants have structures that help them conserve water. In fact, many plants have adaptations that enable them to survive in harsh environments.

Figure 11–12. It's not easy being green.

Spreading Seeds

One way plants can ensure their survival—at least for one more generation—is to successfully spread their seeds within an environment favorable for germination. Plants show remarkable adaptations for spreading seeds, or *seed dispersal.* Some plants enlist the aid of unsuspecting animals for this job. The burs of a burdock, for example, stick to the fur of many animals. The bur, which contains a seed, may be carried a great distance before it falls off.

Animals disperse seeds in other ways as well. Many fruits are eaten, seeds and all, by animals, especially birds. The seeds pass through their bodies undigested.

Humans, too, help with seed dispersal. Have you ever eaten watermelon outside and spit the seeds on the ground or tossed the core of an apple into an empty lot or field? An apple core contains many seeds, some of which might land in an environment suitable for germination. And the same seeds that stick to an animal's fur will also stick to your clothing.

Figure 11–13. Animals often provide "taxi service" for seed dispersal.

Figure 11–14. Some seeds are adapted for flights of fancy.

Water and wind also aid in seed dispersal. A coconut can drop from a tree, travel on an ocean current, and germinate on a beach thousands of kilometers from where it fell. Red mangrove seeds bob along on tidal currents in a coastal marsh until they touch bottom. Then they sprout, growing prop roots to anchor themselves in the shallow water.

Many other seeds are fitted with "wings" and travel on currents of air. In the next activity, you can see how far winged seeds can fly.

DISCOVER BY *Doing*

Gather a cupful of maple seeds or other winged seeds, or make paper "helicopters" to represent seeds. Release some of the seeds or helicopters from a second-story window and observe where they land. Measure how far they fly. Do any of the seeds go farther than the spread of the branches of a large tree? Why do you think that would be important? ✐

A few plants are adapted internally to ensure that their seeds are widely dispersed. The witch-hazel plant, shown in the photograph, launches its seeds from a miniature "catapult." Witch-hazel seeds can be propelled several meters from the parent plant.

When it comes to self-propulsion, however, the witch-hazel is a poor second to the dwarf mistletoe. Dwarf mistletoe fruits absorb water until internal pressure finally bursts the fruits open, throwing the seeds nearly 15 m from the "launch parent."

Figure 11–15. Witch-hazel seeds are catapulted into the air.

What can you learn from a collection of traveling seeds?

MATERIALS

old knee-high socks, envelopes, plant reference books, poster board, wide cellophane tape, marker, ruler

PROCEDURE

1. Pull an old pair of large knee-high socks over your shoes. Then walk through a field or overgrown lot. After your walk, check your socks and other clothing for seeds.
2. Collect as many different seeds as you can, placing each kind in a separate envelope.
3. At home or at school, use reference books to identify the seeds in your collection. The books will also tell you something about the methods of dispersal of the various seeds.
4. Attach the seeds to poster board with clear tape. Use a marker to label each kind of seed. Allow

enough space to write a brief description of the method of dispersal for each kind of seed. Display your collection.

APPLICATION

What unusual methods of seed dispersal did you discover? How is the parent plant adapted for this dispersal method?

 ASK YOURSELF

How does successful seed dispersal help to ensure the survival of a plant species?

Figure 11–16. "Leaves of three—let it be."

Protection Against Animals

A species will soon become extinct if many of its individuals don't live long enough to reproduce. Plants are no exception. Some of the biggest dangers to the survival of plants come from grazing animals and insects. To defend against these attacks, some plants have adapted a kind of "chemical warfare." For example, the "milk" of the milkweed plant must have a bad taste because most insects avoid this plant.

Poison ivy, poison oak, and poison sumac use a slight variation of this adaptation. These plants secrete a chemical that is very irritating to the skin, keeping most animals away and allowing the plants to grow undisturbed.

Figure 11–17. Chemicals produced by the roots of the black walnut tree keep other plants from growing too closely.

Black walnuts seek to avoid a different kind of threat to survival—competition from other plants. The roots of a black walnut secrete a chemical into the soil that keeps other plants from growing too closely.

Some plants have a physical defense against animal agression. Roses, raspberries, black locusts, and wild citruses all have thorns to keep animals away. The spines of many desert plants, although not structurally the same, are equally effective adaptations that keep hungry animals away.

 ASK YOURSELF

In what ways do plants protect themselves from animals?

Figure 11–18. Thorns among roses grow for protection.

Surviving Harsh Climates

Remember, plants grow in almost every environment. However, no matter where a plant grows, one of the biggest dangers to its survival is loss of water. Like all living organisms, plants need water to survive, but it is not always readily available. So plants must conserve the water they do have. In the next activity, you can observe a water-conserving adaptation.

 **BY** *Observing* —————————————

Get a ripe apple and polish it with a cloth. Observe how the skin shines. Now put a drop or two of water on the apple's skin. What happens to the water?

Figure 11–19. Plants have several adaptations to prevent water loss.

Apples are covered with a waxy coating. When you polish it, an apple gleams like a newly waxed car. On a car, wax causes water to bead and run off. But the waterproof skin of an apple keeps it from losing the water it stores inside. Many leaves also have waxy coats. This coating seals the leaves, preventing water from evaporating. Fruits and vegetables sold in supermarkets are sometimes coated with additional wax. How might this be an advantage to people who buy the produce?

There are many other ways in which plants are adapted to conserve water. For example, part of the bark of many trees contains a layer of cork cells. Commercial cork comes from certain species of oak trees. You probably know that cork floats, but did you know it does so because it does not absorb water? The cork cells in tree bark keep water from escaping from the tree's xylem.

Lack of water is the most limiting factor to desert life. However, many desert plants have adaptations to help them collect and store water. For example, the leaves of the cactus have become small, sharp spines. Photosynthesis occurs in the green stems, which are protected from drying by a thick layer of wax.

Figure 11–20. Cactus spines, which are modified leaves, help to reduce the loss of water.

Desert shrubs have small, thick leaves with waxy coverings. During dry spells, these shrubs drop their leaves to conserve water. Photosynthesis continues in the stem cells. Desert plants also have large root systems. Sometimes the roots grow as deep as 30 m in search of water. Others have shallow, wide-spreading roots that absorb as much water as possible during heavy desert thunderstorms.

Shortly after a heavy rain, many of the desert's plants flower. The seeds that develop may lie dormant for many months, or even years, until the next rainy period. When enough moisture is present, they quickly germinate, grow, bloom, and produce new seeds.

Figure 11–21. Many desert plants have shallow roots to absorb the occasional heavy rains.

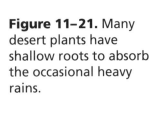

Plants of the tundra are adapted for survival in an equally harsh climate. These plants, too, must grow, flower, and produce seeds in a few short weeks. In addition, most tundra plants are small and are covered by winter snows. Snow actually insulates the plants from the freezing Arctic winds. That is why the Inuit, the natives of northern North America, used to make winter homes from blocks of snow.

Figure 11–22. The growing season is extremely short in the tundra.

It may seem strange, but tundra plants and desert plants face the same problem—lack of water. You might think cold would be the biggest problem for tundra plants. But the tundra is actually much like a desert, since it receives very little precipitation. As with desert shrubs, the leaves of many tundra plants have a thick coating of wax to keep them from drying out. Some tundra plants also have hairs on the leaves to keep water from evaporating.

 ASK YOURSELF

Why do both desert plants and tundra plants have thick, waxy coats on their leaves?

SECTION 2 *REVIEW AND APPLICATION*

Reading Critically

1. Give a specific example for each type of adaptation listed: seed dispersal, protection against animals, protection from climate.
2. Give examples of how seeds are spread by wind, water, and animals.

Thinking Critically

3. How do you think water birds, such as ducks, might spread seeds? What kinds of seeds would they spread?
4. Suppose you and your family moved from southern California to Michigan. Would you be able to grow the same kinds of trees in your Michigan yard as you had in your yard in California? Explain your answer.

INVESTIGATION

Examining Seeds

▶ **MATERIALS**

- bean seeds, dry and soaked ● scalpel ● probes or toothpicks
- hand lens ● pencil ● paper ● corn seeds, dry and soaked

▼ **PROCEDURE**

1. **CAUTION: Scalpels are very sharp. Use extreme care.** Get two soaked bean seeds and peel off their seed coats. This should separate each seed into two halves, making a longitudinal section of the seed.
2. Cut another bean across the seed, as shown below. This makes a cross section of the seed.
3. Use a probe or toothpick on both beans to find the structures of the bean seeds. A hand lens may make it easier for you to see some of the smaller parts.
4. Sketch what you see and label the parts.
5. Prepare a longitudinal section of the soaked corn seed by placing the seed flat on the table, embryo side up. Make a top-to-bottom cut. To prepare the cross section, divide a corn seed in half the other way, at a right angle to the cut you made on the first corn seed.

▶ **ANALYSES AND CONCLUSIONS**

1. Why was it necessary to do a longitudinal section and a cross section of each type of seed?
2. List differences that you found between monocot (corn) and dicot (bean) seeds.
3. Make a table with a list of the parts that you were able to locate. Explain how each part formed.

▶ **APPLICATION**

Compare the soaked seeds with the dry seeds. What changes did soaking make? Why do you think some gardeners soak seeds before planting them?

✳ **Discover More**

Find out what the inside structures of other seeds are like.

The Big Idea

Structural adaptations are important for plants to survive unfavorable conditions. Some adaptations help to ensure reproduction and seed dispersal. Other adaptations help protect plants from animals or harsh climates. Adaptations allow plants to live and grow in nearly every kind of environment on the earth.

For Your Journal

Now how do you think tiny seeds grow into giant trees? How are plants adapted to unfavorable growing conditions? If your ideas have changed, revise your journal entries.

Connecting Ideas

Think about a plant other than a cactus. In your journal, create a concept map similar to the one shown. Relate the plant's adaptations to their specific functions. Be sure to list the name of the plant you chose. Compare your concept map with those made by your classmates.

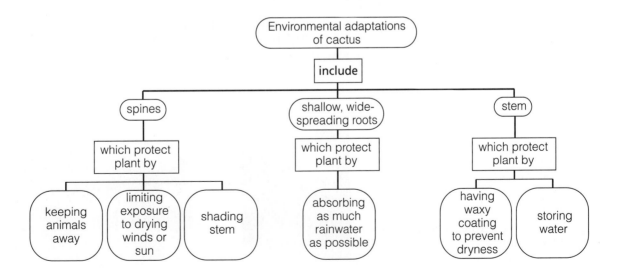

*R*EVIEW

Understanding Vocabulary

For each set of terms, identify the term that does not belong and explain the relationship of the remaining terms.

1. vegetative propagation (292), seeds, stem cuttings, bulbs
2. seed coat, cotyledons (289), mosses, embryos (289)
3. regeneration (294), asexual reproduction, missing parts, seed dispersal (296)
4. phototropism (291), grafting, geotropism (291), tropisms (291)

Understanding Concepts

MULTIPLE CHOICE

5. If a seedling is whitish rather than green, it probably needs
   a) less light.
   b) more light.
   c) less water.
   d) more water.

6. Which of the following plants was *not* grown through vegetative propagation?
   a) a begonia plant grown from a stem
   b) a strawberry plant that broke away from the "mother" plant
   c) a lily plant grown from a seed
   d) an iris plant grown from a bulb

7. The light source of a potted house plant that leans to the south is coming from the
   a) north.
   b) south.
   c) east.
   d) west.

8. A plant with leaves that are not eaten by insects when those of nearby plants are eaten may be protected by
   a) wind-dispersed seeds.
   b) thorns.
   c) chemicals.
   d) flat leaves.

9. Since onions are bulbs, which statement *cannot* be true about onions?
   a) Onions have meristems.
   b) Onions can reproduce asexually.
   c) Onions can only reproduce sexually.
   d) Onions have thick, fleshy leaves surrounding their meristems.

SHORT ANSWER

10. What are tropisms? Provide two examples of plant tropisms.

11. Describe the different ways that seeds can be dispersed.

Interpreting Graphics

12. Look at the photograph below. The bean seed was planted upside down. What tropism is the root displaying?

13. Look at the photographs below. Which plant is receiving light from only one direction? How can you tell?

Reviewing Themes

14. *Environmental Interactions*
Structural adaptations have enabled plants to survive in harsh environments. Explain how the structure of tundra plants has enabled them to survive in their environment.

15. *Environmental Interactions*
Interactions between plant and animal life can be beneficial or harmful to plants. Explain how animal interactions can benefit plants and how they can harm the plants.

Thinking Critically

16. A plant's roots and stems respond to gravity. A plant's stems and leaves respond to light. What would happen to a plant that could <u>not</u> respond to the stimuli of gravity and light?

17. Some plants, such as the one shown below, give off a foul odor similar to that of rotting meat. Explain what advantages and disadvantages this type of odor might have.

18. If a sweet-apple tree branch is grafted to a sour-apple tree, what type of apple will the grafted branch produce? Why?

19. Different types of crocuses bloom in winter, early spring, or fall. Are crocuses long-day, short-day, or day-neutral plants? Explain your answer.

20. Imagine you have planted a plant that has deep roots and big leaves not covered by wax in a desert environment. Would you expect the plant to survive? Explain your answer.

21. Two plants are given the same amount of water and nutrients, but they are exposed to different amounts of light. One plant receives eight hours of light a day, and the other receives four hours of light. Which plant do you think would grow more rapidly? Why?

Discovery Through Reading

Stidworthy, John. *Plants and Seeds.* Gloucester Press, 1990. Text and microscopic photographs introduce various forms of plant life, their methods of reproduction, and their assimilation of nutrients.

Science

PARADE

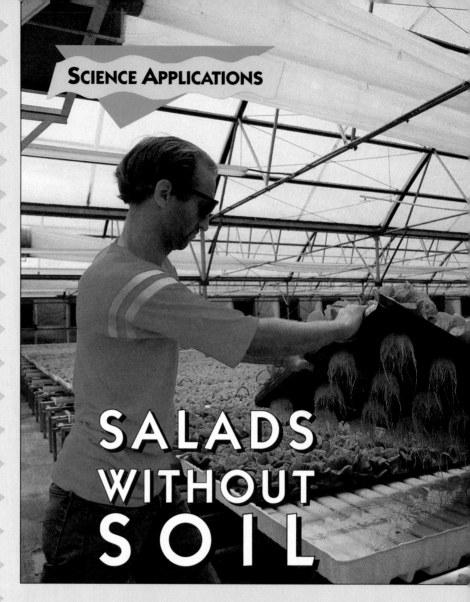

SALADS WITHOUT SOIL

An agricultural planner imagines city apartment dwellers growing fresh vegetables in rooftop gardens. NASA makes plans to grow vegetables in space-station gardens barely larger than closets. Farmers tend giant garden towers in which tons of vegetables are grown under plastic bubbles. All these visions of the future have one thing in common: a special kind of farming called *hydroponics*.

Gets the Dirt Out

On ordinary gardens and farms, plants are grown in soil. The plants' nutrients are provided by the soil or by fertilizers added to the soil. In hydroponic gardens, plants grow in water rather than soil. The plants' roots hang in water that is enriched with the chemicals necessary for growth. Researchers have learned exactly what ingredients must be added to the hydroponic "soup" for plants to thrive. Water, sunlight, and a mixture of about 15 chemical nutrients provide hydroponic crops with everything they need.

Hydroponic gardens

Research on hydroponic lettuce farm

Space-Age Farming

Almost any crop can be grown hydroponically. Seeds are started in a spongelike material and then transplanted to a water tank. Often the base of the plant is anchored in sand. Plants do not grow bigger in hydroponic tanks or taste any different; they just grow faster. Hydroponics also eliminates many harmful pests that live in the soil.

A walk through a hydroponic farm may seem like a space-age dream. Rows of crops grow from giant tanks. Enriched water flows around the plants' roots or is dripped or sprayed from above. Most hydroponic farms are contained in greenhouses or plastic bubbles, which allow crops to be grown year-round.

Hydroponic gardens continue to pop up in the most unexpected places. In some places, restaurant patrons can pick their own fresh salad greens from a hydroponic tank, in the same way they choose a fresh fish or lobster. Rooftops are being transformed into productive gardens for the residents living in apartments below. In the future, fresh fruits might actually be grown inside supermarkets.

Hydroponic farming is not without its drawbacks. Because of the equipment required, hydroponic farms are very expensive to start. A farmer may have to spend $100,000 or more to start a small plot of only $1000m^2$. However, a hydroponics farmer may quickly regain that investment by intensive production. Hydroponic farms can produce as many as five crops a year, whereas traditional farming methods yield only one or two. Hydroponics farmers also suffer fewer losses to insect pests and bad weather.

The Future Is Now

Hydroponics is becoming increasingly popular. More than 40 000 hectares of hydroponic crops are now grown in Europe, where land for farming is less plentiful than in North America. In Japan a national hydroponics research center has helped convert nearly a thousand farmers to hydroponic methods. Although there are presently few hydroponic farms in the United States, the science of hydroponics may some day put gardens in places where nothing could grow before. ◆

Andy Lipkis *and the* TREE PEOPLE

by Mark Wexler
from *Ranger Rick*

"Take a good look now," said the forest ranger, "because in thirty years most of the pine trees in these mountains will be dead. There won't be anything left in some areas but bare hills."

The ranger was talking to a group of summer campers in the San Bernardino National Forest. The pine forest is located just east of Los Angeles in Southern California. "Smog is killing the trees," the ranger said. "Most of this pollution comes from cars and trucks."

The year was 1970. One of the campers was a tenth-grader from Los Angeles named Andy Lipkis. He was stunned by what he heard. "Are *all* of these trees dying?" he asked. "We've got to *do* something!"

"Until we get rid of the smog, there's not much we can do to save the trees," said the ranger. "There are some kinds of trees pollution won't kill. But not many of them grow here."

Andy thought about that, then asked, "Well, why don't we replace the dead trees with ones that can live with the smog?"

The ranger chuckled. He knew how difficult it would be to replace every dead tree.

But Andy wouldn't give up. If the forest couldn't be replanted, then at least he could plant some smog-proof seedlings around the summer camp. With permission from the camp director, Andy and some

308

friends dug up a baseball field and a parking lot. Then they bought two kinds of trees that are not harmed by pollution. They carefully planted them along with some mountain shrubs.

"It took us three weeks, and I really learned a lot," said Andy. "We all did."

That fall, Andy began studying how smog harms plant life. He learned that trees, like people, breathe in the air around them. In the forests of Southern California, however, trees also breathe in smog. The smog is loaded with a kind of gas known as ozone. Ozone destroys the green coloring called *chlorophyll* (KLOR-uh-fil) in the trees' needles. (All green plants need chlorophyll to make their food.) So when a tree breathes in ozone, it slowly loses its ability to make food for itself. It becomes weaker and weaker.

Trees weakened by smog become good targets for insects and disease. A healthy tree closes its wounds with sap and this helps keep pests away. A sick tree cannot defend itself very well. Pests work their way into its bark, and the tree dies.

Andy was angry about the air pollution. He knew it was the cause of the problem and that something would have to be done to stop it. But right now he figured he had to do some-

thing about replacing the dead trees. He thought it would be easy to get people to help. He decided to ask some large companies for money to buy seedlings. Then he would round up volunteers to plant the young trees.

In the next few months the teenager met with im-

portant people in several large California companies. He explained the problem smog was creating in the forests and how they could help. "But because I was just a kid," says Andy, "nobody would listen to me." He was so discouraged he gave up his plan—but only for

309

a time. In 1972 Andy entered college and decided he would *not* abandon his tree planting project. "The forests were dying and I just couldn't let that happen," he said. Andy wrote letters to 25 summer camps in the San Bernardino Mountains. He told them about his idea and asked for their help. Soon he received letters from 20 camp directors saying they would like to help with the tree planting.

Then Andy called the California Division of Forestry, which grows seedlings to sell. He said he needed 20,000 seedlings for his project. "You're just in time," a man told him. "We're ready to plant some new trees. But to make room for them, we have to plow under last year's seedlings. Send us $600 and we'll save some of last year's trees for you." That made Andy really angry! Why couldn't the state just *give* him the trees instead of destroying them?

The more Andy thought about it, the madder he got. He began calling state officials. Finally, the state foresters agreed to donate 8000 smog-proof seedlings for Andy's project. But there was a hitch. The seedlings would be delivered late in the winter in several big boxes. That meant he would have to plant each tree in a pot by itself if it were to survive until summertime. So Andy got a local dairy to donate 8000 milk cartons for the seedlings. Then he and some friends worked day and night until the trees were safely potted.

When Andy learned smog was killing thousands of trees near Los Angeles, he acted! With friends he started planting trees that can live with smog. He planted this taller-than-Andy tree when he was still a teenager.

Andy's troubles, though, weren't over. He had his trees, but he needed money for tools and fertilizer. He turned to a Los Angeles newspaper for help. The newspaper ran a front-page story about Andy and the trees.

"I was amazed," Andy says. "All kinds of people sent money." Before long, Andy had more than $10,000 to use for his tree work.

In the summer of 1973, Andy worked with campers, Scouts, and other volunteers to plant the seedlings. At least 5000 of those 8000 trees are still growing, and Andy and his Tree People are still planting. So far they've planted over 150,000 smog-proof seedlings!

"When we began our planting in 1973," notes Andy, "part of our goal was to save forests. But most important was to teach people that they must help cut down on smog." The Tree People know of a few good ways: You can use automobiles less and use bicycles and public transportation more. Also, you can use fewer electric gadgets. Making electricity makes pollution. "To have healthy forests we have to change some of the ways we live," Andy adds.

Young Tree People scramble along a California hillside planting and watering young trees. They hope a forest will someday stand where grass and weeds now grow.

Pollution is still killing thousands of trees in the Los Angeles area every year. But Andy and his friends are still working very hard to get people to stop smog and to help save the forests. "I get tired," he says," but I have a good time. And I have a good feeling inside, knowing that I'm doing something that counts." ◆

UPDATE

Today Andy Lipkis continues to plant trees. In more than 20 years, Andy Lipkis and his helpers have planted about 200 million trees. Andy's organization, called Tree People, has received attention all over the world for its replanting efforts.

Andy Lipkis and the Tree People are working on projects to teach communities about forest problems. They also have plans to plant 5 million more trees in the Los Angeles area!

GEORGE WASHINGTON CARVER (1864—1943)

One of the greatest scientists of all times, George Washington Carver revitalized and revolutionized farming. He developed over 300 different products made from the oils and other chemicals found in peanuts. New industries sprang up around these products, and the economy of the war-ravaged South prospered. He produced similar results with sweet potatoes and pecans.

Born of slave parents on a plantation near Diamond Grove, Missouri, in 1864, Carver went out on his own at the age of ten to get a formal education. In 1891, after two years at Simpson College, he entered Iowa Agricultural College, now Iowa State University. Because of his outstanding work in botany and chemistry, Carver was appoint-ed assistant instructor in botany and greenhouse director at the agricultural college. After receiving his master's degree in 1896, Carver answered a call for teaching help from Booker T. Washington, head of Tuskegee Institute. Carver spent the rest of his life at Tuskegee Institute as a plant scientist. His research gradually won him the respect of Southern farmers by convincing them of the need for crop rotation. Years of cotton farming had depleted much of the soil of the South. Carver told the farmers that planting peanuts, clover, and peas would help restore the soil, since these crops release nitrogen as they grow. ◆

ELMA GONZALEZ (1942—)

The research of Elma Gonzalez may one day help scientists understand better how plant cells store energy and how seeds germinate.

Elma Gonzalez was born on June 6, 1942, in Mexico. Every spring Gonzalez would leave school to pick crops until late fall. She and her family were migrant farm workers. Through hard work and determination, Gonzalez graduated from high school. She went on to receive her bachelor of science degree in biology and chemistry at the Texas Woman's University in 1965 and then her doctorate in cellular biology at Rutgers University in 1972.

In 1974, Gonzalez became a professor of cellular biology at the University of California, Los Angeles. At the university, she has specialized in plant research. Gonzalez is now the head of her own research laboratory, where she studies the function of various internal bodies of plant cells. ◆

Shirley Mah Kooyman, Plant Scientist

S hirley Mah Kooyman knows about the importance of raising healthy plants. She is a plant scientist in Chanhassen, Minnesota.

What do plant scientists do?

Plant scientists help ensure the survival of plants. They study ways to breed heartier and more productive plants. Our food, many of the things we wear, and even some parts of our homes come from plants. People also get enjoyment from looking at plants and I'm a botanist. I study plants in their natural state, in the wild. For example, I might do research to discover what makes wild wheat grow at the roadside. Then I could share my research results with other plant scientists.

Where do you work?

I work both indoors in a lab and outdoors in an arboretum. An arboretum is a place where many kinds of plants and trees are grown for study. I also work in the classroom, teaching plant identification and

313

gardening. I enjoy teaching people about the plant kingdom.

Why do we need to know what makes plants grow?

They have many important roles. Unfortunately, plants are sometimes taken for granted, and people forget that they are living things. Plants need light, energy, water, and hormones. When the weather is too hot or cold, or if the soil is not properly nourished, plants can die. Research into the vital processes of plants can help make them healthier. This will help protect our food supply for years to come.

What are plant hormones?

Hormones are chemical messengers that tell a plant to produce other chemicals. This, in turn, stimulates plant growth. Hormones are produced in the stem and roots and travel to the rest of the plant. Scientists can now make plant growth regulators, or artificial hormones. If you place them on the root tip, artificial hormones make plants grow. Special types of botanists, called plant physiologists, work on the creation and production of artificial hormones.

Why do we need artificial hormones?

Artificial hormones are used to help produce bigger and better plants. In addition, artificial hormones can also prevent fruit from dropping off trees too early. These hormones are important because plants are used to feed more and more people.

What kind of education does a plant scientist need?

Most plant scientists need to have a four-year college degree in botany. After getting a college degree, plant scientists can work for private companies such as plant nurseries or seed companies. They might choose to work for federal, state, or local governments. For those who are interested in teaching at a college or university, more education is required. ◆

GROWING PLANTS IN SPACE

During the twenty-first century, people may spend months and even years in space. Orbiting space stations and bases on moons and planets must be designed with balanced, self-sustaining life-support systems. These systems must be able to supply air, water, food, and warmth and must dispose of or recycle wastes. Green plants could serve several functions in life-support systems.

The Need for Plants

An adequate food supply for extended missions would take up a large amount of storage space, and its mass would require much fuel. Raising crops in space may help solve the food problem, and fresh food would be a welcome change from prepackaged, freeze-dried fare.

Plants could also recycle wastes, taking up carbon dioxide exhaled by astronauts and releasing oxygen back into the air. Digestive wastes could be used to fertilize the plants.

Plants could also provide a psychological benefit by surrounding people with living reminders of Earth. Plants could create a pleasant oasis in the high-tech environment of a space station or lunar base.

New Growing Techniques

To solve the problem of growing plants in the limited area of a space station, scientists are experimenting with water solutions as substitutes for soil. The solutions contain the ele- ments essential for plant growth. This technology is a type of hydroponics, which you read about earlier.

Another problem with growing plants in space is the lack of gravity. Scientists are designing devices that put plant roots in contact with the hydroponic solutions without letting the solutions escape into the surrounding air.

Moon Dirt

Lunar soil is composed of minerals that are similar to those on Earth. However, they were formed in a different chemical environment. Lunar soil could supply plants with calcium, magnesium, iron, and small amounts of potassium and phosphorus. The two elements that plants require in the largest quantities—carbon from air and nitrogen from soil—are absent on the moon. Scientists must overcome such limitations to grow food away from the familiar soil of Earth. ◆

315

UNIT 5 ANIMALS

I n recent years, many pandas have starved when all the bamboo in their area died. And an even bigger problem for the pandas is poachers. These are people who illegally set traps to catch or kill wild animals. When a panda gets caught in a trap, the poacher can sell its skin for a lot of money.

From "China's Precious Pandas," in *Ranger Rick* magazine

/// CHAPTERS

For millions of years, giant pandas have lived in China. So why are they in trouble now?

Types of Animals

Africa is one place where you can find an amazing variety of animal life. But your own back-yard, the neighborhood in which you live, and the streets you travel to and from school are also all places where animal life can be found. Of course, unless you live in a very exciting neighborhood, you will not see the same types of animals that David Attenborough writes about!

Twenty-five years ago, I went to the tropics for the first time. I still recall, with great clarity, the shock of stepping out of the plane and into the muggy, perfumed air of West Africa. It was like walking into a steam laundry. Moisture hung in the atmosphere so heavily that my skin and shirt were soaked within minutes. A hedge of hibiscus bordered the airport buildings. Sunbirds, glittering with green and blue iridescence, played around it, darting from one scarlet blossom to another, hanging on beating wings as they probed for nectar. Only after I had watched them for some time did I notice, clasping a branch within the hedge, a chameleon, motionless except for its goggling eyes which swivelled to follow every passing insect. Beside the hedge, I trod on what appeared to be grass. To my astonishment, the leaflets immediately folded themselves flat against the stem, transforming green fronds into apparently bare twigs. It was sensitive mimosa. Beyond lay a ditch covered with floating plants. In the spaces between them, the black water wriggled with fish, and over the leaves walked a chestnut-coloured bird, lifting its long-toed feet with the exaggerated care of a man in snow-shoes. Wherever I looked, I found a prodigality of pattern and colour for which I was quite unprepared. It was a revelation of the splendour and fecundity of the natural world from which I have never recovered.

from *Life on Earth: A Natural History*
by David Attenborough

For Your Journal

- ✐ List ten animals that have hair and ten animals that do not have hair.
- ✐ Write helpful *or* harmful *after each animal, depending on whether you think it helps or harms humans.*
- ✐ Think of another way to divide all animals into just two separate groups.

Introduction to Animals

Objectives

Describe the great variety of living animals.

Distinguish between helpful and harmful animals and give examples of each.

Debate the idea that an animal considered harmful might be helpful as well.

Do you suppose David Attenborough liked animals when he was young? How do people grow up to care so much about animals that they build a job, a career, and even a life around animals? Perhaps you feel as if you are one of those people. Perhaps you like some animals and dislike others but are curious about most animals.

How aware are you of the animals around you? You can increase your awareness of the animals in your environment by doing the following activity.

DISCOVER BY *Observing*

Spend about fifteen minutes outdoors looking for animals. Conduct your observations every day for five days. Try to observe at different times of the day or early evening. Write your observations in your journal. Depending on where you live, you may see very small animals or even animals that fly. If you aren't sure whether a living thing is an animal, list it with a question mark after it. Review your list after you have finished this chapter. ✐

Figure 12–1. Marine biologists specialize in the study of plants and animals found in the world's oceans. They study organisms that live in shallow waters as well as those that live in the deep ocean.

Increasing Awareness of Animals

A biology teacher once explained that as a child she liked animals with fur, even mice. But she was frightened of cockroaches and lizards. Then she lived for two years in Asia where she learned to be grateful for lizards because they protected the household from insects whose bite or sting was painful and sometimes dangerous. She explained, "I never thought I would lie in bed at night and be glad to hear the clacking sound of geckos calling to each other. I knew if the geckos were there, other things were not." After a minute she added, "I still haven't found a way to like cockroaches, though."

The more we observe and read about animals, the more respect we have for them. How they move, what they eat, and how they go about their daily lives is a series of amazing facts. Check your animal knowledge by trying to match the facts with the appropriate animals in the following activity.

Figure 12–2. A typical gecko is small and nocturnal with large eyes and expanded toes. Often found in warm countries, it is harmless to humans and useful in destroying insects.

ACTIVITY

Can you match the animals with the facts?

MATERIALS
paper and pencil

PROCEDURE
1. Use the two lists provided. Match the animal from List #1 with the appropriate "amazing fact" from List #2.
2. You are not expected to know all of the facts listed in this activity. To match the lists, use logical reasoning and don't be afraid to guess.
3. Which of the interesting facts might help each animal survive? Give reasons for your answers.
4. What other animal facts do you know that might explain why certain animals are successful at getting food, escaping enemies, or just remaining healthy?

LIST #1 ANIMAL		
1. Ostrich	**2.** Flea	**3.** Emperor Penguin
4. Shrimp	**5.** Blue Whale	**6.** Elephant
7. Tuna	**8.** Giraffe	**9.** Mallard Duck
10. Box Turtle	**11.** Ant	**12.** Monarch Butterfly

LIST #2 AMAZING FACT
a. Has lived to be 129 years old
b. Snacks on 68 kg of hay a day
c. Lays 7-14 eggs at a time
d. Can pick up 50 times its own weight
e. One of its eggs scrambled would feed 12 people
f. Its tongue is 43 cm long
g. Can hold its breath underwater for 18 minutes
h. Can jump 33 cm
i. Travels 69 km per hour
j. Travels 16 km per hour
k. Travels 3 km per hour
l. Drinks 487 L of milk a day as a newborn

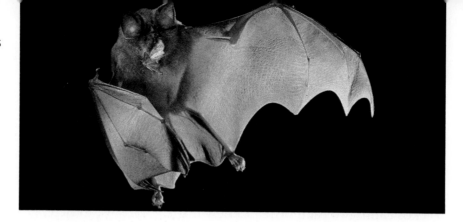

Figure 12–3. The habits of many animals are surprising. Night-flying bats navigate and locate their food by using *echolocation*. The bats give off ultrasonic signals (above human hearing) and then listen for the echoes from the signals.

Awareness of animals and a curiosity about them are what lead most people to pursue a career with animals. What animals are common where you live? Close your eyes and take an imaginary walk around your neighborhood. Are there birds perched on roofs, telephone wires, or tree branches? Is that a spider web attached to a fence? Look closely at an advertising sign. Have wasps built a nest behind it? Look for ants, bees, worms, cats, dogs, and butterflies. Maybe flies have found a food scrap that someone dropped or a garbage can that has not yet been emptied. You don't have to live on a farm or a ranch or in a jungle to find animals. Even large animals such as horses can be found in big cities. The horses of New York City's mounted police attract attention wherever they go in the city.

Sometimes we forget what the word *animal* includes. When we read about animals such as elephants and lions, it is easy to forget that the pigeons roosting on a nearby building are also part of the animal world. Thinking of butterflies and worms as animals may take a little practice. They don't bark, have fur, or walk on four legs like the animals that may share our homes.

 ASK YOURSELF

Name five places where animals may be found in your neighborhood.

Figure 12–4. Often the animals that share our environment go unnoticed. Sometimes we are just so used to seeing them that we don't pay attention.

Contributions of Animals

In at least some way, animals are a part of everyone's life. Many provide food for us and for other animals. Clothing and shoes are sometimes made from animal products. Some people may choose not to eat or wear animal products but are still influenced by animals in other ways. Research studies have shown that often people can be calmed and reassured by holding or petting a cat or dog. Heartbeat and blood pressure rates usually slow when a person is holding a soft, warm animal.

Some groups are experimenting with the idea of lending pets to elderly people or disabled people for a few hours a day. They may be unable to care for a pet 24 hours a day, but they can enjoy and benefit from the period of companionship. The animal is delivered and returned by the workers from the animal-lending group.

Have you ever seen an aquarium in a doctor's office? Physicians and dentists often have a large aquarium in the waiting room because the sight of fish swimming gracefully through the water can relax patients.

Figure 12–5. People of all ages enjoy animal companions. What types of animals would you choose for a Lend-a-Pet program?

DISCOVER BY Researching

Use your library and phone directory to learn about safe ways student volunteers can be involved in organized activities with animals. Ask your parents and teachers for their opinions. Does your community have a rehabilitation area for injured or mistreated animals? Write your findings in your journal.

▼ ASK YOURSELF

Give an example of how animals calm or relax humans.

The Good, the Bad, and the Ugly?

Figure 12–6. You may think of stinging insects as harmful, but often they play an important role in pollinating plants.

Are all animals helpful? What about the scary ones like rats that carry disease and termites that destroy houses? Many animals—particularly insects—are considered pests because they compete with humans for food and living space. Some animals, such as wolves or bears, can actually threaten human life.

All animals—even those that humans consider pests—have a role to play in their environment. For example, bees and other stinging insects are important in pollinating flowers. Removing a supposed "pest" can sometimes cause new problems. Killing the wolves that attack sheep and other livestock has, in some cases, caused deer populations to increase so much that the deer become the new "pests." Although it is important to control the numbers of animals that are destructive, the programs for control must be carefully thought out.

The biology teacher who still dislikes cockroaches is probably not too different from the rest of us. But she and others might be surprised to read some of the sad stories about people imprisoned during wartime. There have been many accounts of the loneliness and depression these prisoners experienced when kept completely alone in cells for months or even years. Several of these stories describe how grateful the prisoner was for the presence of crickets, cockroaches, or other insects in the cell. The prisoner studied the habits of the insects, thus finding a way to stay interested and alert because of the presence of another living thing—even a cockroach.

 ASK YOURSELF

List three animals you think of as pests. For each one, describe some activity of the animal that is actually beneficial.

SECTION 1 *REVIEW AND APPLICATION*

Reading Critically

1. Why could it be reassuring to have a gecko in your house?

2. List three ways animals are helpful to humans.

Thinking Critically

3. In the last two hundred years, what animal do you think has been the most useful to humans? Give reasons for your answer.

4. Would any of the animals mentioned in this section be natural enemies of each other? If so, give an example of such a pair and tell why they would be enemies.

Characteristics of Animals

Objectives

Describe the basic characteristics of animals.

Determine whether or not a living thing is an animal.

Contrast radial and bilateral symmetry in animals.

If somewhere in this chapter you found a question asking you to explain the difference between a rose and a cow, you would probably laugh. The idea sounds funny because we don't usually have trouble seeing the difference between common plants and common animals. But David Attenborough in West Africa might easily have discovered a living thing that mystified him. Is the strange organism a plant or an animal?

Occasionally, we encounter a confusing organism that doesn't match our mind's picture of a common animal or plant. We may find it difficult to classify this organism. That's when it is helpful to think about the characteristics that all animals must have. Applying what we know about animals in general can help us decide whether a new organism is an animal or a plant.

General Characteristics of Animals

What do we know about animals in general? One characteristic of animals is that they must take their foods in from their environment. Animals cannot make their own food. They cannot perform the process of photosynthesis used by plants. If you have ever cared for pets, you know that they must be fed. Even you and your friends are good reminders of this characteristic of animals. A full day of sunshine and fresh air does not take the place of breakfast or supper, and your body complains to you if you delay eating for too long.

Figure 12–7. There are many ways of obtaining food, as can be seen by observing animals such as an American robin (below), a May beetle (bottom left), and a common snapping turtle (top left).

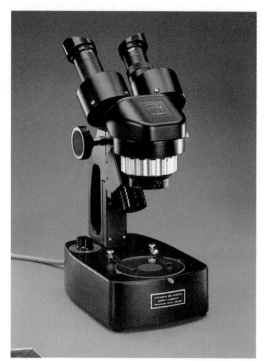

Figure 12–8. Some differences between animals and plants can be seen only on the microscopic level.

Another characteristic of animals is that they usually do not grow beyond a certain adult size. Imagine what would happen if humans, like trees, did not stop growing until they died!

A Closer Look The next two characteristics of animals are things you would not be able to tell if you just walked up to an animal such as the cow mentioned earlier. You would need to examine the animal with a microscope. In order to be called an *animal*, a living organism must be multicellular. Using a microscope, you would be able to see that some organisms have only one cell. Even though these one-celled organisms may look like animals, they are classified as *protists.*

If you used the microscope to look very closely at the cells of the cow, you would find that cow cells, like all animal cells, have cell membranes but no cell walls. Plant cells have both cell membranes and cell walls. So now you have a list of ways the cow and the rose are different, in case anyone ever asks! You can illustrate the differences in animal and plant cells by doing the following activity.

In your journal, sketch a typical animal cell and plant cell as shown in the illustration. Label the cells to show the presence or absence of cell walls, cell membranes, and chloroplasts.

Figure 12–9. One difference between plant and animal cells is that animal cells do not have cell walls. Another difference is the absence of chloroplasts in animal cells. What function can animal cells not perform because they do not have chloroplasts?

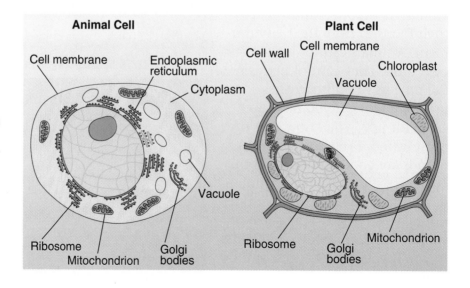

Moving Along

A final characteristic of most animals is that they move from place to place. This type of movement is called *locomotion*. Most of the animals we encounter fly, run, swim, crawl, hop, or slither. Then we come to the animals that confuse us. Some animals move only during early stages of their lives. As adults they anchor onto some solid object and remain there the rest of their lives. Barnacles that attach themselves to piers and boats are an example of animals whose movement is limited. Sponges, which don't seem very animal-like at first glance, also have limited movement. We are so accustomed to watching animals move that it is usually their movement that makes us recognize them as animals. When animals seem to stay in one place most of their life, it can confuse observers.

Figure 12–10. Most animals are *motile,* which means that they move around during their lives. However, this fur seal does not seem to be in a hurry to go anywhere!

 ASK YOURSELF

Animals cannot make their own food. How do animals get the food they need?

Symmetry

All animals except for sponges have some arrangement of their body parts that enables us to describe them in terms of symmetry. **Symmetry** is the balanced arrangement of body parts around a center point or center line. Look at the picture of the starfish on the next page. The starfish has neither a head nor a tail. Its body parts are arranged in a circle around a center point. If you drew imaginary lines through the center of the starfish, they would resemble the spokes of a wheel. This circular body plan is called **radial symmetry.** The jellyfish is another example of radial symmetry.

DISCOVER BY Doing

Create "inkblot animals." First, fold a sheet of paper in half. Then open the sheet and flatten it out. Put ink or paint along the center fold line. Fold the paper in half again. Open the paper. What do you see? What type of symmetry does your inkblot animal have? Make a variety of "inkblot animals."

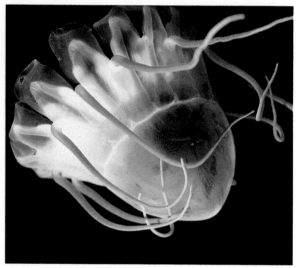

Figure 12–11. Radial symmetry can be seen in this starfish (left) and in this jellyfish (right).

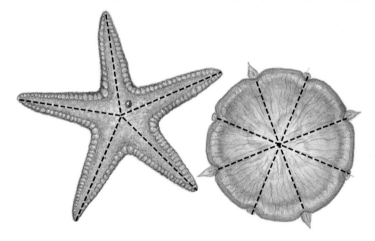

Most animals, including humans, have bilateral symmetry. An animal with **bilateral symmetry** has its exterior body parts arranged the same way on both sides of its body. A dog has two legs on each side, a butterfly has one wing on each side, and humans have one arm and one leg on each side. Nearly every animal with bilateral symmetry has a definite front end and a definite back end. The front end of the animal is called the *anterior*. The back end is called the *posterior*.

To picture bilateral symmetry in your mind, draw an imaginary line from the anterior end of an animal to the posterior end. Then imagine that the animal is divided in half along that line. As you can see from the illustration of the butterfly, the two long halves, or the left and right sides, are mirror images of each other.

Animals with bilateral symmetry also have an upper side and an underside. The upper side is called the *dorsal* side, and the underside is called the *ventral* side. When you rub a puppy's tummy, you are rubbing the ventral side. Petting its back is petting the dorsal side. When the puppy licks your hand and wags its tail, it is using anterior and posterior body parts to tell you that it likes to be petted. The anterior and posterior ends and the dorsal and ventral sides of a dog and a chimpanzee are shown in the illustrations.

Figure 12–12. An animal with bilateral symmetry can be divided by an imaginary line into two equal right and left halves. The halves are mirror images of each other.

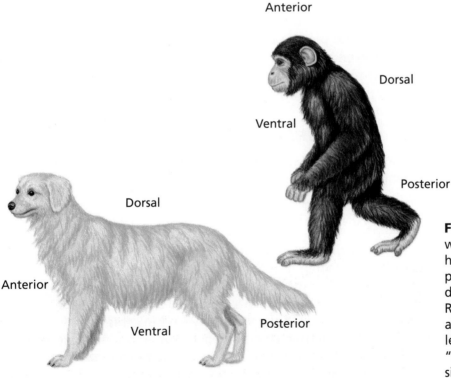

Anterior

Dorsal

Ventral

Posterior

Dorsal

Anterior

Ventral

Posterior

Figure 12–13. Animals with bilateral symmetry have anterior and posterior ends as well as dorsal and ventral sides. Regardless of whether an animal walks on four legs or two legs, the "belly" is the ventral side and the "back" is the dorsal side.

Figure 12–14. The soft skeletons of some types of sponges are often used for bathing. What characteristic of the sponges makes them a good choice for this purpose?

No Front End

A few animals are asymmetrical, or show asymmetry. The prefix *a-* in front of a word sometimes means "not." For example, *atypical* means "not typical." **Asymmetry** means "lacking symmetry"; *asymmetrical* means "not symmetrical." Asymmetrical animals have their body parts arranged in such a way that you cannot divide them into equal halves. Some sponges are examples of asymmetrical animals.

Uses of Symmetry

When artists and manufacturers design homes, automobiles, swimming pools, jewelry, or clothing, they often use symmetry. Cars with sleek lines can be divided down the middle to create mirror images. Domes on buildings and pieces of pottery have a radial design. The word *radial* is even applied to automobile tires. In the Skill activity on the next page, you will have the opportunity to find symmetry in other nonliving objects.

 ASK YOURSELF

Name two types of symmetry, and give examples of animals that exhibit each type.

SECTION 2   *REVIEW AND APPLICATION*

Reading Critically

1. List three characteristics of animals other than symmetry.

2. Name an animal that has no symmetry. What is its type of body arrangement called?

Thinking Critically

3. When humans sense they may be hit by an oncoming object, such as a thrown ball, they usually turn in such a way as to protect their ventral side. Why do you think this is so?

4. Compare the movement of an animal that has bilateral symmetry with the movement of an animal that has radial symmetry.

SKILL

Identifying Symmetry

▶ **MATERIALS**
● ten to fifteen pictures of common objects found at home or school

▼ **PROCEDURE**

1. Sort the objects into three groups: those with radial symmetry, those with bilateral symmetry, and those with asymmetry.

2. Sketch each object.
3. Draw a line or lines through your sketch of each object to show the type of symmetry.

▶ **APPLICATION**

1. What test did you use to determine the type of symmetry for each object?
2. Do any of the objects show more than one type of symmetry?
3. Hold two mirrors at right angles and move so that you can see one image of your face in the mirrors. The two mirrors are reversing your right and left sides. (This is actually how you look to other people!) Is the bilateral symmetry of your face and hair perfect? Explain your answer.

✳ ***Using What You Have Learned***
Select pictures of ten different animals. Divide the animals into groups according to their symmetry.

Animal Classification

Objectives

Compare and contrast *vertebrates and invertebrates.*

List *examples of vertebrates and invertebrates.*

Analyze *the advantage of one type of animal over another.*

If you could be another animal for a day, which would you choose to be? Would you choose one of the strange and colorful animals David Attenborough found in West Africa? Would you pick a more familiar animal—a household pet or an animal that you have seen living in your neighborhood?

Before you answer, think about what your day might be like as the animal you choose. If you choose a bird or a butterfly, you might spend a lot of the day flying over the countryside enjoying the view. Of course, as a bird, you would also have to eat worms! You might think that the cat next door has an easy, pleasant life. Would you choose to be a cat? Since this is turning out to be a little more involved than it sounded at first, maybe we should look closely at the main groups of animals. Finding out more about animal classification might help you in making your decision.

One decision you would have to make is whether or not you wanted a backbone. Any animal that has a backbone is a **vertebrate.** The group gets its name from the small bones called *vertebrae* that make up the spine or backbone. An animal that does not have a backbone is called an **invertebrate.**

Figure 12–15. What type of animal would you choose to be for a day? If you enjoy swimming, you might choose to be a fish.

Vertebrates

In some ways, vertebrates seem to have an advantage over animals without a backbone. The bones of a vertebrate are part of the living tissue of the animal and grow as the animal grows. The bones support the body from within and also protect vital organs. Being squished is not as likely for a vertebrate as it appears to be for the soft, less-protected worm.

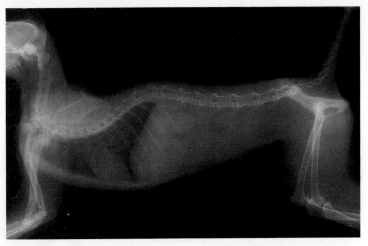

Part of what the vertebrae are supporting and protecting is the *spinal cord*. The spinal cord is a bundle of nerves that is surrounded by the vertebrae. The spinal cord is connected to the brain and carries electrical messages from the brain to all other parts of the body. The messages can also travel the other way.

If you choose to continue being a vertebrate—remember, as a human, you already are a vertebrate—you could choose to be one of 41 700 species of vertebrates. In spite of that large number, vertebrates make up only about 3 percent of the animal kingdom. The other 97 percent of animals are invertebrates.

The following tables give some examples of vertebrates and invertebrates. You may recall that *phylum* is the name given to a large group of animals. *Class* is the term used to further subdivide a phylum. In some cases, a descriptive name is used instead of the more accurate Latin name for a phylum or class. For example, in Table 12–2, instead of using the name *Platyhelminthes*, we will use the descriptive name *Flatworms*.

Figure 12–16. The backbone of a vertebrate forms the central portion of the animal's skeleton. The rest of the skeleton is attached to the backbone.

Table 12-1 **Vertebrates**

Class Name	Example of Animal
Fishes	Trout
Amphibians	Frog
Reptiles	Alligator
Birds	Hummingbird
Mammals	Elephant

 **ASK YOURSELF**

How did vertebrates get their name?

Invertebrates

If you would like to be part of the animal group that was on Earth before animals had backbones, you would choose to be an invertebrate for a day. Fossil evidence indicates that invertebrates appeared on Earth long before vertebrates. Invertebrates may have been able to survive so long because their relatively simple structures enable them to readily adapt to new living conditions.

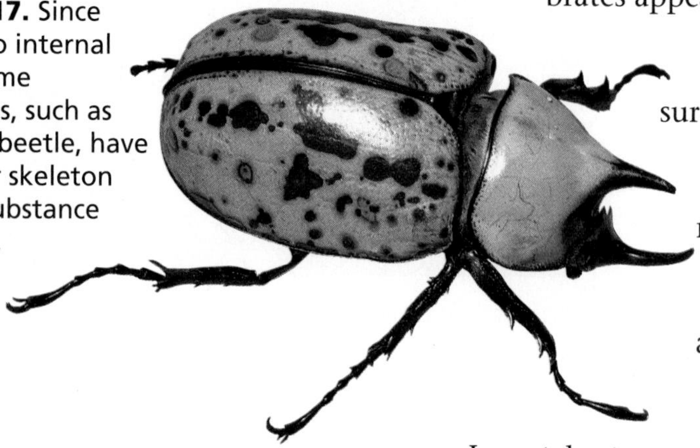

Many invertebrates are well protected by coverings of shell, by spines, or by outside skeletons. Invertebrates are extremely varied, ranging from butterflies and corals to worms and spiders. They include starfish, jellyfish, grasshoppers, clams, and even octopuses.

Certain invertebrates that live in water have radial symmetry. The jellyfish and the starfish are invertebrates with radial symmetry. Most invertebrates and all vertebrates have bilateral symmetry. Remember the asymmetrical sponge? That sponge is also an invertebrate.

Figure 12–17. Since they have no internal skeleton, some invertebrates, such as the unicorn beetle, have a hard outer skeleton made of a substance called *chitin.*

Table 12-2	Invertebrates	
Phylum Name	**Example of Animal**
Poriferans	Sponge
Coelenterates	Jellyfish
Flatworms	Tapeworm
Roundworms	Hookworm
Segmented Worms	Earthworm
Mollusks	Snail
Echinoderms	Starfish
Arthropods	
Arachnids	Spider
Crustaceans	Lobster
Myriapods	Centipede
Insects	Grasshopper

DISCOVER BY Researching

List twenty-five animals that are as different from one another as possible. Using your observational skills as well as reference books, list two advantages and two disadvantages of being each of the animals. Record the information in your journal. Then choose the animal you would most like to be for a day. ✎

Figure 12–18. Invertebrates come in many different forms. The tiger swallowtail butterfly (left), the cowrie (top right), and the marbled orb-weaver spider (bottom right) are all invertebrates.

What factors did you use in making your decision? In the next two chapters you will read more about invertebrates and vertebrates. You may change your mind when you know more about both types of animals!

▼ ASK YOURSELF

List three invertebrates and describe any protective covering each has.

SECTION 3   REVIEW AND APPLICATION

Reading Critically
1. How are vertebrates and invertebrates different?
2. Name two functions of vertebrae.
3. Why is the spinal cord important enough to be enclosed by protective bone?

Thinking Critically
4. According to the classification system used by biologists, why is a whale more closely related to a tiger than to a jellyfish?
5. Grasshoppers and turtles both have hard outer coverings. Can you conclude that both are invertebrates? Explain.

INVESTIGATION

Comparing Vertebrates and Invertebrates

▶ **MATERIALS**
- preserved frog ● preserved earthworm ● paper towel

▼ **PROCEDURE**

Part A
1. Place the frog and the earthworm side by side on the paper towel so that you can observe them at the same time.
2. List five ways in which the frog is like the earthworm and five ways in which the two animals differ.
3. Which animal seems more complex to you? Give reasons for your answer.

Part B
1. Run your finger down the middle of the frog's dorsal side. (The dorsal side is the one with the frog's eyes.) Move your fingers from the anterior end to the posterior end of the frog's body. Can you feel a backbone? Record your answer.
2. On a sheet of paper draw the outline of the frog's body. Draw a dotted line from anterior to posterior, dividing the frog's body down the middle. Do the two halves resemble each other? What kind of body symmetry does a frog have? Record your answers.

Part C
1. Now examine the earthworm. The dorsal side of the earthworm is darker in color than the ventral side. Be sure the earthworm is lying on its ventral side. The anterior end of the earthworm is larger than the posterior end.
2. Run your finger down the middle of the dorsal side of the earthworm. Move your fingers from the anterior end to the posterior end of the earthworm. Can you feel a backbone? Record your answer.

3. On a sheet of paper draw the outline of the earthworm's body. Draw a dotted line from anterior to posterior, dividing the earthworm's body down the middle. Do the two halves resemble each other? What kind of body symmetry does an earthworm have? Record your answers.

▶ **ANALYSES AND CONCLUSIONS**
1. Is the frog a vertebrate or an invertebrate? Explain.
2. Is the earthworm a vertebrate or an invertebrate? Explain.

▶ **APPLICATION**
Have you changed your mind about which animal is more complex? If so, explain why. If not, give new reasons to support your opinion.

✳ *Discover More*
Make a list that includes one animal with radial symmetry, one with bilateral symmetry, and one with asymmetry. Rank the animals in order from the least advanced to the most advanced. Explain your reasons for ranking the animals as you did.

The Big Idea

Animals are found everywhere, and they vary greatly in size. They range from tiny insects to enormous elephants and whales. But although they come in many different sizes, all animals have common structural traits.

Some animals have a backbone that protects the bundle of nerves called the spinal cord. These animals are called *vertebrates*. Animals without a backbone are called *invertebrates*.

All animals share the same general characteristics. Animals take in food from their environment, they are multicellular, they have cells without cell walls, and most have some pattern of body symmetry.

For Your Journal

Add to the journal entries you made before you read the chapter. Write the word vertebrate *or* invertebrate *next to each animal you listed. Were most of your animals vertebrates? If so, add some more invertebrates to your list. For those animals you labeled as harmful, add one positive trait for each animal.*

Connecting Ideas

Animals make many contributions to humans and to the environment. Sometimes, however, animals are considered pests, or even dangerous. The confusing thing is that many animals are considered both helpful *and* harmful. Make a list of animals you consider to be both helpful and harmful. Try to show why your opinions could both be true. For each animal, illustrate your point with a diagram like the one shown.

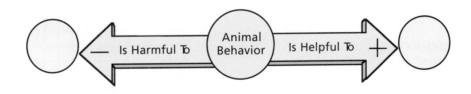

REVIEW

Understanding Vocabulary

For each pair of terms, complete this statement:
_____ is different from _____ because _____.

1. A vertebrate (332); an invertebrate (332)

2. Asymmetry (330); symmetry (327)

3. Bilateral symmetry (328); radial symmetry (327)

Understanding Concepts

MULTIPLE CHOICE

4. A characteristic of animals is that
 a) they take in food from their environment.
 b) they grow throughout their lifetimes.
 c) their cells are protected by a cell wall.
 d) their cells have no nuclei.

5. Some sponges are asymmetrical because
 a) they live in the water.
 b) they are invertebrates.
 c) their body parts are mirror images of each other.
 d) their body parts do not have a balanced arrangement around a center point or line.

6. All animals are
 a) symmetrical.
 b) multicellular.
 c) vertebrates.
 d) invertebrates.

7. An animal with radial symmetry
 a) has anterior and posterior ends.
 b) has a circular arrangement of body parts.
 c) has difficulty moving quickly.
 d) has more than one spinal cord.

SHORT ANSWER

8. Justify the classification of a mosquito as an animal.

9. Explain the function of the spinal cord.

10. Give two examples of ways in which invertebrates protect themselves.

Interpreting Graphics

11. Copy the "animal" below. What type of symmetry does it have? Use the labels *anterior, posterior, dorsal,* and *ventral* to mark the different parts of the animal.

12. Explain how the diagram below shows the ratio of vertebrates to invertebrates in the animal kingdom. Hint: Recall that *percent* is a ratio of 100.

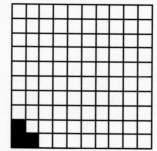

Reviewing Themes

13. *Systems and Structures*
A brown tarantula (a very large spider) and a brown squirrel might both be about 25 cm long. In what ways would the animals be different?

14. *Energy*
How do animals differ from plants in the way they get the energy they need to live?

Thinking Critically

15. Why do environmental groups working to save an endangered habitat find it necessary to choose a "cute" animal as their mascot? For example, why might a small, fuzzy monkey be chosen instead of a colorful snake?

16. A "broad-spectrum" pesticide is one that kills a wide variety of insects. Explain why these pesticides have, in some cases, created new pest problems for farmers.

17. Discuss why growing to a predetermined size is a characteristic of animals rather than a characteristic of living things in general.

18. Your friend claims there are no animals living in her house because her family does not have house pets. Explain why she might be wrong in her claim.

19. In this chapter you read about both vertebrates and invertebrates. A vertebrate, like the cat shown here, has a skeleton inside of its body. Since invertebrates do not have a backbone, they must have some other means of body support. The grasshopper shown here has its skeleton outside its body. What are the advantages and disadvantages of an internal skeleton? What are the advantages and disadvantages of an external skeleton?

Discovery Through Reading

Kerrod, Robin. *Mammals: Primates, Insect-Eaters and Baleen Whales.* Facts on File, 1989. Brief essays highlight the characteristics of the various mammals depicted to explain how each is unique.

Invertebrates

Invertebrates have been around for a very long time. We know what invertebrates look like today. However, since all organisms undergo changes over the course of time, what were the invertebrates that lived long ago like? Author Rachel Carson offers her view.

Five hundred million years ago life in the sea, the fossils tell us, had already progressed very far. All the main groups of animals without a backbone had been developed.

*H*ave you ever heard someone described as spineless? Such a description is negative—it means that the person is thought to be weak and cowardly. But scientists use the word spineless to refer to invertebrates, or animals without a backbone. The vast majority of animals on Earth, including some of the most beautiful, some of the most unusual, and some of the most frightening-looking creatures, are invertebrates.

But there were no animals with backbones, no insects or spiders. And there was still no plant or animal that could venture onto the forbidding land. So for at least three-fourths of the time that life has existed on the planet, the continents lay silent and uninhabited, while the sea prepared the life that was later to invade them. Meanwhile, with violent tremblings of the earth and with the fire and smoke of roaring volcanoes, mountains rose and wore away, glaciers moved to and fro over the earth, and the sea crept over the continents and again withdrew.

It was not until some 350 million years ago that the first pioneer of land life crept out on the shore. It was an arthropod, one of the great tribe from which, later on, crabs and lobsters and insects came. It must have been something like a modern scorpion, but it lived a strange life, half on land, half in water....

from The Sea Around Us
by Rachel Carson

For Your Journal

- ✎ *What do you think the first invertebrates on Earth might have looked like?*
- ✎ *Name as many invertebrates as you can.*
- ✎ *Imagine what life is like for an invertebrate. Write about it.*

Sponges, Coelenterates, and Worms

Objectives

Explain regeneration.

Distinguish between the sexual and asexual stages of coelenterate reproduction.

Summarize the effects that different worms have on humans.

The underwater environment contains many fantastic sights and sounds. Imagine scuba diving, not in Earth's ancient ocean, but in the crystal-clear water of a modern-day tropical paradise, and seeing beds of gigantic seaweed and many beautiful fish. It would be obvious to you that the seaweeds are plants and the fish are animals. Then you might see sponges like the ones in the illustration and ask yourself, "Are they plants or are they animals?" Other people have asked the same question. In fact, scientists debated the answer to that question for more than 2000 years!

Characteristics of Sponges

Today scientists know that sponges are animals. Sponges are considered to be the simplest invertebrates because their bodies are only two cell layers thick. They have no tissues and, therefore, no organs or organ systems.

Sponges can be found in salt water and in fresh water. In its adult form, a sponge is attached to an object in the water and does not move from that spot. Animals that do not move from place to place are called *sessile* (SEHS ihl).

The body of a sponge is covered with many small openings called *pores*. Water enters the pores of a sponge, carrying algae, protozoa, and other one-celled organisms. Inside the sponge, these organisms are filtered out of the water and digested as food. The filtered water is then pushed out of the sponge through an opening in the top part of its body.

Figure 13–1. Pores allow water and nutrients to flow constantly into and out of a sponge.

The body of a sponge is supported by a type of "skeleton" made of either spongin or spicules. *Spongin* is a flexible protein. Most *spicules* (SPIHK yoolz) are very sharp spikes that look like crystals when viewed under a microscope. Which type of sponge would you rather bathe with? You will discover more about sponges in the next activity.

Problem Solving

Artificial sponges are less expensive to buy than natural sponges. Since natural sponges are available in great numbers, why would a sponge manufactured in a factory be less expensive? Develop a hypothesis to explain this difference in cost. Write the hypothesis in your journal.

Figure 13–2. Can you tell which sponges are natural and which are synthetic?

Synthetic sponges are made in factories and are packaged for use in household cleaning. You've probably seen many of these sponges, which can be purchased in a wide array of colors. Occasionally, natural sponges may also be brightly colored. The natural coloring is a result of algae and bacteria that live with the sponges.

▼ ASK YOURSELF

Why is a sponge classified as a simple invertebrate?

The Life Cycle of Sponges

Sponges can reproduce both sexually and asexually. A sponge that reproduces sexually produces both sperm cells and egg cells. A sperm and egg combine to produce a fertilized cell. The fertilized cell then develops into a *larva* (plural, *larvae*) or young sponge. The sponge larva shown in the illustration has many flagella. The motion of the flagella allows the larva to swim around in the water. After about 24 hours, the larva settles on a rock or another object and attaches itself. This location becomes its permanent home. The larva then grows into the mature form we recognize as a sponge.

Figure 13–3. Although adult sponges are sessile, larvae, like this one, are free swimming.

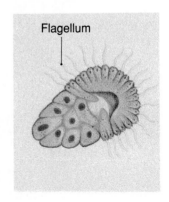

Flagellum

Sponges also have the ability to reproduce asexually. This can occur in two ways. *Fragmentation* occurs when a small branch or piece of a mature sponge breaks off and forms a completely new sponge. This process is called **regeneration.** You would have the ability to regenerate if, for example, your ear could break off and then grow into a whole brand-new you!

A less common form of asexual reproduction may occur in sponges that live in fresh water. During harsh conditions, a sponge may produce gemmules (JEHM yoolz). A *gemmule* is a small group of sponge cells with a supply of food and a hard protective covering. Gemmules can live through unfavorable conditions, such as the drying up of a lake or stream. When the conditions are again right for survival, the gemmules open, and the cells escape to form new sponges.

 ASK YOURSELF

Describe how sponges reproduce.

Characteristics of Coelenterates

On your scuba-diving expedition, you see an unusual animal called a jellyfish. Despite its name, a jellyfish is not a fish but a *coelenterate* (sih LEHN tuh rayt), a baglike animal that has tentacles around its mouth. Other examples of coelenterates are the hydra, Portuguese man-of-war, sea anemone, and coral. Although coelenterates are more complex than sponges, they are still considered "simple" because even though they have tissues, they have no organs.

Figure 13–4. A Portuguese man-of-war may look beautiful, but its tentacles can be deadly to small animals and sometimes even to humans.

The tentacles around the mouth of a coelenterate have stinging cells, which paralyze or kill prey. These tentacles also hold prey and carry it into the coelenterate's mouth, where the prey enters the body cavity. Any waste material from the digestive process leaves the body cavity through the animal's mouth.

The tentacles and stinging cells of a coelenterate can cause serious allergic reactions and pain in humans. Any swimmer or person walking on a beach who is stung by a coelenterate should contact a physician immediately. The following activity will help you compare the body plans of coelenterates and sponges.

Make a poster showing the body plan of a sponge and that of a coelenterate. Identify all the parts you can. Below each diagram, list the differences between the two animals.

The Portuguese man-of-war shown in the photograph on page 344 is a type of coelenterate that is made up of several animals living together in a group called a *colony*. These animals function in a way similar to workers in a factory—everyone works at different tasks to reach a common goal. Some animals in a colony gather food, others are stingers, and still others are used during reproduction.

Another type of coelenterate is the jellyfish. While some jellyfish are small, some have tentacles that may be 9–12 m long and a bell 2 m across. One jellyfish was reported to have tentacles so long that they would touch the ground if the jellyfish were placed on the roof of a 20-story building!

Figure 13–5. Sea anemones use stinging cells in their tentacles to capture small animals for food.

Sea anemones and corals are often compared to flowers. The photographs show that these coelenterates can be very beautiful. However, the brightly colored tentacles that look like harmless flower petals are used for stinging and capturing crabs and small fish for food.

Corals also live in colonies. Unlike the Portuguese man-of-war, however, each coral is a single animal. Corals have hard skeletons that are often made of limestone. Coral reefs are created over hundreds of years as living corals build their skeletons on top of the skeletons of dead corals. The following activity offers you the opportunity to think about a certain use of coral.

Observing

If possible, examine pieces of coral jewelry. Research the use of coral in jewelry-making. In your journal, explain how this use of coral might negatively impact the ocean environment. Find out whether there is such a thing as "fake" coral jewelry for the environmentally conscious consumer. ✎

▼ ASK YOURSELF

Explain how coelenterates are more complex than sponges.

The Life Cycle of Coelenterates

If you look at photographs of yourself taken at different times of your life, what changes would you notice? Even though you have gone through many changes, there has never been a time in your life when you did not have the form that we recognize as that of a human being. However, some animals look very different from one stage of their growth to the next. In either case, the different stages of an animal's life make up its *life cycle*. In the following activity, you can reflect on the changes in an animal's development.

Figure 13–6. The life cycle of a jellyfish is shown on the right. Although the adult jellyfish (left) is free swimming, the larval stage is sessile.

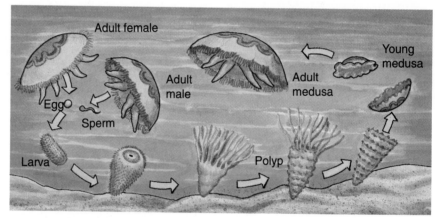

Writing

Crawling and learning to walk are two examples of significant stages in your development. You have experienced many others. Personalize your journal by recording the significant stages in your development. You might wish to illustrate the various stages you have experienced. ✎

Most coelenterates are unlike you—their life cycles occur in two stages, the polyp and the medusa. In the *polyp* stage, the animal's body is vase-shaped. In the *medusa* stage, it is bell-shaped. The illustration on page 346 shows how the jellyfish experiences both the polyp stage and the medusa stage in its life cycle.

Just as the life cycles of different coelenterates vary, the ways in which coelenterates reproduce also vary. The hydra can reproduce sexually, with sperm cells swimming to nearby egg cells that are attached to the hydra for fertilization. The hydra can also reproduce asexually by developing buds that grow into new hydras. The following activity will help you learn more about hydras.

ACTIVITY

How can you directly observe the characteristics of a typical coelenterate?

MATERIALS
living hydras, Petri dish, water, microscope or magnifying glass, millimeter ruler, pencil, medicine dropper, *Daphnia*

PROCEDURE
1. Place one or two hydras in a Petri dish and add enough water to cover them. Using a microscope or a magnifying glass, observe a hydra carefully.
2. Measure the length of the hydra when it is stretched out. What happens when you touch the hydra gently with a pencil point?
3. Use a medicine dropper to create a current around the hydra. How does it react?
4. Add two or three live *Daphnia* to the Petri dish. If the hydra is hungry, it will swallow and digest the *Daphnia*.

APPLICATION
1. What evidence indicates that a hydra has cells that contract and cells that sting?
2. Why might early observers have thought that the hydra was a plant?
3. What does careful observation reveal that indicates the hydra is an animal?
4. Imagine that you are the first person ever to see a hydra. How would you describe it to everyone else?

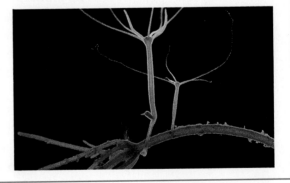

It is easy to see that hydras have a fairly simple life. Not all coelenterates, however, lead such a simple life, especially in terms of reproduction. Jellyfish reproduction is more sophisticated than that of the hydra. In jellyfish reproduction, the female medusa releases egg cells into the water, and the male medusa releases sperm cells near the egg cells. The larvae that develop after fertilization swim around until they attach themselves somewhere on the bottom of the ocean. After the larvae develop into polyps and then mature, they release themselves to become free-swimming young jellyfish.

 ASK YOURSELF

How do the polyp and the medusa stages of development differ?

Flatworms

The name *flatworm* gives a very accurate description of the animals in this group—all have bodies that are completely flat. Even though flatworms are considered to be "simple," they are more complex than sponges and coelenterates. The bodies of flatworms have three tissue layers; coelenterates have two. Unlike sponges and coelenterates, flatworms have definite head and tail regions. Some flatworms are parasites.

Figure 13–7.
Planarians are sometimes called "cross-eyed worms." An injured planarian can regenerate, or regrow, body parts.

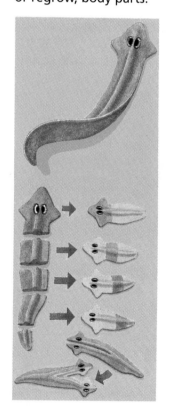

Flat as a Plane   The planarian shown in the illustration is an example of a flatworm that is not a parasite. Most planarians live in water, but some live on moist land. They are usually about 1 cm long, although some planarians grow to about 60 cm long! A planarian's head is shaped like a triangle, and it has two *eyespots*. The eyespots are not really eyes at all; they are tissue that can only sense changes in light. The eyespots help the planarian move about its environment.

The planarian has a mouth on its bottom side. From its mouth it can extend a feeding tube called a *pharynx* (FAR inks). The pharynx draws in food in the form of microscopic organisms. Waste matter also passes out of the mouth, so the mouth is similar in function to that of coelenterates.

Planarians reproduce both sexually and asexually. However, as is true of most flatworms, there are no male or female planarians. Each individual worm produces both egg cells and sperm cells. Also, a small body part can form a completely new organism through the process of regeneration. The illustration shows how planarians regenerate.

It's a Fluke   Flukes are parasitic flatworms that must find several different hosts to continue their life cycle. The liver fluke, shown in the next illustration, must live in the digestive tract of a sheep during one stage of its development and in a snail during another stage!

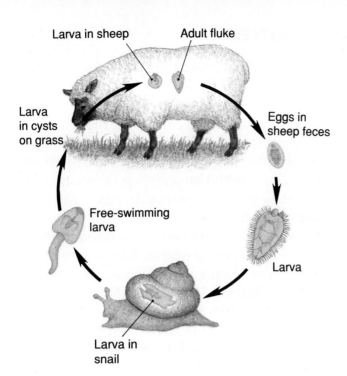

Larva in sheep   Adult fluke

Larva in cysts on grass

Eggs in sheep feces

Free-swimming larva

Larva

Larva in snail

Figure 13–8. Liver flukes require a different host for each part of their life cycle.

Sticks Like Tape, Too

Tapeworms are probably the best known of the flatworms. You may have heard about tapeworms. They look like long tapes or ribbons. These parasites enter hosts that graze on grasses containing tapeworm eggs. Once inside the host, the tapeworm egg develops into a larva. The larva burrows into the flesh of the host and lives there. Humans can get tapeworms if they eat raw or undercooked meat that contains tapeworm larvae. Once a tapeworm larva is in the body, it attaches itself to the intestine and lives on the digested food of its host. Its eggs may then pass out with the feces of the host. A tapeworm is dangerous because it uses the nutrients that should feed the host. Tapeworms, like other parasites that are found in food, can be killed by the proper cooking of food before it is eaten and by the proper disposal of human waste. Careful inspection of meat has made tapeworm infections in humans rare today.

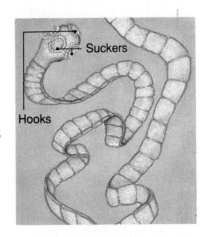

Suckers

Hooks

Figure 13–9. The hooks on the head of a tapeworm are used by the animal to attach itself to the intestine of its host.

 ASK YOURSELF

Explain how parasitic worms pose a threat to their host organisms.

Roundworms

Roundworms are more advanced than all of the other worms you have met so far. They have a fairly complex digestive tract and separate sexes.

While roundworms do have a more complex digestive system than flatworms, the system is still considered to be "simple." The roundworm's digestive tract consists of a long tube with two openings: a mouth and an anus. Food enters through the mouth, and waste is removed through the anus.

The occurrence of two sexes in roundworms is a great advance over flatworms. Roundworms only reproduce sexually. During reproduction, the sperm cell and egg cell combine to form a larva. Since many roundworms are parasites, a host is often required to complete the development of the larva.

The hookworm is an example of a typical parasitic roundworm. During the larval stage, the hookworm enters a human host. The larva can enter a human being when the person walks barefoot in areas where hookworm larvae live. The larva is so small that it can pass painlessly through the skin on the soles of the feet. Once in the body, the larva makes its way into a blood vessel and is carried into the lungs. The human host then coughs up the larva and swallows it into the digestive system. In the intestine, the worm continues its development and feeds on the blood of the human host. This blood loss can be very dangerous to the host. A hookworm can also damage the host's small intestine. Hookworms are a problem in most warm, damp areas of the world, including the southeastern United States.

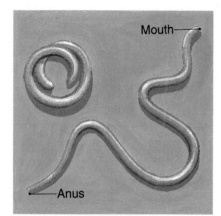

Figure 13–10. Most roundworms are parasites and cause serious illness in humans and other animals. The anatomy of a roundworm (top) is very simple.

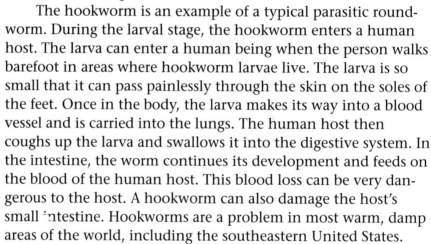

ASK YOURSELF

How are roundworms more advanced than flatworms?

Segmented Worms

Segmented worms are the most advanced variety of worms. Like other types of worms, they have three tissue layers. In addition, segmented worms have a "tube-within-a-tube" body plan. This means that their internal organs are separated from the other parts of their body. Segmented worms also have simple circulatory, muscular, and nervous systems. The bodies of these worms have segments that look like a series of rings running the length of the body.

Earthworms
by Valerie Worth

Garden soil,
Spaded up,
Gleams with
Gravel-glints,
Mica-sparks, and
Bright wet
Glimpses of
Earthworms
Stirring beneath:
Put on the palm,
Still rough
With crumbs,
They roll and
Glisten in the sun
As fresh
As new rubies
Dug out of
Deepest earth

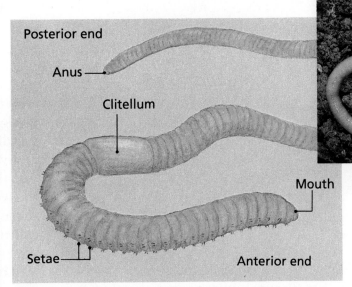

Figure 13–11. The best-known segmented worm is the earthworm. Segmented worms are more complex than flatworms or roundworms.

Gone Fishing Earthworms are probably the best known of the segmented worms. You may have seen many earthworms. If you look closely at the outside of an earthworm, you can see a distinct feature—a large band of tissue that surrounds several of the segments near the anterior end of the worm. The band, called the *clitellum* (klih TEHL uhm), is involved in the reproduction of the worm.

If you gently rubbed your finger over an earthworm, you would find that it is not completely smooth. *Setae* (SEET ee), or tiny bristles, are located on the ventral side of the worm. The setae are attached to muscles that help the earthworm move efficiently through soil. Does the poem help you picture earthworms in a garden? Use your observations from the next activity to help you write your own poem about earthworms.

DISCOVER BY *Observing*

Carefully examine an earthworm. Feel its setae, count its segments, and find its mouth and anus. Then place it in a glass jar partially filled with soil. Observe the earthworm for a few days. How has it affected the soil? How could this kind of activity help gardeners? Write your conclusions in your journal. Then compose your own poem about earthworms. Release the earthworm in the soil outside. ✎

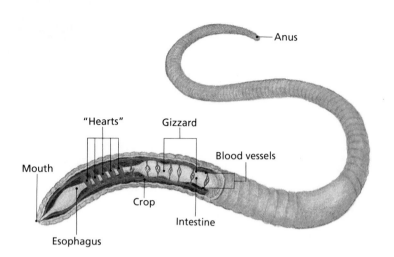

The illustration shows that the digestive system of segmented worms consists of a tubelike structure called the *esophagus*, a thin-walled crop for storage, and a thick, muscular gizzard for grinding. The rest of the digestive system is an intestine that runs the length of the worm and ends with the anus. Solid wastes, called *castings,* pass through the anus. Castings are rich in minerals and improve the soil in which they are deposited. In what other ways are earthworms good for the soil?

Figure 13–12. The body plan of an earthworm is a tube within a tube. Segmented worms have digestive, muscular, circulatory, respiratory, nervous, excretory, and reproductive systems.

Have a Heart (Hearts?)

Segmented worms are the first animals you have met in this chapter that have blood vessels. Earthworms have five pairs of "hearts." Actually, the "hearts" are very simple muscles that pump blood into the blood vessels.

Earthworms reproduce sexually, but each earthworm produces both sperm and eggs. Although both sexes are found in the same individual, the worms do not fertilize themselves. During reproduction, two worms exchange sperm cells, and both are fertilized. The fertilized eggs are covered with a sticky substance secreted by the clitellum and then deposited. The sticky substance forms a kind of cocoon, which protects the eggs.

 ASK YOURSELF

In what ways are earthworms more advanced than roundworms or flatworms?

SECTION 1 *REVIEW AND APPLICATION*

Reading Critically

1. Give two examples of worms that are parasites.
2. Why should people be cautious when they are near some coelenterates?
3. Describe how hookworms enter the human digestive system.

Thinking Critically

4. If a liver fluke cannot find a snail host, what will happen to the next generation of liver flukes?
5. Hookworm infections occur more frequently in rural areas than in cities. Explain why.
6. Knowing that adult sponges are sessile, how would you answer scientists who say locomotion is a necessary characteristic of animals?

INVESTIGATION

Tracing the Life Cycle of a Liver Fluke

▶ MATERIALS
- die ● pencil ● paper

▼ PROCEDURE

The procedure in this investigation is similar to a classification key. At each step, you will throw the die and follow the directions for an odd number or an even number. Record your results for each try. Continue the investigation until you have completed the life cycle.

1. Odd number
The sheep eats grass on which a cyst has been deposited. Move to step 2.
Even number
The grass is mowed, and the cyst is destroyed before the sheep eats it. Start over.

2. Even number
The cyst dissolves in the sheep, and the larva develops into an adult fluke. Move to step 3.
Odd number
The cyst dissolves in the sheep, but the larva dies. Start over.

Odd number
Eggs are passed out of the sheep in feces and fall into water. Move to step 4.
Even number
Eggs are passed out of the sheep in feces but never reach water. Start over.

4. Even number
Eggs develop into larvae, and the larvae enter a snail host. Move to step 5.
Odd number
Eggs develop into larvae, but the larvae cannot find a snail host. Start over.

5. Odd number
Larvae reproduce asexually in the snail and then leave the snail host. Move to step 6.
Even number
The snail dies before the larvae reproduce asexually. Start over.

6. Even number
The larvae swim to grass and form cysts. Life cycle is completed.
Odd number
The larvae have been carried too far away by the snail and cannot reach a grassy area on which to settle. Start over.

▶ ANALYSES AND CONCLUSIONS
1. How many attempts did it take for your liver fluke to complete its life cycle?
2. Do you think there would be this many problems in real life? Explain.

▶ APPLICATION
Liver flukes are commonly found in Asia, Africa, and South America. How does the presence of liver flukes affect humans? Why are diseases caused by liver flukes uncommon in the United States? Where in the United States might diseases caused by liver flukes be a problem?

✳ Discover More
Change the outcomes for each roll of the die to reflect the life cycle of a tapeworm. Then repeat the investigation. Compare your results from each investigation. How do you account for the similarities and differences in your results?

Mollusks and Echinoderms

"Happy as a clam!" Have you ever heard this expression? What does it mean—can clams be happy or sad? Clams are one example of mollusks. Were clams among the first organisms to emerge from the sea? Perhaps they were, but we are not sure. Regardless of which organisms first crawled from the ancient ocean, mollusks have evolved into many diverse life forms. If you like eating clams, oysters, scallops, or squid, your taste buds relish the mollusk phylum!

Types of Mollusks

Mollusks (MAHL uhsks) are soft-bodied animals that may have shells to protect them. Some mollusks live in the ocean and in fresh water, while others live on land. You probably have seen different kinds of mollusks—oysters, clams, and snails are mollusks with shells; slugs and octopuses are mollusks without shells. Does it surprise you that these animals are in the same phylum? The differences between them remind us that animals don't have to look alike for their tissue and organ development to be similar.

Mollusks are more complex than any of the worms. Mollusks have all of the advanced characteristics of the segmented worms, and they also have a true heart and are either male or female. They also have body parts—three main parts, in fact: the head-foot, the visceral mass, and the mantle.

The *head-foot* is the part of the mollusk that contains the mouth and sensory organs. It is also the body part that helps the animal move from place to place. The *visceral mass* contains all the other organs of the body, including the heart. Digestion occurs within the visceral mass. The *mantle* is a thin membrane covering the visceral mass. The mantle produces the shell of the mollusk.

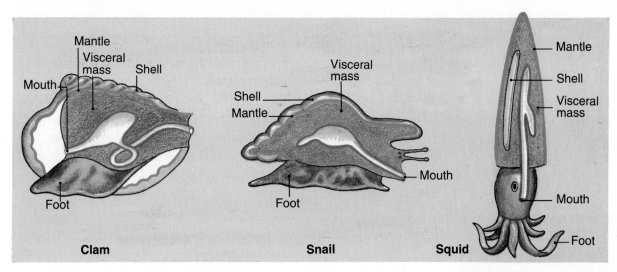

Clam   Snail   Squid

Most mollusks reproduce sexually, creating larvae. Each larva swims freely until it becomes an adult.

Some mollusks such as clams and oysters have a shell that is made of two parts, or valves. A clam, an oyster, or a mussel has two valves and moves from place to place on a hatchet-shaped foot extended from its shell. For that reason, such mollusks are called hatchet-footed mollusks.

The snail is an example of a mollusk with a shell consisting of only one valve. Snails are found both in water and on land. On land, a snail moves by gliding across a layer of mucus that is secreted by its foot. Because the snail appears to glide on its stomach, it is called a stomach-footed mollusk. The following activity will provide you with the opportunity to directly observe how a snail moves.

Figure 13–13. Although these animals look very different from one another, they are all mollusks. Note that all have the body parts of a typical mollusk.

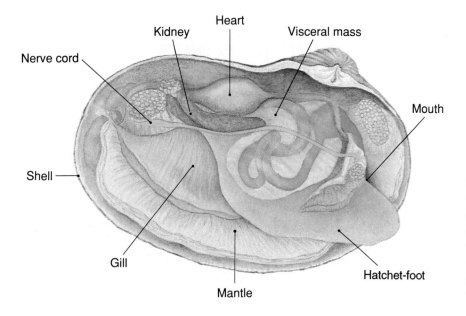

Figure 13–14. As water flows through the body of a clam, nutrients and oxygen are removed from the water, and waste and carbon dioxide are removed from the animal.

Section 2 Mollusks and Echinoderms   **355**

Observing

Your teacher will supply you with a live snail. Put the snail on a smooth surface, such as a piece of glass or a ceramic tile. How does the snail move? Describe the secretion left by the snail. Now put the snail on a rough surface such as sandpaper. Compare the movement of the snail on the smooth surface and the rough surface. Was the reaction of the snail to sandpaper the same as your reaction to walking barefoot across gravel? Explain.

Ogden Nash, a poet who often wrote nonsensical verses, had this question for one of the mollusks.

> *Tell me, O Octopus, I begs,*
> *Is those things arms, or is they legs?*
> *I marvel at thee, Octopus;*
> *If I were thou, I'd call me us.*

This silly rhyme focuses on the octopus—an advanced mollusk. The octopus and the squid are the most advanced mollusks. Both are male or female head-footed mollusks. A head-footed mollusk has a very distinct head, and its foot is a set of tentacles with suckers (the arms or legs of the octopus, for example). An advanced mollusk may have an external or internal shell or no shell at all.

Figure 13–15. The octopus may look like a frightening animal, but it is really rather timid.

▼ **ASK YOURSELF**

Identify some of the characteristics of mollusks.

Animals with Spiny Skin

An *echinoderm* (ih KY nuh durm) is an animal with hardened bumpy or spiny plates under its skin. Starfish, sea urchins, and sand dollars are common echinoderms. The body of an echinoderm usually has five parts that look like rays coming from a center point.

Coats of Many Colors   Starfish are found in colorful shades of deep purple and red as well as in the yellowish sand color of most dried starfish. They are quite strong. Their rays can pry open the shells of clams and oysters. On the underside of these rays are rows of tube feet, which are hollow cylinders tipped with suckers. Starfish use the tube feet for movement. They can even climb up the support poles of piers and docks.

Can You Stomach This? Starfish are meat eaters that use their great strength to pry open mollusk shells. Starfish can be very persistent and patient—some will struggle for as long as two days and nights to open the shell of an oyster or a clam! Would you be able to wait that long to eat? Once the starfish has pried a shell open, it inserts its stomach into the shell. Its stomach secretes digestive juices that slowly turn the soft part of the mollusk into a liquid, which the starfish draws up through a tube. Then it moves its stomach back into its own body.

Even though starfish eat meat, they would never want you for dinner—you're much too large! Besides, you're able to move too fast. A starfish moves only about 15 cm per minute.

Figure 13–16. Mollusks provide food for starfish.

DISCOVER BY *Calculating*

Using meters or kilometers, estimate the distance from your home to your school. Then estimate the length of time it would take you to walk that distance. Given the fact that starfish move at a rate of about 15 cm per minute, determine how long it would take a starfish to travel from your home to your school. ✏️

And Baby Makes Three or More Starfish reproduce sexually. A female starfish may release as many as 2.5 million eggs at one time. She expels the eggs into the water at the same time that the male releases sperm. Starfish can also reproduce asexually by regeneration. Some starfish can grow and develop a whole new individual from one detached ray.

▼ **ASK YOURSELF**

List characteristics of starfish.

SECTION 2   *REVIEW AND APPLICATION*

Reading Critically

1. How do mollusks move?
2. Is a clam stronger than a starfish, or is a starfish stronger than a clam? Explain.
3. How does a starfish digest a meal?

Thinking Critically

4. Since mollusks have rather heavy shells to drag around with them, how have they managed to become so widespread in the ocean?
5. Gatherers of clams and oysters used to destroy starfish by cutting them up and throwing the pieces back into the water. Was this a good way to keep the starfish from damaging clam and oyster beds? Explain.

Insects—The Most Common Arthropods

Objectives

List characteristics of arthropods.

Identify the economic importance of certain insects.

Distinguish between complete and incomplete metamorphosis.

What were the first insects on Earth like? Were they anything like the insects that we have today? Were the pesky mosquitoes or ugly earwigs around then? We can't be sure, but we do know that today there are nearly one million different insects. When people talk about insects, they often refer to insects as "bugs." Correctly speaking, bugs are insects, but not all insects are bugs. This harmless mistake doesn't seem to bother most people—does it "bug" you?

Characteristics of Arthropods

Insects belong to the phylum of arthropods. The word *arthropod* means "jointed leg." Arthropods have several pairs of jointed legs, or appendages, which extend out from their bodies. Arthropods may also have other jointed appendages such as claws, jaws, fangs, egg depositors, and pincers.

The body of an arthropod is divided into body sections, or segments, and is covered by an outer skeleton called an **exoskeleton.** The exoskeleton protects soft body parts like a suit of armor. Because the exoskeleton does not grow, the arthropod sheds it occasionally and forms another. This shedding process is called *molting.* After molting, an arthropod is defenseless until the new exoskeleton forms.

Figure 13–17. The red Hawaiian lobster (far right) and the molting crab (above) are both arthropods.

Examples of arthropods include lobsters, crabs, scorpions, spiders, and insects. Did you know there are about 900 000 known species of insects? Of course, they cannot all be listed here! A short list of insects, however, includes ants, bees, grasshoppers, beetles, mosquitoes, butterflies, cockroaches, fleas, and termites. Do you have a most favorite or least favorite arthropod?

Arthropods have complex circulatory and nervous systems. Those that live in water use gills to breathe, and those that live on land breathe through air ducts or through structures called *book lungs*. Spiders are arthropods with book lungs—their lung tissue is layered and looks something like the pages of a book.

 ASK YOURSELF

What are the characteristics of arthropods?

Characteristics of Insects

You are probably very familiar with many different kinds of insects. If you looked closely at an insect, you would see three distinct body sections—the head, the thorax, and the abdomen. In addition, you would see three pairs of legs. Although most insects have wings, some are wingless. Insects also have antennae and compound eyes. An insect seeing through compound eyes will see a single mosaic image of an object.

Insects, as well as other arthropods, have specialized mouthparts. Each part performs a different function, depending on the needs of the animal. As the illustrations show, the mouthparts of insects are adapted to their feeding habits.

 ASK YOURSELF

What are the three body parts of insects?

Classification of Insects

You have probably classified, or ordered, various things in your life. Maybe you've ordered your music collection alphabetically.

One of the many characteristics used to classify insects is wings. A close examination of insects shows that they display a great variety of wings. Some insects have one pair of wings, others have two pairs, while still others have no wings at all. Wings may be thin and delicate, covered with scales, straight, or curved. Wings may be out to the side, straight down the back, or overlapping each other.

 ASK YOURSELF

How do insect wings differ?

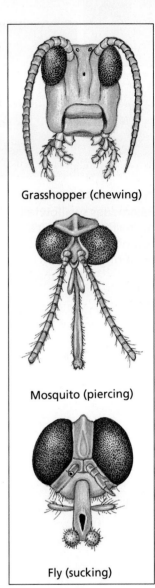

Grasshopper (chewing)

Mosquito (piercing)

Fly (sucking)

Figure 13–18. The structure of a compound eye causes an insect to see multiple images. Insects also have various types of mouthparts.

Figure 13–19. The Palamedes swallowtail butterfly experiences complete metamorphosis during its life cycle.

Insect Development

Many insects change form through the various stages of their life cycle. Often the immature form does not resemble the adult form. *Metamorphosis* is a series of changes that an insect experiences during its life cycle. Metamorphosis can be complete or incomplete.

What Am I—Identity Crisis
An insect that goes through **complete metamorphosis** has four stages of development—egg, larva, pupa, and adult. The butterfly experiences complete metamorphosis. When a butterfly egg hatches, a very young caterpillar emerges. A caterpillar is the larval stage of a butterfly. After a time, the caterpillar enters an inactive pupa stage. During this stage, the caterpillar forms a chrysalis (KRIHS uh lihs), a protective covering, around itself. While in this stage, the caterpillar gradually changes into a butterfly. When the butterfly is completely developed, it leaves the chrysalis as an adult.

Child or Adult?
The grasshopper experiences **incomplete metamorphosis,** which consists of only three stages of development—egg, nymph, and adult. A young grasshopper, called a *nymph,* looks much like an adult grasshopper. Although a nymph looks like an adult, the nymph does not have wings. The wings develop as the insect matures. Since the grasshopper does not go through an inactive pupa stage, its metamorphosis is incomplete.

Some insects, such as the silverfish, do not experience metamorphosis. Like you, these insects just grow bigger as they become adults.

 ASK YOURSELF

Which stage of development is not present in incomplete metamorphosis but is present in complete metamorphosis?

Economic Importance of Insects

Insects are everywhere in the world around you, and they affect many things that you do. Have ants ever ruined a picnic for you? Have mosquitoes ever made your camping trip miserable? For better or for worse, insects have an impact on the environment of Earth.

The destruction caused by an insect, while a serious problem, is told with humor in the following poem.

The Termite
by Ogden Nash

Some primal termite knocked on wood
And tasted it and found it good,
And that is why your Cousin May
Fell through the parlor floor today.

Although some insects cause damage, many others can benefit the environment. Some insects pollinate crops. Bees provide honey, butterflies provide beauty, and most insects are a food source for other animals. The following activity will help you discover some positive and negative ways that insects are economically important.

  **Researching**

After thinking over what you have learned about insects and after interviewing friends and relatives, compile lists of both positive and negative ways insects are economically important. Record the lists in your journal. ✎

◤ ASK YOURSELF

What benefits do insects provide?

SECTION 3 *REVIEW AND APPLICATION*

Reading Critically
1. Describe the characteristics of insects.
2. List four characteristics of arthropods.
3. Describe the four stages of complete metamorphosis.

Thinking Critically
4. A grasshopper nymph has a tough exoskeleton. What happens to the exoskeleton as the nymph becomes an adult?
5. In what ways do you think the behavior of an arthropod might change when it is molting?

Other Arthropods

Organize *the large number of arthropods and* **identify** *criteria for their classification.*

Compare and contrast *arthropod classes.*

Explain *why spiders are not insects.*

Imagine what it must have been like for the first arthropod that emerged from the sea. What did this creature find? What did it look like? Why was it drawn to the land? Rachel Carson thought that the first arthropod might have looked something like a modern scorpion and that it probably lived its life half on land and half in water. If they were similar to scorpions, then these arthropods were ancestors of present-day arachnids.

Arachnids

Spiders

by Mary Ann Hoberman

Spiders seldom see too well.
Spiders have no sense of smell.
Spiders spin out silken threads.
Spiders don't have separate heads.
Spider bodies are two-part.
Spider webs are works of art.
Spiders don't have any wings.
Spiders live on living things.
Spiders always have eight legs.
Spiders hatch straight out of eggs.
Since all these facts are surely so,
Spiders are not insects, no!

The poem says spiders are not insects. Many people, however, incorrectly think of them as insects. As you read, you will discover differences between spiders, one type of arachnid (uh RAK nihd), and insects.

What's the Difference? *Arachnids* have eight legs, while insects have only six. Insects have three body segments, but arachnids have only two—the *cephalothorax* (sef uh luh THAWR aks) and the abdomen. The cephalothorax is the arachnid's head and chest region. The abdomen contains most of its internal organs. An arachnid can have as many as 12 simple eyes. These eyes can

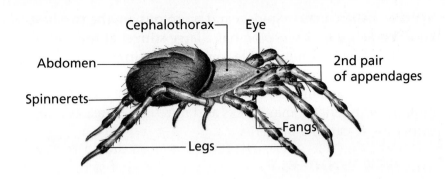

Cephalothorax — Eye

Abdomen —

2nd pair
of appendages

Spinnerets —

Fangs

Legs

Figure 13–20. Spiders can be identified by their eight legs and two body segments.

sense light and dark, but they cannot form images. You can see arachnids much better than they can see you!

In addition to having an extra pair of legs, arachnids have more appendages than insects. Arachnids have sharp, fanglike structures that are used to inject poison into their prey. The poison creates a kind of "milkshake" by liquefying the soft inner part of the prey. The arachnid then sucks out this soft material and swallows it.

Oh, the Web You Weave   Spiders have structures that enable them to make webs. At the rear of the spider's abdomen are several small structures called *spinnerets*. Spinnerets are tiny tubes through which a liquid protein passes. This protein hardens into the familiar silk of spider webs. In the following activity, you will observe a spider web.

DISCOVER BY *Doing*

Locate and observe a spider web. In your journal, draw what you see. If the spider is in the web, observe it for several minutes. Write a description of the spider and sketch it. Describe its behavior. Use a classification guide to help you identify the type of spider that made the web. ✏

How Many in a Batch?   Reproduction in spiders begins when the male deposits sperm cells into the female. The female spider can store sperm cells for as long as 18 months. As a result, she can fertilize several batches of eggs with sperm from one deposit. Some spiders can lay as many as 2000 eggs in a batch, depending upon the size of the spider. The average female lays about 100 eggs. Why do you think it is an advantage for some spiders to lay so many eggs?

Spider Relatives   Spiders and scorpions are the two best-known arachnids. A scorpion has a large stinger at the end of its abdomen. The scorpion is able to whip the stinger over its back and stab its prey, injecting poison into the victim. Ticks and mites are also arachnids—the tiniest ones. Usually less than one millimeter in length, many ticks and mites are parasites and carry disease-causing bacteria.

ASK YOURSELF

Compare the characteristics of insects and arachnids.

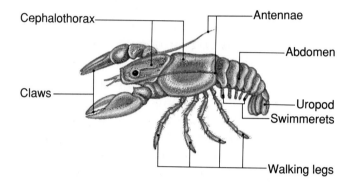

Figure 13–21. Crayfish live in fresh water and look like lobsters.

Crustaceans

Crustaceans have more appendages than arachnids. Crustaceans have two pairs of appendages called *antennae* (an TEHN ee). The antennae are sense organs that provide information about the environment. Crustaceans also have structures called *mandibles,* or jaws, which are used for chewing, crushing, and grinding food.

Figure 13–22. The eyes of a crustacean are usually located on stalks, allowing it to see a larger area.

Most crustaceans live in water. Examples include lobsters and crayfish—lobsters live in salt water and crayfish live in fresh water. There are a few crustaceans, however, that live in damp places on land. The pill bug is one such crustacean.

A crustacean such as a crayfish uses its small claws to help hold food and its large claws to protect itself and to capture food. The crayfish's next four pairs of appendages are "walking legs." Behind these are five pairs of swimmerets. Swimmerets are much shorter than the other appendages and are used in swimming. The uropod, a flipperlike structure at the posterior of the animal, is also used for swimming. A crayfish can move backward very quickly by snapping its tail forward.

Some crustaceans are male; others are female. In most cases, fertilized eggs are attached to the swimmerets and carried by the female until they hatch.

 ASK YOURSELF

Which animal has more appendages—a spider or a crayfish?

Myriapods

Centipedes and millipedes are myriapods. *Myriapod* means "having many legs." If you ever see millipedes or centipedes, you might be able to tell them apart by watching how they move. Centipedes wriggle because the legs on either side of the body move alternately. Millipedes have a smoother motion because the legs on either side of the body move at the same time. The reason for this difference in motion is the number of legs each myriapod has. Centipedes have one leg on each side of every body segment. Millipedes have two legs on each side of every body segment.

All myriapods need a moist environment in which to live. They may be found in moist soil, in decaying logs, and under objects such as rocks.

Millipedes have rounded, tube-shaped bodies. They are vegetarians that live in damp, dark places and eat dead plants. Millipedes often give off an unpleasant odor when they are frightened. Centipedes have bodies that look like flattened tubes. They often have fangs that they use to attack other invertebrates, including earthworms. Both of these arthropods care for their eggs.

Figure 13–23. A centipede (top) may have from 15 to 170 pairs of legs. A millipede (bottom) may have up to 115 pairs of legs.

 ASK YOURSELF

How is a centipede different from a millipede?

SECTION 4   *REVIEW AND APPLICATION*

Reading Critically
1. Name four types of arthropods and give an example of each.
2. State two ways arachnids are different from crustaceans.
3. List several ways in which centipedes and millipedes differ from each other.

Thinking Critically
4. Could millipedes and centipedes live in the same environment? Explain.
5. Why don't spiders have mouthparts?
6. What characteristics of arthropods point to the supposition that these animals may have been the first to adapt to a land-based lifestyle?

SKILL

Classifying Insects

▶ **BACKGROUND**

All insects are classified as follows:

Kingdom: Animal;   **Phylum:** Arthropod;   **Class:** Insect

The next level of classification is **Order.** To determine order, it is necessary to examine the insect's wings, legs, and mouth parts.

▼ **PROCEDURE**

Imagine you have returned from a field trip on which you observed various insects. Back in your laboratory, you read the characteristics recorded in your journal. Use the table to help you identify the insects in these three entries.

1. "Small insect resting on green leaves, feeding on plant with a piercing, sucking motion, has two pairs of wings that make a rooflike structure over the insect's body when it is at rest."
2. "Insect about as long as the first joint of my forefinger. Brownish color, two pairs of wings, legs powerful enough for the insect to jump a meter."
3. "Small insect with one pair of thin wings. Landed on my wrist. Bit or stung me with a piercing needlelike structure."

TABLE 1: INSECT ORDERS AND CHARACTERISTICS

Order		Examples	Characteristics
Orthoptera ("straight-winged")		Grasshoppers, crickets	Two pairs of straight wings; legs modified for jumping
Homoptera ("uniformly winged")		Cicadas, aphids, scale insects	One or two pairs of wings that form a roof over body when at rest (some species are wingless); mouthparts adapted for piercing-sucking; feed on plants
Diptera ("two-winged")		Flies, gnats, mosquitoes	One pair of wings; mouthparts adapted for lapping or piercing-sucking
Coleoptera ("shield-winged")		Beetles (includes fireflies, ladybugs)	Two pairs of wings; hard forewings join to form a straight line down the back
Hymenoptera ("membrane-winged")		Ants, bees, wasps	Two pairs of wings; chewing mouthparts, social insect

▶ **APPLICATION**

1. To which order does each recorded insect belong?
2. In the first journal entry, what characteristic helped you decide on a certain order?
3. Name three members of the order you chose for the third journal entry.
4. If the insect described in the third journal entry had two pairs of wings instead of one, what order would you have thought about? What other characteristic might have caused you difficulty in classifying this insect?

 Using What You Have Learned

Observe some insects in their natural habitat. To avoid bites and stings, do not touch or otherwise disturb the insects. Using the procedure outlined in this activity, classify the insects into their proper orders.

The Big Idea

Invertebrates have evolved and successfully adapted to become the largest group of animals in our world. Invertebrates are also diverse. Land and marine organisms with a diameter of several millimeters and giant squids with lengths of almost 20 m are invertebrates. All invertebrates occupy a significant role in their surrounding habitat. The food chain, for example, would be affected if various orders of invertebrates suddenly disappeared. Invertebrates also make important economic contributions to the lives of people.

For Your Journal

You have just finished studying a group of animals that make up almost all of the animals in the world. Look back at the ideas you wrote at the beginning of the chapter. Thinking of the different animals you studied, add to your list of beautiful and frightening-looking invertebrates. Also include the name of your favorite invertebrate and a brief explanation of why it is your favorite.

Connecting Ideas

Copy the following concept map into your journal. Complete the concept map by adding appropriate terms in the blanks.

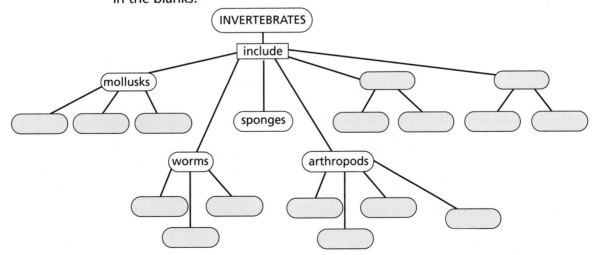

REVIEW

Understanding Vocabulary

For each set of terms, explain the similarities and differences in their meanings.

1. pharynx (348), clitellum (351) 2. setae (351), sessile (342)
3. webs, spinnerets (363)   4. exoskeleton (358), regeneration (344)
5. incomplete metamorphosis (360), complete metamorphosis (360)

Understanding Concepts

MULTIPLE CHOICE

6. Why is it important for sponges to live in water that moves?
   a) Moving water offers sponges a greater and more constant variety of nutrients.
   b) Moving water gives sponges the ability to move from a less desirable location to a more desirable location.
   c) Moving water helps protect sponges by keeping predators away.
   d) Moving water is not an advantage to sponges.

7. How do earthworms benefit soil?
   a) Earthworm castings provide nutrients for soil.
   b) Earthworm pathways allow oxygen to penetrate soil.
   c) Earthworm pathways help rainfall penetrate soil.
   d) All of these choices are correct.

8. When an arthropod molts, it takes between several hours and several days for the new exoskeleton to harden. What is the most likely behavior of an arthropod during this time?
   a) The arthropod will hide because it is especially vulnerable to predators.
   b) The arthropod will become more active than usual.
   c) The arthropod will lie in the sun to help speed the hardening process.
   d) The arthropod will carry on normal life activities.

9. Which type of mollusk could move great distances in the shortest period of time?
   a) hatchet-footed
   b) stomach-footed
   c) head-footed
   d) All mollusks move at the same speed.

10. Which of the following invertebrates is least likely to be eaten by another animal?
    a) clam
    b) starfish
    c) worm
    d) sponge

11. The setae of an earthworm are similar to what part of your body?
    a) hair
    b) heart
    c) eyes
    d) legs

SHORT ANSWER

12. Explain the term *sessile* and give two examples of a sessile animal.

13. Explain how a starfish obtains and digests food.

14. Explain the difference between complete metamorphosis and incomplete metamorphosis. Give an example of an insect that experiences each type of metamorphosis.

15. Explain the difference between a colonial animal, such as a Portuguese man-of-war, and animals that live in colonies, such as corals.

Reviewing Themes

16. Systems and Structures
Using examples other than those you read about, list five ways that invertebrates benefit the world economically.

17. Diversity
Choose an invertebrate and describe a situation that shows the invertebrate adapting to its environment.

Interpreting Graphics

18. How does the skin of a sea urchin help protect it from predators?

19. A parrot fish will eat a sea urchin. Explain how you think this might occur.

Thinking Critically

20. A female starfish will lay about 2.5 million eggs at one time. Why aren't the world's oceans overcrowded with starfish?

21. How do the members of the colony in a colonial organism benefit one another?

22. Make a list of the problems that might occur if all of the insects of the world were to die simultaneously.

23. Discuss the advantages and disadvantages of having an exoskeleton.

24. Many animals that begin as larvae live in water. Using what you know about larvae, explain why this is true.

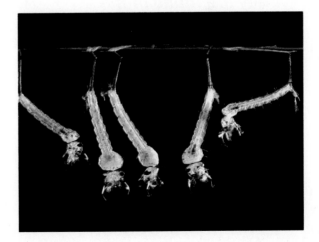

Discovery Through Reading

Wiessinger, John R. *Bugs, Slugs, and Crayfish—Right Before Your Eyes*. Enslow Publishers, Inc., 1989. Text and drawings describe the physical characteristics, habits, and habitats of insects and other animals without a backbone.

25. Why is it advantageous for parasitic flatworms to have both sexes in the same animal? Explain in detail.

VERTEBRATES

*F*ishers are fur-bearing animals that can reach a length of three feet. They are found in the forests of eastern North America and are extremely rare. In his novel **The Winter of the Fisher,** Cameron Langford describes one year in the life of a fisher. Notice in this part of the story how the fisher learns from and about other animals. All of them are vertebrates—animals with a backbone.

*H*e followed and spied on many animals in the triangle, unconsciously filling in the gaps in his education left by his mother's death. From the weasels he learned the use of cover, from the foxes the way to circle a rabbit until it ran itself out; the deer showed him how to blend motionlessly with any background, and the squirrels and jays taught him the meaning of their danger calls....

He learned the language of the tracks, written fleetingly across the impermanent face of the snow. The tiny marks of deer mice crisscrossed everywhere, small and birdlike, the thin streak of the pendant tail cutting like a knife edge between each set of tightly grouped prints.

Occasionally the paired footmarks of a weasel paralleled them, measuring a foot in each bound, and ending often in a large hole beside a small hole in a drift where the mouse had dived for safety, only to find that the weasel could burrow as well as he. The martens left soft, oval two-inch tracks as they moved from tree to tree, usually in tight pursuit of the crescent-shaped groups of prints put down by fleeing squirrels. The deer punched their strange, split-hooved signature across the open patches, and it was some time before the fisher realized that the two-legged tracks had been left by a four-legged beast, for as they threaded through the trees, the deer carefully placed each hoof on the mark of the hoof before, as if the front and rear halves of the animals were two distinct creatures, the one treading carefully in the other's steps. Surprisingly, the wolves following them walked in much the same way, placing each paw almost directly over the print of the paw in front. He learned to recognize the huge, fuzzy pads of the lynx, the chunky, triangular mark of the mink, and the strange, paw-studded, eight-inch gutters left lining the steep stream banks by the sliding clowns of the north, the river otters.

from *The Winter of the Fisher*
by Cameron Langford

For Your Journal

✐ **What characteristics do you think all vertebrates share?**

✐ **Name two animals you think are vertebrates. How do you think vertebrates reproduce?**

✐ **How are all the animals described alike?**

Fishes, Amphibians, and Reptiles

Objectives

Organize *fishes according to major characteristics.*

Summarize *frog metamorphosis.*

Describe *the characteristics of reptiles.*

Imagine being a fish for a day! Where would you most like to live—in a stream, a lake, or an ocean? Or, would you like to be a fish in a large aquarium where you would not have to search for food? Look at the two fishes in the illustration. Would you prefer to be a very large fish like the shark or a colorful fish like the betta? Perhaps you would have enemies to worry about. Which of the animals described at the beginning of the chapter might be your enemy?

Fishes

As a fish, you would notice several things—first of all, you are very wet! Second, your body temperature is constantly changing. All fishes, as well as amphibians and reptiles, are cold-blooded vertebrates. **Coldblooded** animals do not maintain a constant temperature. Their temperature changes when the temperature of the environment changes.

With or Without Bones   If you decided to be a fish for a day, you would have another decision to make. Would you want to be a jawless fish, a cartilaginous fish, or a bony fish? Before you make such a decision, you may want to find out more about each type of fish.

Figure 14–1. The seas abound in strange and beautiful creatures, such as the hammerhead shark (above) and the betta fish (right).

Observe a fish in a fishbowl or an aquarium. Describe how the fish swims. How many fins does the fish have? How does it use its fins in swimming? What other characteristics or behaviors can you identify? Sketch the fish in your journal, and write notes based on your observations of the fish. ✐

Jawless Fishes

Jawless fishes are the least developed vertebrates living today. This group once had many members but now includes only two types of fishes—lampreys and hagfish.

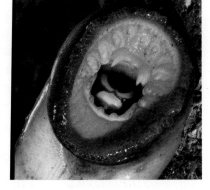

Jawless fishes could also be called *boneless fishes* because they have no real bones. Their skeletons are made of cartilage, and as a result, they are very flexible. Both the lamprey and the hagfish have long, round, tubelike bodies. Each fish attaches itself to another fish and sucks out the blood and body fluids of the host fish. Hagfish also eat the tissue of the host. While lampreys attack healthy, living fishes, hagfish usually feed on dead or dying animals.

Figure 14–2. The mouth of the lamprey (left) is perfectly suited to its feeding behavior. The hagfish (right) has no bony skeleton.

Cartilaginous Fishes

Like jawless fishes, *cartilaginous fishes* have skeletons made of cartilage. Unlike jawless fishes, however, they have jaws, which are also made of cartilage. Instead of the small, round scales that cover most fishes, cartilaginous fishes have sharply pointed scales. These scales make the fishes' skin feel rough, like sandpaper.

Sharks, rays, and skates are examples of cartilaginous fishes. Sharks have a fish shape with pectoral, pelvic, dorsal, and caudal fins. Skates and rays do not look like fish at all. Instead, their bodies are flattened. Their shape allows skates and rays to glide through the water in a graceful manner. They almost look as if they are flying through the water.

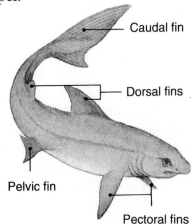

Caudal fin

Dorsal fins

Pelvic fin

Pectoral fins

Figure 14–3. The positions of its fins enable a shark to change directions quickly as it pursues its prey.

Figure 14–4. A shark's powerful jaws contain several rows of razor-sharp teeth.

Guess Who's Coming to Dinner?

Have you ever seen a movie about shark attacks on humans? At first, all seems fine as swimmers enjoy the water. Then threatening music builds. Suddenly, a shark attacks! Contrary to popular belief, however, most sharks avoid humans and are not a threat to them. In fact, the chance of a person being attacked by a shark is less than the chance of a person being struck by a car while crossing a street. Even so, swimmers should leave the water if a shark is spotted. The diet of most sharks includes crustaceans, mollusks, and other fishes. In the following poem, E. V. Rieu takes a light-hearted look at the feeding habits of a shark.

The Flattered Flying Fish

Said the shark to the Flying Fish over the phone:
"Will you join me tonight? I am dining alone.
Let me order a nice little dinner for two!
And come as you are, in your shimmering blue."

Said the Flying Fish: "Fancy remembering me,
And the dress that I wore at the Porpoises' tea!"
"How could I forget?" said the shark in his guile:
"I expect you at eight!" and rang off with a smile.

She has powdered her nose; she has put on her things;
She is off with one flap of her luminous wings.
O little one, lovely, light-hearted and vain,
The Moon will not shine on your beauty again!

Figure 14–5. Despite a threatening appearance, the manta ray is actually a rather shy and gentle animal. Estimate the size of the manta ray shown.

Rays of the Sea

The flying fish became a meal for the shark, but would she have been a likely meal for a ray? Rays usually feed on creatures such as clams, fishes, and even plankton. Like most sharks, most rays are harmless to humans. For example, the large manta ray shown in the picture is not dangerous to humans even though it weighs as much as 1360 kg. The stingray, however, produces painful wounds if its spiny tail strikes a swimmer. The spine at the base of the tail gives off a venom that can be fatal to humans. Other rays can produce an electric shock of as much as 220 volts.

Bony Fishes   When you hear the word *fish,* you probably do not first think of lampreys, sharks, or rays. You are more likely to think of trout, perch, salmon, codfish, catfish, and flounder. These are examples of *bony fishes.*

Most bony fishes are ray-finned fishes. *Ray-finned* simply means that the paired fins of the fish have long, bony rays. These rays allow the fins to fan out and to hold their shape. If you have ever held a bony fish, you may have noticed that you could lift the fins up and fold them back down, much as you might open and close a fan.

The illustration shows the external features of bony fishes, including the fins. Find the *operculum.* This is a bony plate that covers and protects the gills. Note that there are no eyelids. Since fishes are in water, they have no need for eyelids and tear ducts, which protect and moisten the eyes of land animals. Also notice the openings on the snout of the fish. These openings are like your nostrils. The sense of smell seems to be a fish's strongest sense.

A fish can control its movements and receive messages about its environment by using two unusual structures. A fish has a swim bladder that fills with or loses air. This structure allows the fish to rise and sink in the water. A fish also has a lateral line extending the length of the side of the body. The lateral line is sensitive to vibrations that occur in the water and permits the fish to sense nearby movement and the presence of food.

When a fish swallows a worm, an insect, or some other piece of food, the food goes through a digestive system. The system consists of an esophagus, a stomach, and an intestine. In addition, there is a liver that helps digest fats and produce proteins. Waste matter from digestion passes out of the intestine through the anus.

Figure 14–6.
Although bony fishes vary greatly in appearance, they all have the same basic internal and external structures.

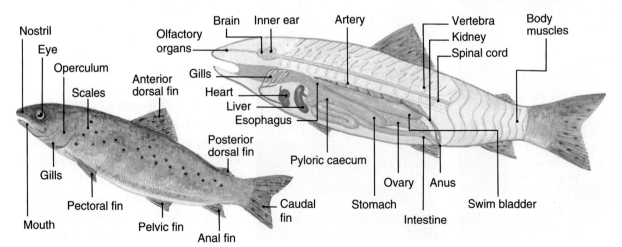

Figure 14–7. Fish eggs are often fertilized externally. As this male grunion swims over the eggs, he sprays them with milt.

A fish obtains oxygen from the water that continually passes over its *gills*. The gills contain many blood vessels that absorb the oxygen from the water. The oxygen passes through the thin cell membranes of the blood vessels and is carried by the blood to all parts of the fish's body. The blood is pumped by a simple two-chambered heart.

Bony fishes reproduce sexually. The female lays eggs, and the male fish sprays them with a fluid called *milt*. Milt contains sperm. Fishes lay many more eggs than most other vertebrates because fishes' eggs are usually unprotected. Some fishes do guard their eggs, but most do not protect them. The only other vertebrates that lay such large numbers of unprotected eggs are the amphibians.

 ASK YOURSELF

How do bony fishes differ from cartilaginous fishes?

Amphibians

Figure 14–8. Frogs, such as the poison-arrow frog, the wood frog, and the green tree frog, are found in many different parts of the world. What adaptations does each frog display?

Frogs are probably the best-known *amphibians*. Like all amphibians, frogs have successfully made the adjustment to living on land, but they still need a water environment.

Although adult frogs have lungs and are able to breathe oxygen from the air, they spend a great deal of time submerged in water. The skin of a frog is very thin, and oxygen from water can pass directly through the skin cells. Frogs and other amphibians can also absorb oxygen through the linings of their mouths.

You have seen that fishes are specifically designed to function in water. Over time, some sea creatures developed characteristics that enabled them to survive partly on land. What changes were needed for them to move from the water to the land and back again? The ancestors of today's amphibians developed some, but not all, of the adaptations needed to survive partly on land. What kind of adaptations do you think these animals had to make? The following activity may help you decide on the types of changes that were necessary.

ACTIVITY

How would you redesign a fish for life on land?

MATERIALS
paper, pencil, aquarium, fishes

PROCEDURE
1. List the ways that fishes are specifically designed for life in the water.
2. Imagine that you must create a method for a fish to enter the air-breathing world but still spend most of its life in the water. Design ways to overcome problems resulting from fish structure. For example, fishes have no eyelids or tear glands because these structures are not needed in an underwater environment. Fishes have a protective covering of slime on their scales that would quickly dry out in the air. Fishes are also coldblooded.

APPLICATION
1. List some ways in which humans have prepared themselves for entering and exploring the underwater environment.

2. Humans can move from a very hot environment to a very cold one without experiencing much discomfort. Why do fishes have problems with rapid changes in temperature?

What changes did you make in your redesigned fish? Your fish needed a way to take in oxygen from the air, but it still had to be able to take in oxygen when it returned to the water. After the changes you made, could your fish return to water without drowning? Your fish would probably need legs or some sort of muscle development so that it could move on land. Your re-designed fish may resemble amphibians in some ways. Amphibians are adapted for life on land and in water.

Leapfrog You are probably familiar with many of the external features of frogs. A frog has four legs, the back legs being longer and stronger than the front legs. Its large, round eyes bulge from the top of its head. Because its nostrils are also on the top of its head, a frog can breathe even when almost totally submerged in water. Look at the three frogs in the photographs on the previous page. How are they alike and how are they different?

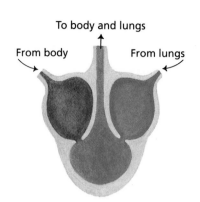

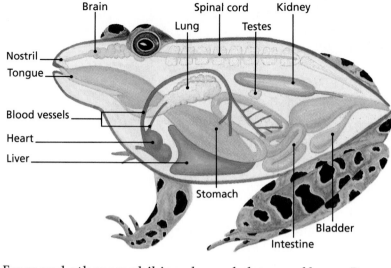

Figure 14-9. The internal anatomy of the frog is shown. Note the three-chambered heart, which is more sophisticated than the two-chambered heart of the fish.

Frogs and other amphibians have skeletons of bone. Because they are vertebrates, they have a backbone. Their small brain is protected by a bony skull, and their limbs are supported by bones. Hinged joints enable frogs to bend and move their limbs.

The internal organs of the frog are similar to those of more complex vertebrates. The liver is the largest of these organs. It produces bile and aids in digestion. Blood is pumped through the frog's body by a three-chambered heart, which is more advanced than the two-chambered heart of the fishes, but not as advanced as the hearts of birds and mammals.

Frogs and other amphibians reproduce sexually. The eggs are fertilized outside the female's body. The female produces large numbers of eggs that fill her body cavity before they are released. During mating season, the male frog grasps the female from the back. The female releases the eggs, and the male sprays them with sperm.

Figure 14-10. Toads such as this *Bufo* do not live in the water, but they must return to the water to breed.

A Bumpy Frog?
Toads, while they appear similar to frogs, are quite different. Toads are a little fatter, and their skin is bumpier and drier. Toads do not spend as much time in water as frogs do. In fact, many toads have adaptations that allow them to live in dry areas where a frog would die. Some toads even live in the desert. However, toads must always return to water to reproduce.

Frogs are the subject of many stories, poems, and cartoons. Kermit the Frog and the Frog Prince may have been childhood favorites of yours. In the following activity you can create some of your own frog characters.

Make a list in your journal of frog characters you remember. Tell some fact about each frog that made it special. For example, Kermit's eyes were once made of Ping-Pong balls. Create a frog character of your own. Name your frog and tell facts that make it special. Draw a picture of your frog character or write a nonsensical poem about it. ✐

Just a Tad Different   Recall that a butterfly undergoes great changes on its way to the adult stage. From egg, then caterpillar, then pupa, the butterfly emerges. This type of dramatic change is called *metamorphosis*. Frogs and toads also develop through metamorphosis. Frog and toad eggs hatch at various times from a few days to several weeks after fertilization. Water temperature has some influence on the length of time—warmer water apparently speeds up the hatching process somewhat.

The young frogs and toads that emerge from the eggs do not look at all like adults. They look a great deal like small fish. The young are called *tadpoles*. A tadpole has gills and must stay in the water. Gradually, a tadpole develops legs, its tail is absorbed, its lungs develop, and the animal becomes an adult frog or toad. In some cases, metamorphosis can take several years to complete. As the tadpole develops into the adult form, it changes its diet from plant life to small animals, such as insects. Using the illustration below as a guide, describe the life cycle of the frog.

Figure 14–11. A tadpole may look like a small fish, but it will soon develop legs and lose its tail.

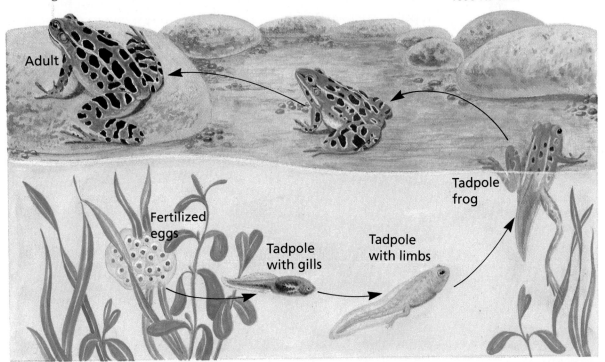

Adult

Fertilized eggs

Tadpole with gills

Tadpole with limbs

Tadpole frog

Figure 14–12. A salamander (top) and a caecilian (bottom) are amphibians. Yet salamanders are often mistaken for lizards, and caecilians are mistaken for earthworms.

Cousins to the Frog　Salamanders are quite different from frogs and toads. In fact, a salamander looks more like a lizard than a frog or a toad. Sometimes it is difficult to remember that these lizardlike creatures are amphibians. Unlike lizards, salamanders have smooth skin with no scales, and they have no claws on their feet. As the photograph shows, salamanders have a tail, and their legs grow from the sides of the body. Most salamanders live near water, and most return to the water to reproduce. Some species do lay their eggs on the ground, but the eggs must always be kept moist. Unlike the eggs of frogs and toads, salamander eggs are fertilized in the body of the female.

Caecilians are even less like frogs and toads than salamanders are. In fact, caecilians resemble earthworms more than they do frogs, toads, or salamanders. Compare the photograph of a caecilian with those of a salamander and a frog.

 ASK YOURSELF

Name two structures found in an adult frog but not found in a tadpole.

Reptiles

The fossil record suggests that the earliest organisms on the earth lived in water. Later, amphibians appeared, spending part of their lives in water and part on land. The first vertebrates to be true land dwellers were the *reptiles*.

Reptiles do not have to protect their skin from drying out, and they do not have to return to water to reproduce. Many reptiles, such as lizards, snakes, and tortoises, live in dry areas. However, crocodiles, alligators, and many turtles prefer watery environments.

Figure 14–13. The name *dinosaur* means "terrible lizard," yet many dinosaurs were small plant eaters that ran from danger.

In a While, Crocodile

Reptiles have hard, leathery scales covering their bodies. The scales of alligators and crocodiles are so hard that they are more like plates. If you have ever spent time looking at alligators and crocodiles at the zoo, you have probably heard other visitors trying to decide which of the animals were alligators and which were crocodiles. Experts say you can tell the difference by observing the snouts or the teeth. Observing snouts seems like the safer method. Alligators have broad, somewhat flat snouts. The snout of a crocodile is much narrower. Sometimes you can see two bottom teeth of the crocodile even when the mouth is closed.

Alligator

Crocodile

Figure 14–14. Describe the observable differences between the alligator and the crocodile.

In a poem by Ogden Nash, Professor Twist insists on distinguishing alligators and crocodiles despite unusual conditions.

> #### The Purist
>
> *I give you now Professor Twist,*
> *A conscientious scientist.*
> *Trustees exclaimed, "He never bungles!"*
> *And sent him off to distant jungles.*
> *Camped on a tropic riverside,*
> *One day he missed his loving bride.*
> *She had, the guide informed him later,*
> *Been eaten by an alligator.*
> *Professor Twist could not but smile.*
> *"You mean," he said, "a crocodile."*

Figure 14–15. Komodo dragons are the largest living lizards. They are native to islands of Indonesia. They are aggressive animals and can be dangerous to humans.

Eggs, Leathery Side Up

Many reptiles, including turtles, alligators, and crocodiles, lay eggs covered with shells. Snakes are unusual reptiles because, although some lay eggs, others bear live young. Reptile eggs are fertilized inside the female's body. The shells are usually not as breakable as those of bird eggs. Since the shells of most reptile eggs are somewhat leathery, they bend and tear rather than shatter during hatching.

Get That Bug

Of the many species of lizards, most are harmless, and, in fact, do a good job of controlling small, pesky insects. The Gila (HEE luh) monster and the beaded lizard are the only venomous lizards. *Venomous* animals are able to poison their victims by biting them.

No Need for Stockings

Snakes are reptiles without legs. The jaws of most snakes can unhinge, or separate, from one another. An unhinged jaw allows the snake to swallow animals that are actually larger around than itself. The photograph below shows a snake eating an egg with the roof of its mouth. The skin of a snake is not wet and slimy; it is dry and patterned with scales.

Snakes frighten many people—possibly because there is a lot of false information about snakes. Although there are several venomous snakes in the United States, most snakes are harmless. The water moccasin, copperhead, rattlesnake, and coral snake are the best-known venomous snakes in the United States.

Figure 14–17. An egg-eating snake does not actually swallow an egg whole. It cracks the egg open with its upper jaw and swallows the contents.

ASK YOURSELF

How are reptiles different from amphibians?

SECTION 1 *REVIEW AND APPLICATION*

Reading Critically

1. Give an example of a jawless fish, a cartilaginous fish, and a bony fish. Tell how they are alike and how they are different.

2. Give two reasons why amphibians need to live near water.

3. Describe four different types of reptiles.

Thinking Critically

4. Poisonous amphibians are usually brightly colored. How does this coloration benefit the amphibian and other animals?

5. Explain why reptiles do not produce as many eggs as amphibians do.

6. Describe the adaptations that probably occurred in the evolution of fishes, amphibians, and reptiles.

INVESTIGATION

Identifying Foods That Tadpoles Need

▶ MATERIALS
- aquaria (2) • pond water or dechlorinated water • rocks • tadpoles (4–6 per liter of water)
- aquarium plants or boiled lettuce • air pump (optional) • ground beef or mealworms

▼ PROCEDURE

1. Pour pond water or dechlorinated water into both containers until the water is about 10 cm deep. Tap water may be used if it is allowed to stand at least 24 hours. Make sure both containers are kept in exactly the same type of environment.
2. Place several rocks in each container so that part of each rock is above the water level.
3. Place an equal number of tadpoles in each container.
4. To both containers, add three or four sprigs of an aquarium plant, such as *Elodea,* or a few lettuce leaves.
5. If an air pump is used, force air into the water at a slow pace. If an air pump is not used, change the water every few days by removing 1/4 to 1/3 of the water and replacing it with fresh pond water or dechlorinated water.
6. Mark one container "A" and the other one "B." To container A, add a few pinches of raw hamburger meat or a few mealworms each day for 10 days. Do not add any meat or mealworms to container B. Before adding fresh food to container A, remove any uneaten food left from the day before.
7. Check the containers each day. Observe the tadpoles closely. Record your observations in your journal.

▶ ANALYSES AND CONCLUSIONS
1. Do any of the tadpoles grow in size or change their shape?
2. Do you find any differences between the tadpoles in container A and those in container B? Describe the differences in your journal.
3. What is the variable in the experimental setup?

▶ APPLICATION
1. What happens to the tadpoles that do not have any meat in their diets?
2. Would you describe a frog as a meat-eating animal? Explain your answer.

✳ *Discover More*
Why is it important for human beings to eat a well-balanced diet? Give examples of foods that should be included in a person's daily diet.

Birds and Mammals

Imagine hiking in a forested area on a warm spring morning. You come to a clearing, and you see a doe and her fawn grazing quietly. You hear birds calling to one another. You see chipmunks racing across the path ahead of you. As you reach a pond, you notice a duck nesting in the reeds. You marvel at the beauty of spring as you walk quietly down the path. You wonder what the lives of the animals you see must be like, and you want to know more about them.

You know that birds and mammals are warmblooded animals. A **warmblooded** animal keeps a nearly constant body temperature regardless of the temperature of its environment. You also know that you are a mammal, and as a mammal, your body temperature stays approximately 37°C. This is true whether you are playing volleyball outdoors or shopping in an air-conditioned store. Because birds use a large amount of energy for flight, their body temperature is slightly higher than yours. A sparrow's average body temperature is 42°C, while that of a thrush is as high as 45°C.

Birds

Birds are easy to identify. They have wings, a bill or beak, two legs, and a covering of feathers all over their bodies. Feathers, besides being pretty, help birds in many ways. They provide birds with a covering that is strong, lightweight, and flexible. The movement of some feathers, called *flight feathers,* helps birds fly. Feathers also help hold in body heat and, in some cases, repel water.

Figure 14–18.
Bald eagle and mallard duck in flight

Light as a Feather   Feathers grow from the outer layer of a bird's skin and can be compared to the scales on a reptile. A feather has three parts: the vane, the rachis, and the quill. The *vane* is the flat part of a feather, made up of many threadlike structures that are zipped together. The *rachis,* or shaft, supports the vane. The *quill* is the bottom section of the rachis and is attached to the skin of a bird.

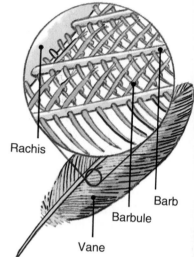

Figure 14–19. Depending on the species, a bird may have up to 25 000 feathers. A bird's feathers help it fly and keep warm.

Rachis

Barb

Barbule

Vane

Bird for a Day If you were a bird for a day, what color feathers would you choose? Would bright feathers attract enemies? Would dull feathers be boring? Feather color is the result of *pigment,* the coloring matter in cells and tissues. Each species of bird has characteristic colors. Although it is rare for a bird's color to be affected by its diet, flamingo feathers are an exception. Zoo keepers have discovered that flamingos kept in captivity and away from their natural diets lose their pink color. However, adding shrimp and carrot juice to their diet restores their bright color.

Colored feathers are legally used in craft projects as well as in the fashion industry. However, to protect certain birds, especially songbirds, laws have been passed limiting the right to use or even to own certain feathers.

DISCOVER BY *Researching*

Research the laws that control what feathers can be used and under what conditions. Check which feathers individuals are able to possess legally. Were special exceptions made for Native Americans who use certain feathers in cultural activities? Write your findings in your journal. ✎

▼ ASK YOURSELF

What are the functions of feathers?

Structure and Reproduction of Birds

All birds have very efficient bodies, and most are adapted to flying. Even a bird's skeleton has special features to help it fly. For example, most birds have lightweight bones that contain air spaces.

Wonderful Wings   The wings of most birds are highly specialized for powered flight. Even so, the bone structure of a bird's wing is similar to that of a human arm and hand. Use the illustration to compare the structure of a bird's wing and the structure of a human arm.

Figure 14–20. The bones of a bird's wings are similar to those of the human arm. For most birds, however, the shape and function of the wings are modified for flying.

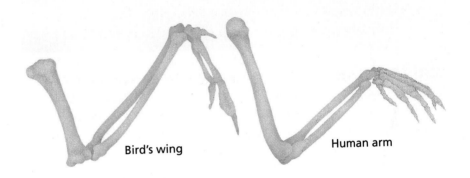

Bird's wing   Human arm

Let's Eat   The digestive system of a bird is similar to that of other vertebrates. It includes an esophagus, a stomach, and an intestine. However, birds have unique structures—a crop and a gizzard—that other vertebrates do not have. The esophagus connects the bird's mouth to its crop, which is used to store food. The gizzard grinds the food. This organ is especially important because birds do not have teeth. Waste materials travel through the intestine and leave the bird's body through an opening called the *cloaca* (kloh AY kuh). Use the illustration to trace the passage of food through a bird's digestive system.

Figure 14–21. The digestive system of a bird is extremely efficient. Almost all of the nutritious components that can be obtained from food are absorbed into the blood.

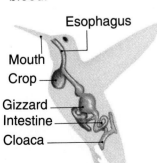

Esophagus
Mouth
Crop
Gizzard
Intestine
Cloaca

The diets of birds vary greatly. Many birds are *herbivores*, eating seeds and other plant parts. However, some are *carnivores*, or meat eaters, and still others are *omnivores*, which eat both plants and animals. Regardless of what a bird eats, it eats large amounts for its size. People who nibble or pick at their food are often accused of "eating like a bird," while people who eat large amounts of food are often accused of "eating like a horse." Actually, in relation to the size of each animal, a bird eats far more food than a horse does. Birds must eat large amounts of food to provide the energy needed for flying.

With Every Breath

The illustration shows the respiratory system and heart of a bird. The respiratory system takes in oxygen through the nostrils. The oxygen passes down the trachea to the branched bronchi. Located in front of the bronchi is the syrinx (SIHR ihngks), or voice box, of the bird. From the bronchi, the air enters small spongy lungs. Branching off from the lungs are air sacs that fill with air and make the bird lighter.

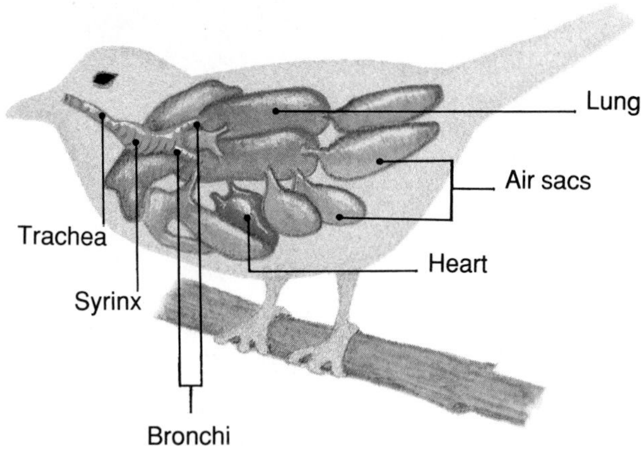

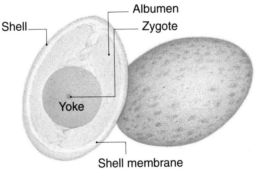

Figure 14–22. A bird's respiratory system is designed to help the bird in flight. Air sacs attached to the lungs make the bird lighter.

And the Beat Goes On

A bird's heart has four chambers and therefore is the most advanced of any vertebrate heart that you have studied so far. The four-chambered heart allows blood to flow through the lungs for a fresh supply of oxygen before it circulates through the body again.

Reproduction

Even the reproductive organs of birds help make the body lighter for flying. In females, a single *ovary* develops as the bird matures. This ovary does not always stay the same size; it gets larger at mating time. In this way, the weight of the female bird is kept to a minimum most of the time.

The male bird produces sperm in reproductive organs called *testes*. After they are fertilized in the female, the eggs move down a tube called the *oviduct. Albumen,* which is the white part of the egg, serves as food for the developing bird. The yolk is also a food source. The egg passes through a shell gland where it is covered with a shell. The egg with its shell leaves the female's body through the cloaca. Egg production is the same whether or not the egg is fertilized. Most eggs sold in stores are not fertilized.

Figure 14–23. The bird's egg is attached to the inside of the shell. The yolk and albumen feed the developing bird until it hatches.

Eggs of Many Colors and Shapes

Bird eggs are found in many different colors. Generally, dark eggs or eggs with spots and blotches are laid by birds whose nests may be easily seen. The colors and spotting are a type of camouflage. White eggs are usually laid by birds whose nests are well hidden or by birds whose mates share the responsibility of staying with the eggs.

Figure 14–24. You are probably most familiar with the oval, white eggs of chickens. Yet bird eggs come in a variety of colors and shapes.

Bird eggs come in many shapes and sizes. Birds that have nests on bare rocks or cliffs produce eggs that are very pointed at one end. This shape seems to protect the eggs if they are disturbed. Instead of falling off the rock or cliff, the eggs roll in small circles. If these eggs were shaped like chicken eggs, they would fall as easily as a chicken egg can roll off your kitchen counter. In the next activity, you will find out about other dangers to bird eggs.

DISCOVER BY *Observing*

In addition to natural hazards, such as enemies and weather, bird eggs are also threatened by human-caused pollution. Your teacher will demonstrate the effect of vinegar on an eggshell. If an eggshell can be destroyed so easily through contact with a common household substance, what effect do you think environmental conditions and a bird's diet might have on its shell? ✎

Figure 14–25. Both Canada goose parents—male and female—care for the eggs. As one parent keeps the eggs warm, the other searches for food.

Warm Up Unlike fishes, amphibians, or reptiles, birds are warmblooded. As a result, it is necessary for bird eggs to be kept warm. Keeping eggs warm until they hatch is called **incubation.** Depending on the type of bird, incubating eggs is the job of the female, the male, or both parents. Which sex incubates the eggs of Canada geese?

While in the egg, the developing bird is nourished by the yolk and albumen and receives oxygen through the pores in the eggshell. Incubation times vary and may be as long as 80 days for the large white egg of the royal albatross or as short as 12 days for the yellow-and-brown egg of the lark. The green egg of the mallard duck must incubate approximately 26 days, whereas the chalky white egg of the flamingo needs about 32 days. A chicken egg develops in 21 days.

 ASK YOURSELF

Describe the reproduction of birds.

Mammals

Spending a day at a zoo, such as the San Diego zoo, shows that humans are interested in other mammals. Which of the animals in the photographs would you expect to see in a zoo?

Who's Who at the Zoo Even though other classes of vertebrates also live in zoos, mammals are often the leading attractions. Close your eyes for a moment and take an imaginary walk around a zoo. Which mammals would you probably stop to look at for the longest time?

Figure 14–26. Mammals include a wide variety of animals, including humans. Shown here are the tiger, opossum, hump-back whale, and giraffe.

If you could safely and appropriately care for mammals in your own private zoo, which ones would you select? In your journal, list 15 mammals you would allow to live in your zoo. Tell why you selected each of the animals. Did the possibility of extinction influence your selection? Draw a layout of your zoo. Write a description of the conditions in your zoo. ✎

Figure 14–27. Experienced trackers can identify what tracks belong to which animal.

In the Human World   All animals contribute to the environment in some way. Mammals, perhaps because they are so visible, seem to make major contributions. In some cases, animals greatly improve the quality of human life. Guide dogs for the visually impaired and dogs trained to assist hearing-impaired people provide their owners with the opportunity to enjoy independent living.

Mammals are often present in our lives without letting us see them. Have you ever wandered into wooded areas and seen the tracks left by animals? The tracks tell stories without words about the activities of the animals. In the passage at the beginning of the chapter, the fisher learned much from observing animal tracks. Give some examples of what the fisher learned. Then do the following activity to become more of an expert yourself!

Using a book such as *Field Guide to Animal Tracks,* draw and label 20 animal tracks in your journal. Study the tracks often until you become an expert at recognizing them. ✎

Sometimes, when you are riding in a car at night, you might catch the reflected glow of a mammal's eyes. Many mammalian eyes reflect light. The eyes of bears and opossums reflect red light. The eyes of house cats, cougars, and other cats reflect green light. Deer eyes reflect amber light.

Mammal or Not? Two main characteristics are used to classify animals as mammals. One of these is the presence of hair somewhere on the body at some time during the animal's life. The other characteristic is the presence of *mammary glands* that produce milk for the young. The mammary glands give this class of vertebrates its name. Most young mammals need parental care for longer periods of time than other vertebrate young do. Feeding the young from milk-producing glands is part of that parental care. The milk contains water; butterfat; lactose, a sugar; proteins; and various salts.

In some cases, the amount of hair a mammal has on its body has been reduced to only a few bristles. Whales and dolphins are mammals that have just a few bristles of hair around the chin and snout early in their lives. Dolphins actually have no more than eight of these bristles.

Mammals whose hair covering has been reduced usually have thickened skin or even armored plates to protect them. Elephants, for example, have little hair but very thick skin. Armadillos have armorlike skin that protects them from the weather and from enemies. In some animals, hair is associated with the senses, especially with the sense of touch. Whiskers help animals recognize the presence of objects, especially in burrows or other dark places.

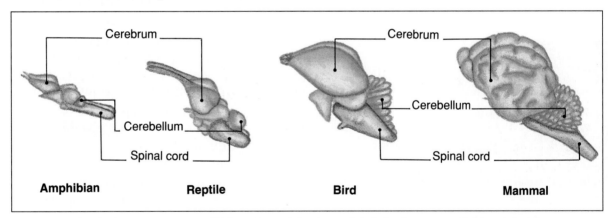

Amphibian **Reptile** **Bird** **Mammal**

Mammals have a very well-developed brain, which is the center of their nervous system. The *cerebrum,* which is the largest part of the mammalian brain, is involved with intelligence and the organizing of information. As you can see from the illustration, the brains of amphibians, reptiles, and birds are organized differently. The exceptional learning ability and memory of mammals are attributed to their well-developed cerebrum.

Figure 14–28.
The cerebrum is associated with intelligence. A large, furrowed cerebrum is characteristic of mammals.

 ASK YOURSELF

By what characteristics are mammals classified?

Types of Mammals

There are three main types of mammals. They are the monotremes (MAHN uh treems), the marsupials (mahr SOO pee uhlz), and the placentals (pluh SEHN tuhlz). These animals are grouped according to how their young develop.

Figure 14–29. The duck-billed platypus is one of only two mammals that lay eggs. What is the other mammal?

Mammal Eggs? We don't usually think about mammals laying eggs. Most bear live young. If someone tried to sell us an elephant egg, we would know right away it was a joke. But there really are two mammals that lay eggs. The duck-billed platypus and the spiny anteater are members of a primitive group of mammals called *monotremes*. The platypus actually lays eggs in a nest. Can you imagine finding a nest of platypus eggs? Unless you live in Australia or Tasmania, you will never find such a nest. The spiny anteater keeps her eggs in a pouch on the side of her body. After monotreme eggs are hatched, the newborns feed on milk from their mother's mammary glands.

In the Pocket *Marsupials* are mammals with pouches, such as the kangaroo and the opossum. The marsupials are the main group of mammals found in Australia. Marsupials give birth to their young long before the young are fully developed. The tiny, underdeveloped animals must then find their way to the mother's pouch, where the mammary glands are located. The young continue their development in the pouch, nourished by the mother's milk.

Figure 14–30. Marsupials, such as this kangaroo, give birth to premature young. The young climb into the mother's pouch, where they attach themselves to the mammary glands and remain until fully developed.

On the Life Line *Placental* mammals are those in which the young develop completely within the mother's body. More than 95 percent of all mammals are placental mammals. This group gets its name from the placenta. The *placenta* is an organ that attaches the young unborn animal to the mother. The young mammal is attached to the placenta by the umbilical cord. During development of the young, nutrients and oxygen pass from the mother to the young. At the same time, carbon dioxide and wastes move from the young to the mother.

Figure 14–31. The well-developed young of placental mammals can feed on the milk provided by the mother. Some placental animals, such as humans, are dependent on the mother for a longer time than others, such as dogs.

Placental mammals give birth to relatively well-developed young. Once an egg is fertilized by a sperm in the body of a female, a zygote is formed. The zygote attaches itself to the wall of an organ called the *uterus*. This attachment begins the period of **gestation** (jehs TAY shuhn), or the time when a young mammal is developing within the mother's body. As the zygote grows, it is called an *embryo*.

The young grows inside the mother's body until it is time for birth. The muscular uterus begins to contract and relax, firmly pushing the new mammal out of the mother's body. This process is the same for all placental mammals, including bats, whales, deer, gorillas, and humans.

Another word for gestation is *pregnancy*. The period of gestation varies for different mammals, just as the incubation time of birds does. The gestation period of a mouse is as short as three weeks, while that of an elephant is nearly two years! Gestation time for humans is approximately 40 weeks.

 ASK YOURSELF

How are development and birth different in the three types of mammals?

SECTION 2   *REVIEW AND APPLICATION*

Reading Critically
1. Describe how birds are adapted for flight.
2. Describe the relationships between humans and other mammals.
3. What is the purpose of the placenta?

Thinking Critically
4. What are the advantages of bird parents sharing duties during the incubation period?
5. Most mammal eggs contain a small amount of yolk in comparison to bird eggs. Why do you think this is so?

SKILL Making and Interpreting a Graph

► MATERIALS
- graph paper
- pencils in 2 different colors

▼ PROCEDURE

From 10:00 to 10:30, a science student watched a bird feeder. As part of his observations, he listed the times and numbers of birds at the feeder every time the number changed. He recorded his data in a table like the one shown.

BIRD-FEEDING TIMES

Times	Number of Birds	Times	Number of Birds
10:02	3	10:18	7
10:08	5	10:20	0
10:10	0	10:22	1
10:14	2	10:23	2
10:15	0	10:29	9

1. Use a ruler to draw a vertical line down the left side of a sheet of graph paper. Leave space to the left of the line to write a title and numbers. This line is called the *Y axis*.
2. At the bottom of the graph paper, draw a horizontal line that intersects with the bottom of the vertical line. Make sure that there is enough space under the line to write a title and numbers. This line is called the *X axis*.
3. Starting at the bottom of the *Y axis*, number from 1 to 10 going up the axis.

There should be an equal amount of space between each two numbers. This is the *range* of the *Y axis*—in this case, the number of birds.

4. Starting at the left of the *X axis*, write the times in 2-minute intervals starting with 10:00. There should be an equal amount of space in each interval. This is the *range* of the *X axis*—in this case, time.
5. Plot the data from the table on the graph. For example, the first data point is directly over the time 10:02 and directly to the right of the number 3 (for 3 birds). Then, from that point, draw a horizontal line segment from 10:02 across to a point directly above 10:08. Repeat this process with the rest of the data.

► APPLICATION
1. During what time period was there the greatest number of birds at the feeder?
2. What was the longest period of time during which the number of birds stayed the same? When was this?
3. How is a graph an advantage over a table when you are reading certain types of information?

※ *Using What You Have Learned*
You can compare data on a graph by using different colors. Suppose you wanted to see which of two bird feeders was used most during specific times of the day. Make up data giving the number of bird visitors for a second bird feeder. Using a different colored pencil, chart the information with a second set of line segments on the same graph.

The Big Idea

Classification schemes for living organisms help organize the great quantity of information collected about life. In this chapter, the vertebrates are classified into five groups: fishes, amphibians, reptiles, birds, and mammals. The groups help to emphasize common characteristics and also point to evolutionary links among the species.

Early fossil records show the appearance of ocean-dwelling creatures. Gradually, over long periods of time, some of the fishlike creatures began to adapt for life on land. The five groups of vertebrates parallel the development of life in Earth's history. Each class of vertebrates is more advanced than the previous one. Mammals are the most advanced. Brain and nervous system advances as well as lengthy and responsible care of young are seen in human beings, the most advanced of all the mammals.

For Your Journal

Review the questions you answered in your journal before you read the chapter. Add to your journal entries by describing ten more vertebrates. Include two animals from each of the five classes covered in this chapter: fishes, amphibians, reptiles, birds, and mammals. Sketch each animal and describe a few of its characteristics. How are these animals alike? How are they different?

Connecting Ideas

Copy this unfinished concept map into your journal. Complete the concept map by writing in the appropriate terms.

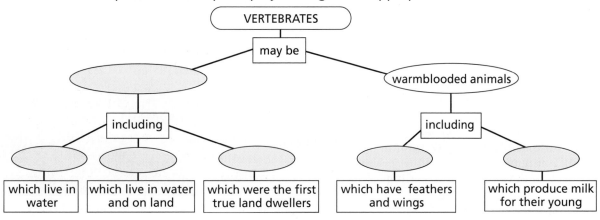

REVIEW

Understanding Vocabulary

For each pair of terms, explain how they are alike and how they differ.

1. coldblooded (372) and warmblooded (384)
2. amphibians (376) and reptiles (380)
3. jawless fishes (373) and cartilaginous fishes (373)
4. incubation (388) and gestation (393)

Understanding Concepts

MULTIPLE CHOICE

5. The first vertebrates to be true land dwellers were the
   a) fishes.   b) mammals.
   c) amphibians.   d) reptiles.

6. Tadpoles develop into frogs through a process called
   a) incubation.   b) gestation.
   c) metamorphosis.   d) reproduction.

7. Birds that nest on cliffs may lay eggs that are
   a) round.   b) very large.
   c) pointed at one end.   d) colored like rocks.

8. One characteristic that distinguishes mammals from other vertebrates is that
   a) they are warmblooded.
   b) their young are fully developed when born.
   c) they do not lay eggs.
   d) they have body hair at some time during their lives.

9. Which process is described by the term *gestation*?
   a) A young mammal is developing within its mother's body.
   b) Eggs are kept warm until they hatch.
   c) A new mammal is created by means of internal fertilization.
   d) Young mammals continue their development by getting nourishment from the mother's milk.

SHORT ANSWER

10. How might changes in weather affect the body temperature of a coldblooded animal?

11. What might happen if fish eggs had shells?

12. One example of a warmblooded vertebrate is the mammal. Give a second example.

13. Why are some bird eggs colored?

14. Explain why birds eat large quantities of food.

15. What similarity exists between the food provided for developing birds and that provided for young mammals?

Interpreting Graphics

16. The mud puppy shown here is a type of amphibian that resembles a tadpole. Note the mud puppy's external gills, which are never lost, even as an adult. Are mud puppies more or less advanced than other amphibians? Justify your answer.

Reviewing Themes

17. *Changes Over Time*
List the characteristics of fishes, amphibians, and reptiles that you think are the most valuable for their survival.

18. *Systems and Structures*
Compare and contrast how birds and humans care for their young. Include as many different facts as you can.

Thinking Critically

19. There are fewer amphibians on Earth than any other type of vertebrate. Explain why you think this is so.

20. If you wanted to encourage birds to visit your yard, what steps would you take? Why are birds discouraged from visiting some places?

21. Why do you think placental mammals are more numerous than marsupial mammals?

22. There are several laws that prohibit owning certain types of feathers. Why do you think these laws were passed?

23. Fishes produce many young. They actually produce even more eggs than those that hatch. Is this true of amphibians? What about reptiles? Do birds lay as many eggs as frogs do? Do elephants have as many baby elephants as ducks have baby ducks? Over a lifetime, what animal probably has the fewest offspring? Develop a hypothesis to explain why numbers of eggs and numbers of young are reduced as the animal type becomes more advanced.

24. If a coldblooded animal is to remain living, it must learn to adapt itself to changes in the temperature of its environment. Dinosaurs had some very unusual adaptations, such as huge fins, that helped to regulate their body temperature. What are some of the things that modern reptiles do to keep from becoming too hot or too cold? Do you think being too hot would be more dangerous than being too cold? Explain.

Discovery Through Reading

Darling, Kathy. *Manatee On Location*. Lothrop, Lee & Shepard Books, 1991. Text and photographs describe the life history, physical characteristics, behavior, and underwater activities of the Florida manatee and how scientists and others are trying to save this endangered sea mammal.

Animal Behavior

My stop! I whistled to the driver. The air brakes wheezed and puffed as the bus slowed and stopped on a steep hill. I hauled my backpack down the bus steps and jumped out into brilliant sunshine.

If you have a pet, you can probably think of quite a few animal behaviors. But wild animals and those in zoos have specific behaviors, too. Most behaviors help ensure an animal's survival. Now travel through the rain forest of Costa Rica with Dr. Adrian Forsyth, studying the behaviors of some of the animals that live there.

I was standing at Kilometer 149 on the Pan American Highway in Costa Rica. A sleepy brown pig was there to greet me. It looked up from its siesta, snorted and closed its eyes again. "Welcome to Lagarto," it seemed to say. Lizardville, you might call it in English. Lagarto was a small collection of houses, shacks, and a general store named after the Rio Lagarto, the Lizard River, which ran nearby.

It looked shady and cool under the huge umbrella-shaped trees down by the river, so I wandered over for a closer look. This river was a kind of signpost for me. The Rio Lagarto wound its way in the same direction as the first leg of my journey. Upstream were the Tilaran Mountains, part of the backbone of mountains extending from the Canadian Rockies all the way down through Central America. Another bus would take me up through the dry foothills on the Pacific slope of these mountains until I reached their wet, windy top. Then my trail would go downhill through the lush rain forests coating the Atlantic slopes of the mountains. Along the way I hoped to discover many strange and beautiful tropical plants and animals.

As I strolled along the riverbank I heard loud quacking noises overhead. It sounded as though the treetops were filled with ducks. I had found a group of boat-billed herons. All the quacking was because I had disturbed their daily sleep. These herons feed at night.

The quacking disturbed another animal sleeping in the tree. I saw an iguana lizard, as large as a ctenosaur, scrambling along a branch reaching over the river. The iguana had been so well camouflaged by his greenish skin that I didn't see him until he reached the end of the branch. The lizard fell off like a stone, belly-flopping with a huge splash into the river, and then swam away under water. It wasn't a graceful exit but it was effective.

from *Journey Through a Tropical Jungle* by Adrian Forsyth

For Your Journal

- What animal behaviors do you know of?
- How do wild animals behave?
- Why would people be interested in studying the behavior of animals?

Defense

Many animals have markings or colorings that help them hide from other animals. Such markings or colorings are called **camouflage** (KAM uh flahj). Camouflage may save an animal from a predator or help a predator hunt its prey. As Dr. Forsyth found out, even large animals can hide.

One of my goals on this trip was to collect flies for a friend of mine, a scientist who was studying a special kind of fly. These flies are fast, and the only way I could catch them was to walk slowly along the trails and net the flies when they were sunning themselves on leaves. This calm, quiet way of walking helped me discover some of the cleverly camouflaged animals of the jungle.

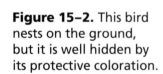
Figure 15–1. A careless mouse might become the next meal for this camouflaged boa.

Figure 15–2. This bird nests on the ground, but it is well hidden by its protective coloration.

One morning on my favorite fly-catching route, a path that ran down through the forest to a coffee farm, I spotted a large boa constrictor on the ground. The wavy pattern of the boa's skin made it blend in with the leaves, helping the boa to catch a mouse or some other small mammal that was moving too fast and too carelessly to notice it.

Lower down in the valley I almost stepped on an even better mimic of dead leaves. A long winged bird fluttered up from under my foot. It was a pauraque (PAHR uh kay), a kind of whippoorwill that nests on the ground during the day and hunts at night. Any bird that nests on the ground has to be very skilled at hiding from its enemies. I had seen for myself how difficult it is to spot one of these birds standing still.

Figure 15–3. The arctic fox blends in with the snow of winter and the bare ground of summer.

Camouflage

Camouflage is important for the survival of many animals. An animal that lives in tall grasses might have stripes to confuse a predator. The stripes of zebras and certain antelopes, for example, help them to blend into their surroundings. Tigers also have stripes, but for slightly different reasons. Its stripes help a tiger to sneak up on its prey without being seen.

The best-camouflaged animals are those whose colors match their surroundings. Its white fur helps a polar bear blend into its icy environment.

The coats of some animals even change color to keep up with the changing seasons. An arctic fox's gray fur blends in with the bare ground of the summer tundra. In the fall, its fur changes to white, blending in with the snow of the long tundra winter.

Animals such as certain lizards, fishes, and invertebrates have the ability to change color as needed. Chameleons are so good at changing color that the word *chameleon* has a second meaning in the dictionary. In addition to a type of lizard, *chameleon* is also defined as "a person who is changeable or moody."

DISCOVER BY Researching

Use your library to learn more about the lizards called *chameleons*. Find out about their environment and the causes of their color changes. Write your findings in your journal and report them to your classmates.

Another type of camouflage, seen often in fishes and other marine animals, is countershading. Countershaded animals are darker on top than they are on their undersides. The killer whale, for example, has a black upper surface and a white underside. The underside blends with the water surface when seen from below, while the topside blends with the ocean depths when seen from above. This camouflage allows the killer whale to hunt with better results.

Figure 15–4. The dark top and light bottom of killer whales help them to blend in with their ocean surroundings.

Some animals use bright colors instead of camouflage for protection. Bright colors "advertise" the danger of close encounters. Dr. Forsyth describes one close encounter.

Figure 15–5. Bright colors are also a kind of protective coloration.

One day I was walking along, looking for flies, when I suddenly stopped and jumped back without really knowing why. I looked down, and at my feet was an unusually beautiful snake. Its markings were bold and bright, the exact opposite of camouflage. When I realized the snake was banded with shiny rings of black and red, I jumped back even farther. I was glad I was wearing my tall rubber boots. It was a dazzling but deadly coral snake, with venom strong enough to paralyze and kill a human.

But the snake had no interest in harming me. Snakes, after all, have no arms or legs to help them capture food, so this coral snake couldn't have been considering anything as big as me for dinner. Their venom is simply a special saliva used to subdue more suitable prey and then help digest it. But they also use their poisonous bite to defend themselves from animals that step on them. So the bright colors of the coral snake had done us both a favor. They provided a warning that had stopped me from injuring the snake, and my retreat had stopped the snake from having to bite me in self-defense.

▼ **ASK YOURSELF**

In what ways is camouflage useful for an animal?

Mimicry

Mimicry (MIHM ihk ree) is another type of protective coloration. But mimicry is different from camouflage. **Mimicry** makes an animal look like something else. That is, the animal mimics, or copies, the appearance of another organism.

Some animals mimic parts of their environment. For example, the pink flower mantis, shown in the photograph, looks like a beautiful orchid. Certain other insects look so much like small tree branches that they are called *walking sticks*. Dr. Forsyth found animals that even copy the behavior of the object they are mimicking.

Figure 15–6. The coloration of the pink flower mantis is an example of mimicry.

> There were moths that looked just like dead sticks and old leaves. Their wings were gray and brown, with fine lines and spots looking like tree bark or leaf veins, and they were twisted in crinkled, wrinkled shapes. They even had blotchy patches on them that looked like mold and fungus. The moths that looked like dead twigs stayed completely still, and the ones that looked like dead leaves let go of the branches they were resting on and fluttered slowly to the ground.

Animals can also mimic other animals. For example, certain nonpoisonous frogs have the same bright colors as those frogs that do have poison glands. Predators cannot tell the nonpoisonous frogs from the poisonous ones. As a result, predators usually pass both types of frogs by for a safer meal. The viceroy butterfly, which birds find tasty, has the same colors and markings as the monarch butterfly, which will actually make birds sick if eaten. Once a bird eats a monarch, it avoids eating another one. This benefits the viceroy, because birds will not eat it either. In the next activity, you can test the effectiveness of some types of camouflage and mimicry.

Figure 15–7. The monarch butterfly (left) and the viceroy butterfly (right) look alike to birds.

ACTIVITY

How effective is protective coloration?

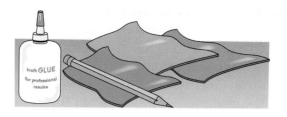

MATERIALS
yellow pencil, materials for camouflaging pencil

PROCEDURE

1. Using whatever materials are available, camouflage or disguise your pencil so that it can be hidden in the open somewhere in your room.

2. Have another team try to find your pencil while you try to find their "hidden" pencil.

APPLICATION
What do you think was the most successful example of protective coloration of a yellow pencil in your class? Explain why you think it was successful.

 ASK YOURSELF

What is the difference between protective coloration and mimicry?

Other Means of Defense

Camouflage and mimicry are often used as methods of defense. These methods involve tricking predators or sneaking up on prey. Animals also have other adaptations for protection. Some animals are able to swim, fly, or run faster than most of their predators. The Thompson's gazelle, for example, has been clocked at 50 km/h, while one of its predators, the lion, can run only about 35 km/h.

Have you ever heard the saying "There is safety in numbers"? Animals often travel in large groups for protection. Predators have a hard time attacking large numbers of animals. However, any animal that becomes separated from the herd is in danger of attack. In the activity at the end of this section, you can see how large numbers of offspring help to ensure a species' survival.

Figure 15–8. Animals can often protect themselves from predators by forming herds.

Sometimes size is the best defense. Very few predators will attack an elephant simply because elephants are so large. However, young or sick elephants sometimes become prey. Stronger and larger elephants try to protect weaker elephants whenever possible. Elephants have been known to free other elephants from traps.

Figure 15–9. The belly of the seemingly well-protected porcupine is open to attack by the fisher.

Outer coverings can also offer protection for animals. Armadillos have tough, leathery plates that protect their internal organs by making it difficult for predators to bite or claw them. Armadillos try to flee from attack, but if given no choice, they can roll themselves into balls, making it even harder for predators to find an unprotected place to attack. However, no animal is immune to attack. For example, the fisher, a relative of the weasel, preys on well-armored porcupines by rolling them over and attacking their soft bellies.

Skunks and other animals with scent glands protect themselves by spraying attacking animals with an offensive-smelling liquid. Skunks are usually left alone by would-be predators that have had unpleasant experiences with this little striped animal.

 ASK YOURSELF

Explain how traveling in large groups can be a benefit to both predators and prey.

SECTION 1 *REVIEW AND APPLICATION*

Reading Critically
1. Which types of defense depend on an animal deceiving its predator?
2. Give three examples of animal defenses that do not depend on deceiving a predator.

Thinking Critically
3. Some toads blend so well with the ground that they are almost invisible. Is this camouflage or mimicry? Explain your answer.
4. Some fish are able to puff themselves up to look much larger than they really are. Would this be an example of camouflage, mimicry, or another defense? Explain your answer.

SKILL

Interpreting Data

▶ MATERIALS
- pencil • graph paper

▼ PROCEDURE

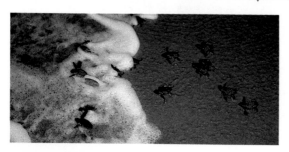

1. Loggerhead sea turtles lay great numbers of eggs about three times each year. They lay their eggs in nests dug into sandy beaches. The baby turtles, called *hatchlings,* crawl from their nest and scurry down the beach toward the sea. If all the hatchlings survived, the population of loggerhead sea turtles would be very large.

2. Study the table showing the survival rate of loggerhead hatchlings. The table shows what happened to a sample population of turtles over the course of 3 years. Assume that in the original population, all 500 000 eggs hatched, and all the hatchlings survived. Now interpret the data. How many turtles would there be after 4 months?

3. Hungry predators, such as gulls, ospreys, and crabs, grab many hatchlings from the sand. Since not all hatchlings survive, the overall population of sea turtles does not change much. Use the data table to construct a graph of hatchling survival.

SURVIVAL RATE OF LOGGERHEAD TURTLE HATCHLINGS	
Number of Months	**Remaining Hatchlings Surviving**
4	6% of original population
8	11% of population at 4 months
12	19% of population at 8 months
16	28% of population at 12 months
20	35% of population at 16 months
24	41% of population at 20 months
28	51% of population at 24 months
32	57% of population at 28 months
36	59% of population at 32 months

▶ APPLICATION
Interpret the data on your graph. How many turtles would there be after one year? After two years? After three years?

✳ *Using What You Have Learned*
Get together with a group of classmates. List your ideas about the ways in which interpreting data might be important to scientists in their work.

Group Behavior

Objectives

Describe four types of social behaviors among animals.

Explain several ways in which animals communicate.

Compare and contrast seasonal behaviors.

*"A*aaaarrrrroooooooo-oooo-oooo-gaaaahhhh!"

I almost jumped out of my rubber boots. It sounded as if there was a huge and fierce creature right above me. Branches were rattling, sticks and leaves rained down, along with an avalanche of noise.

I had startled a large male howler monkey feeding at the end of a branch not far above my head. And he had startled me! The howler glared down, shaking the shaggy mane and beard around his face. He thrust out his strong lower jaw and began bellowing out more threats at me.

Social Behavior

Their behavior is one way the howler monkeys defend their territory. Dr. Forsyth was interested in territorial behaviors. His story continues:

I wasn't worried. Howlers are big monkeys, but they are harmless in spite of their fearsome calls. In fact the animal's annoyance was my good luck. I had hoped to get a good look at the special equipment the male howler has for producing his deep roars. I could see clearly that his throat was massive, with a giant voice box. He was using this sound system for producing deep gravelly bass notes.

Figure 15–10. A troop of howler monkeys claims a territory.

Figure 15–11. Male howler monkeys have a massive voice box. Howler monkeys spend the night in the treetops.

The howler's roaring is an important social behavior directed at other howler monkeys. A troop normally roars together, led by the largest dominant male. These howling sessions begin each morning as the monkeys wake up. Often they repeat the chorus just before bedding down in the treetops for the night. You can hear the calls echoing down the vast valleys. The clamor advertises their location and discourages other troops from using the same area of forest and fighting over the same feeding trees.

The roaring of the howler monkeys is an example of group behavior. Animals often live or work together in groups. For example, white pelicans fish in groups. All the pelicans in the group dip their bills into the water at the same time. Any fish that escapes the bill of one pelican is likely to swim right into the open bill of another pelican. Few fish escape this cooperative effort, and all the pelicans get enough to eat.

Some animals, such as dolphins, form social groups rather than work groups. During the day, dolphins swim and play in groups that may number nearly 100. At night they split up and feed on their own.

Figure 15–12. Animals often feed together.

Figure 15–13.
Dolphins socialize only part of the time.

Figure 15–14.
Schooling is a form of group behavior for some fish.

Protection is another benefit of group living. Fish, for example, often live in large groups called *schools*. An attacking predator is likely to catch only the fish from the fringe of the school. The fish in the center are protected.

Did you ever wonder why the word *school* is used for a group of fish? All kinds of names are used for different animal groups. For instance, a group of whales is a *pod*, while a group of baboons is a *troop*. What other unusual names do you know? In the next activity, you can test your knowledge of animal group names.

DISCOVER BY *Researching*

Here is a list of animals. Next to the list are terms that refer to groups of these animals. See how many animals you can match with their group names. Add any other names for groups of animals that you know.

a _____ of frogs   bed
a _____ of wolves   army
a _____ of lions   flock
a _____ of oysters   pack
a _____ of crows   pride
a _____ of sheep   murder
a _____ of geese   town
a _____ of prairie dogs   gaggle

ASK YOURSELF

What are some of the reasons for group behaviors among animals?

Communication

Animals communicate with one another in many ways. Perhaps you have watched birds and listened to them sing or call to one another. Although animal communications can sometimes be seen or heard, humans often fail to understand them. The most interesting communication occurs as animals make one another aware of boundaries, predators, food, or nesting sites.

For example, when they are ready to mate, pigeons signal by clapping their wings, bowing, and sometimes dragging their tail feathers. Nest sites are almost always chosen by the male. He communicates his choice to his mate by nodding repeatedly toward the ground. The female joins him and also makes nodding motions if she agrees with his selection.

Although they do protect their nests, pigeons do not seem concerned about territory. Perhaps this is because pigeons often live in crowded, busy places.

Many interesting behaviors can also be observed among animals that almost never come in contact with humans. Dr. Forsyth has observed and photographed the fascinating behavior of rain-forest lizards as they communicate about territory.

Figure 15–15. Pigeons make use of many structures built by humans.

I saw something strange by the river. A huge lizard, as big as a large dog, waddled along the ground. I thought, "Perhaps this is how the river (Lagarto) got its name." It was a male ctenosaur (TEHN uh sahr) lizard looking like something out of the Dinosaur Age. He had scaly skin banded with stripes of blue-gray and black and a huge spiny crested head with a very wide mouth. He was tilting his head up and down again and again. The cause of all this head-bobbing was another ctenosaur nearby, who was also head-bobbing. Each was signaling that this was his territory: "Keep out, no trespassing." But it soon became clear that both lizards would rather feed than fight. As I watched, they put their heads down and began to pull up large mouthfuls of weeds.

Figure 15–16. Aggression between lizards is a form of communication.

Complex communication occurs among bees. Bees use "dances" to tell other bees where food is located. If the food is closer than 90 m, a bee does a "round dance." If the food is farther away, the dance becomes a more complicated "waggle dance." In the next activity, you can observe other insects.

Figure 15–17. Communication among animals can help to ensure the survival of the entire group.

ACTIVITY

What behavior can you observe in an ant colony?

MATERIALS
ant colony; shovel; newspaper; small trowel; plastic bag; small paper tube; large, clear glass jar with tiny holes in the lid; cornmeal, grass seeds, or sugar; moist sponge; tape; black paper

PROCEDURE
1. **CAUTION: Some ants may bite. Do this activity only under the supervision of your teacher.** Look for an ant colony. Dig down with a shovel, and place the soil on the newspaper.
2. Use the trowel to sort through the soil. Tiny white objects that look like grains of rice are ant cocoons and larvae. Place ants, cocoons, and larvae—about 30 in all—in a plastic bag along with some soil. Seal the plastic bag, and place it in the refrigerator for 5 to 10 minutes to slow down the ants.
3. Stand the paper tube in the center of the jar. Then transfer the ants and the soil into the jar around the outside of the tube. Cover the jar with the lid.
4. Feed the ants by placing a pinch of cornmeal, grass seeds, or sugar on top of the soil. Place a small piece of moist sponge on the top of the soil, too. Keep the sponge moist, and give your ants fresh food every few days.
5. Tape the black paper around the jar to keep out the light. Remove the paper to observe the ants every day for two weeks. Each time, re-cover the jar with the paper after making your observations.
6. When you have finished your observations, return the ants to their colony by carefully emptying the jar near the place where you found them.

APPLICATION
How might you compare the ants' behavior to the behavior of people?

The types of communications seen so far have involved sight or sound. Animals also communicate using chemicals called **pheromones** (FEHR uh mohnz). These chemicals serve as signals that are picked up by animals of the same species. Depending on the animal, the signals are picked up by antennae, nostrils, or tongues.

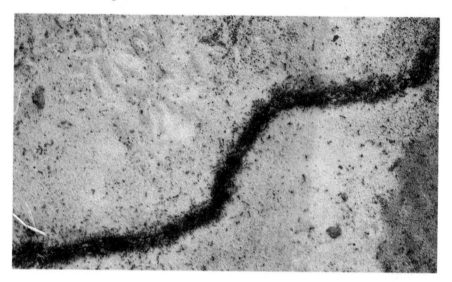

Figure 15–18. Communication by pheromones enables ants to create and follow trails.

You may have seen ants trailing after one another on the way to some food scraps found by ant scouts. Because of the pheromones being deposited by the ant scouts, the rest of the colony knows where to gather and can easily follow the trail.

Attracting mates is another use for chemical communications. Moths use pheromones in this way. The female usually sends out the chemical, and the male picks up the scent with his antennae. Butterflies also attract mates, but they appear to rely on their bright colors and patterns more than on scents. This may be due to the fact that butterflies mate in the daytime, while moths mate at night.

DISCOVER BY Researching

Scientists are investigating the use of pheromones as a way to control insect pests. Use your library resources to learn more about the biological control of insects. Write your findings in your journal and be sure to present the pros and cons of this method of pest control. ✎

ASK YOURSELF

How might animals use pheromones to find mates?

Seasonal Behavior

Animals often change their behaviors as the seasons change. Animals that could be in danger of freezing or starving during a harsh winter sometimes survive by going into *hibernation*. Actually, the term for this inactive period is torpor. **Torpor** is a state in which an animal's body functions, or metabolism, actually slows down. Body temperature drops and breathing and heartbeat rates slow considerably.

Contrary to what most people think, bears do not hibernate. Although they sleep long and often during cold months, bears wake periodically and even venture outside their dens on warm days. More importantly, their metabolism does not slow down.

The ground squirrel, a small North American rodent, is an animal that truly hibernates. Before its winter torpor, a ground squirrel eats extra food, which is stored as fat, to supply the energy its body will need to survive. Then it finds a cozy place to sleep for the entire winter. The warm days of spring awaken the squirrel. By this time a new supply of food will await the hungry animal, who may have lost 50 percent of its normal body mass.

Some desert animals go into a kind of torpor, called *estivation,* to escape the heat of summer. The spadefoot toad, for example, digs deep into the bottom of a desert water hole. Even if the water hole dries up, the toad is protected by cool mud until the late summer rains and the cooler days of fall return.

Figure 15–19. The ground squirrel hibernates in winter, while the spadefoot toad estivates in summer.

Not all animals escape unfavorable conditions by hibernating or estivating. Some simply move to different locations. This seasonal movement of animals from one location to another is called *migration.*

Monarch butterflies, for example, spend the summer in Canada, then fly to Mexico for the winter. Humpback whales mate and feed off the coast of Alaska. In winter they migrate to the waters off Baja California, where the females give birth before heading north again in the spring. Many African animals, such as zebras, antelopes, and wildebeests, migrate across the veldt, not to seek warmer climates, but to find food and water. Even fishes, such as eels, leave the open ocean to lay their eggs in coastal rivers and streams.

 ASK YOURSELF

How are hibernation and estivation similar? How are they different?

Parental Behavior

In the animal kingdom, there are few examples of parents caring for their young. Most animals simply mate, lay their eggs, and leave their young to fend for themselves. Mammals are the notable exception. A young mammal is unable to survive in the world on its own. Mammals need a certain amount of parental care while learning to find food and defend themselves from predators. Dr. Forsyth observed that howler monkeys spend a great deal of time with their young. Notice the similarities in behavior between howler monkeys and human children.

> Soon I came upon the rest of the howler troop, quietly feeding on leaf buds. Small babies were riding on their mothers' backs or clinging to their bellies as the females wandered slowly from one branch to another, looking for fresh new growth. They seemed slow and sleepy. Sometimes they would stop feeding to scratch, groom each other or do a little sunbathing, dozing peacefully with their legs draped over the tree limbs. One of their common gestures was to stick their tongues out at one another. It was just like watching a family on a picnic.

Figure 15–20.
Monarch butterflies fly to Mexico in the fall to avoid the harsh winter weather of Canada.

Figure 15–21. Howler monkey families are similar to human families.

Although parental care is most common among mammals, a few other animals, even some invertebrates, care for their young. Dr. Forsyth describes an encounter he had with a scorpion mother:

I knew it was a mother because her back was covered with baby scorpions. The mother scorpion carries her young around so that she can keep them out of harm's reach. She is better at catching food than her youngsters, so they stay with her until they are large enough to catch supper by themselves. Until then the mother grabs large insects such as cockroaches with her strong pincers, then prepares a meal by chewing open the skin of the roach and adding her saliva, which is full of digestive enzymes. These enzymes begin to pre-digest the food. The young scorpions can then climb down and drink their dinner of liquid cockroach.

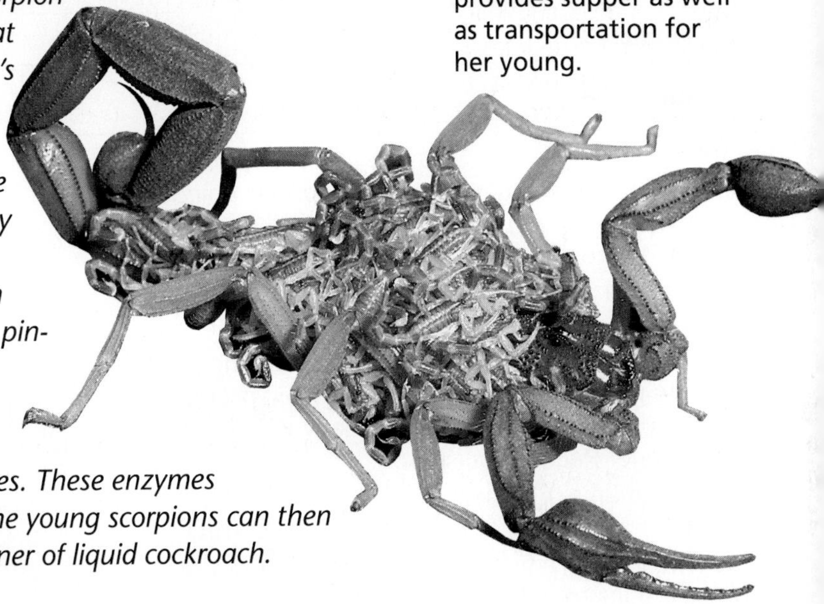

Figure 15–22. This mother scorpion provides supper as well as transportation for her young.

 ASK YOURSELF

Why do mammals take good care of their young?

SECTION 2 *REVIEW AND APPLICATION*

Reading Critically

1. Describe three reasons why animals need to communicate with each other.
2. What are pheromones? Explain how they are used by ants to find food.

Thinking Critically

3. Explain how an understanding of animal behavior might help a performer who works with animals.
4. When might humans migrate?
5. Why is parenting more common among mammals than among other animal groups?

INVESTIGATION

Predicting Animal Responses

▶ MATERIALS
- scissors ● shoe box with lid ● 10 sow bugs ● sponge ● water

▼ PROCEDURE

1. Copy this table for your data.

<table>
<tr><th colspan="5">SOW BUG RESPONSES</th></tr>
<tr><th>Sow
Bugs</th><th>Number
in Light</th><th>Number
in Dark</th><th>Number with
No Sponge</th><th>Number with
Wet Sponge</th></tr>
<tr><td>Predicted</td><td></td><td></td><td></td><td></td></tr>
<tr><td>Observed</td><td></td><td></td><td></td><td></td></tr>
</table>

2. Predict how many sow bugs will prefer a light environment and how many will prefer a dark one. Write your predictions in the table.

3. Design a simple experiment using the shoe box to see if your prediction is correct.

4. Gently remove the sow bugs from the shoe box. Now predict how many sow bugs will prefer a damp environment and how many will prefer a dry one. Record your predictions.

5. Design an experiment to see if your prediction about a damp or dry environment is correct.

6. When you are finished with them, return the sow bugs to their natural environment.

▶ ANALYSES AND CONCLUSIONS
Were your predictions correct? How did the sow bugs respond to the changes in their environment?

▶ APPLICATION
Using your data, hypothesize where you might find sow bugs.

✳ Discover More
Find out the kind of environment another animal, such as an earthworm, prefers.

HIGHLIGHTS

The Big Idea

Scientists observe animal behavior and how it changes as conditions change. Analysis of observations helps scientists make generalizations about relationships in nature and patterns of change. These generalizations in turn help scientists predict what will happen under different circumstances.

For Your Journal

Look back at the list of animal behaviors you wrote in your journal. What new behaviors can you add to your list now that you have completed this chapter? You may wish to do more research and continue to add to your list.

Connecting Ideas

The boxes below contain generalizations about animal behavior. Write the number of each generalization in your journal. After each number, name an animal whose behavior is described by the generalization.

1 Some animals live together because they feed together in the same area.

2 Some animals live in colonies and work together for survival.

3 Some animals live in large groups for protection.

4 Some animals enter an inactive state to survive cold winters.

5 Some animals enter an inactive state to escape summer heat.

6 Some animals move from one location to another to avoid harsh conditions.

7 Some animals spend a lot of time caring for their young.

Understanding Vocabulary

Explain how the terms in each set are related to each other.

1. mimicry (403), camouflage (400)
2. sounds, movements, pheromones (412)
3. torpor (413), hibernation (413), estivation (413)
4. herd, school

Understanding Concepts

MULTIPLE CHOICE

5. An example of mimicry is
a) the stripes of a tiger.
b) the white coat of a polar bear.
c) the markings on a viceroy butterfly.
d) the light underbelly of a killer whale.

6. Pigeons use body movements to communicate with each other about
a) when to migrate.
b) where to build their nest.
c) how many eggs to lay.
d) how to find food sources.

7. Ants know when a food supply is gone because
a) the other ants are no longer leaving pheromones.
b) they see other ants returning without food.
c) they no longer smell the food.
d) other ants signal to them with their antennae.

8. Caribou herds moving south in the winter is an example of
a) hibernation.
b) migration.
c) estivation.
d) communication.

9. Copy and complete the concept map.

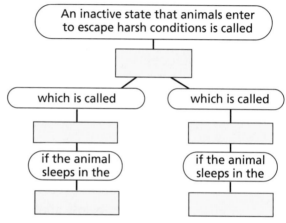

SHORT ANSWER

10. What reason have scientists given to explain why moths use pheromones more than butterflies do?

11. Why do mammal parents have to take good care of their offspring?

Interpreting Graphics

12. Look at the photographs below. Which shows an example of camouflage? Which shows an example of mimicry? Explain the difference.

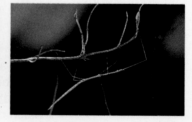

Reviewing Themes

13. Environmental Interactions
Explain why some animals migrate each year. Give at least two examples.

14. Nature of Science
How does the journey of Dr. Forsyth demonstrate a scientific method? If you were to make a similar journey, where would you go? What methods would you use in your studies?

Thinking Critically

15. Make a list of the nonverbal communications exchanged by human beings. Explain what you think each communicates.

16. Scientists have hypothesized that humans use pheromones, just as other animals do. In fact, humans often use the pheromones of other animals as perfume. Musk is an example of this. If humans do have pheromones, what do you think their functions are?

17. Do humans ever use camouflage? List examples and explain the purpose of each kind of camouflage.

18. Zebras' black-and-white stripes help them blend in with each other. How might this blending help them avoid predators?

19. A military aircraft may use countershading as camouflage. What would the airplane look like? How would countershading give the airplane an advantage when flying?

20. The venomous coral snake has red, yellow, and black bands. The nonvenomous scarlet king snake also has red, yellow, and black bands, but they are arranged in a different pattern from the coral snake's. Yet the scarlet king snake is still avoided by predators and humans. Why?

Discovery Through Reading

Patent, D. H. *How Smart Are Animals?* Harcourt Brace Jovanovich,1990. Discusses recent research on levels of intelligence in both wild and domestic animals.

Science PARADE

Romancing the Panda

Ling-Ling and Hsing-Hsing at the National Zoo

Probably no pair of animals had ever been so closely watched as Ling-Ling and Hsing-Hsing, the giant pandas at the National Zoo in Washington, D.C. Human observers and television cameras recorded the couple's every move, waiting for the pandas to mate and give the American people one of the world's rarest zoo animals—a panda born in captivity.

Bear Went a Courtin'

Getting pandas to reproduce outside their natural habitat is not easy. The reproductive cycle of a female panda is, at best, unpredictable. Panda watchers keep an eye on the female for signs that she is in heat, or ready to mate. Then zookeepers must decide how long to keep the male and female together. Pandas are solitary creatures and will fight with each other if left together too long. Even in their native mountains, female pandas often go through heat without meeting a likely male.

Every time Ling-Ling went into heat, panda fans hoped Hsing-Hsing would show some interest in fatherhood. Observers were ecstatic when the pair successfully mated for the first time in 1983, after a two-day courtship. However, biologists were not sure that Ling-Ling was pregnant, so they arranged for her to be artificially inseminated.

In artificial insemination, sperm from a male is injected into the female. The procedure is commonly used in farm animals but is often unsuccessful with wild animals.

After the mating and insemination, zoo biologists kept an eye on Ling-Ling for any sign that she might be pregnant. With pandas it is not easy to tell. A panda's pregnancy is impossible to detect until very near birth. One problem is the difference in size between mother and fetus. Ling-Ling weighed as much as a football lineman, while a panda cub at birth weighs only as much as a stick of butter. Another problem is that the time from fertilization to birth can vary from 97 to 175 days. Biologists hypothesize that a female panda's reproductive system actually times the birth to occur in early fall. That way the cub will be weaned in early spring, when the weather is good and food is plentiful.

To the joy of panda watchers, Ling-Ling had cubs several times. But each time, the joy was short-lived because the cubs died within days of birth.

Pandas use a sixth finger to grasp bamboo.

The Need to Breed

Breeding pandas in captivity is important because so few pandas are left in the wild. Pandas live only in remote mountain regions of southern China. Their numbers have dwindled to less than 1000 because humans have destroyed their natural habitat.

Pandas have also been the victims of natural disasters. Pandas feed almost exclusively on bamboo. Once every 100 years or so, bamboo plants flower and die off in great numbers. This happened in the 1970s and left wild pandas with little to eat. In the past, they might have moved to other locations and eaten bamboo that was on a different cycle. But today much of China's bamboo has been cut down to provide farmland for its growing population.

Another reason for their decreasing numbers is that pandas are not very adaptable. They have the simple stomachs and the short intestines typical of meat-eating animals, but they are too slow to catch prey. So they live by eating up to 24 kg of bamboo shoots a day. They have large grinding teeth and a sixth "finger" for grasping bamboo. Because they are such specialized feeders, pandas cannot easily switch to another food supply. When the bamboo dies off, pandas starve. Nearly 140 pandas died during the most recent bamboo die-off.

The Only Hope

Captive breeding may be the only way to save the giant panda. Biologists hope that one day they will have as much success at breeding pandas as they have had with other endangered species, such as tamarin monkeys. These monkeys live in the rapidly disappearing rain forests of coastal Brazil. By studying the few remaining wild tamarin monkeys, biologists learned that they are monogamous—that is, each has only one mate. When zoo biologists imitated this arrangement, the monkeys bred successfully.

Ling-Ling died unexpectedly on December 23, 1992. She was 23 years old. Scientists continue to attempt captive breeding with mating pairs in other locations. Will scientists ever be as successful at breeding pandas? So far, only a handful of cubs have survived in zoos outside China. Without giant pandas, the world would be a poorer place. ◆

Giant panda at the Metro Zoo in Toronto, Canada

Watch Your Step

by Adrian Forsyth
from *Journey Through a Tropical Jungle*

Michael and Patricia Fogden

A rooster that wouldn't quit crowing had us moving as soon as the first light hit the treetops. It was one of those rare mornings with just a few drifting clouds and the promise of sunshine—not to be wasted. After a quick breakfast of bananas, I reorganized my insect-collecting gear. Eladio worked on his machete, filing a fresh, razor-sharp edge on it, and we headed off into the forest with Eladio leading the way.

Eladio was the perfect person to walk with in the jungle. He was silent, alert and skillful. He was able to find and cut a path with no apparent effort. His machete strokes were neat and economical, disposing of a vine here or a fallen branch there. It seemed as easy as walking. But I knew that his skill was based on years of practice and that it would take me twice as much time to cut a trail half as long.

Eladio opened up a trail into some of the wettest rain forest I had ever seen. Small streams poured down the steep slopes. Mushrooms and molds were sprouting on every fallen log. I even found insects coated with fungus. One of them was a large speckled weevil with a long snout. The weevil had actually been invaded by a fungus that spreads through its body, feeding

Michael and Patricia Fogden

Eladio 🐾

on its tissues. Before the weevil dies this remarkable fungus somehow causes the weevil, normally a low-lying creature, to crawl to the very top of a palm leaf and clamp on in a tight death grip. Then the fungus bursts right out of the weevil, growing into a mushroom-like structure that opens to spread its infective spores on the breeze.

When we reached the Peñas Blancas River, Eladio headed upriver to cut more trail for another day's hike. Before I turned back I sat down for a rest on the river edge. The river was swollen with yesterday's rain and ripping along at high speed. I could hear the groaning of large rocks and boulders grinding and rolling under water along the riverbed. The water and gravity were working at the slow and endless task of wearing the mountain down, turning rock into gravel and gravel into sand, and carrying the sand to the lowlands and the ocean. I looked at the sand and gravel bar and thought how long it had taken to turn a huge boulder into fine sand. I thought about how long it had taken the river to carve the valley out of the mountain and how much time had gone into growing the great forest that blanketed the slopes.

And I had so little time to see it all. I splashed cool clean river water in my face and headed back along the trail alone.

Along the streams, where the sun broke through, sun-loving plants like the heliconia grew in large clumps and produced tremendous stalks of flowers. I stood under one that reached from just above my head all the way down to my ankles. It was easy to be entranced by the almost magical vegetation. I kept staring up, guessing at the height of tremendous trees, and I had to remind myself that this was a place where one had to walk very, very carefully.

The warm lowland rain forest is home to a great variety of venomous snakes and stinging insects. Pit vipers, for example, thrive in warm wet jungles, and I soon saw a mottled green one coiled and camouflaged in some vegetation along the trail. It was warm enough here for the fer-de-lance, a large camouflaged snake that waits coiled beside game trails to ambush small mammals. One thing I wanted to avoid was a close encounter with one of these snakes.

Actually, I had to watch out for ants even more than for snakes. I knew that Eladio had lived down here for many years without once being bitten by a snake. But he had told me that he was always getting stung by ants and wasps.

Some of these ants showed up in strange places. Along one section of trail near the river there were many cecropia trees with hollow stems that looked a bit like bamboo. When I tapped one stem, small azteca ants came boiling out of holes all along the trunk. They ran around furiously with their pointed back ends held straight up. When they reached my finger they swarmed all over it and began biting. Luckily their bite was irritating only on the soft skin between

Fer-de-lance, 2 m (6 ft.) long 🐾

Adrian Forsyth

423

Cecropia tree swarming with ants ⮴

Michael and Patricia Fogden

cecropia leaves are usually seen in the largest cecropia trees, the ones that have lost the ant colonies.

At one point the trail skirted around a large fallen tree. The fallen tree had opened up a gap in the foliage, allowing sunlight to reach the forest floor, encouraging a dense growth of weedy plants. On these plants I found some green treehopper insects and some black and yellow polka-dotted ones. They had also developed a relationship with the ants. The hoppers were sucking sap out of plant stems, and the ants were walking all over the hoppers, milking the sweet sap from them. But the ants were working for their supper: when I poked at the treehoppers the ants rushed at my finger, biting wildly. The arrangement was something like dairy farming, with the treehoppers getting protection as they fed on the plant, and the ants getting sugary rewards for tending the hoppers.

But while I was watching all this, I found that I had stepped where I shouldn't have. Swarming up my pant leg, both inside and out, was a cloud of huge ants.

I was standing in the midst of an army ant raid! I jumped back along the trail, kicking off my boot and hiking up my pants. Only one large soldier ant, as big as a paperclip, had made its way onto my skin, but that was enough. It locked its great sickle-shaped mandibles into my calf and

my fingers. But I could see that climbing a cecropia tree would be almost impossible because hundreds of these ants would swarm over my face and body.

The cecropia tree provides a home for the aztecas, and they aggressively attack any insect or animal they encounter on their tree. With the ants on duty, the tree stays free from grazing and leaf damage. To encourage this relationship, cecropia trees have hollow stems where the ants live. They also have special glandular areas that produce tiny white bits of a special protein-rich food for the ants.

For some reason large cecropias have few ants or no ants at all. Perhaps that is why the three-toed sloths that love to eat

Army ants carry off a wasp larva. ❧

Adrian Forsyth

plunged its sting repeatedly into the skin. The fiery hot sting made me jump even higher.

When I yanked the army ant off my leg, I saw how determined its grip was. Its head remained firmly attached to my skin even when the rest of its body was broken off. I could see why South American Indians use these gripping army ant heads to stitch up open wounds.

I hastily brushed the remaining ants off my pants and boots and stepped back carefully to take a closer look at the raid. It was easy to do as long as I kept an eye on the shifting columns of ants spreading through the forest. At the head of the raid, scouts were dragging their abdomens along the ground and along branches, laying down a chemical signal to direct the other army ants to new areas and food sources. Sometimes to cross a gap they strung their bodies together in a chain so that other ants could run across from tree to tree. The endless columns of ants divided and criss-crossed and blended together again, always advancing, weaving their way up and across the forest.

Careful to keep out of their path (one army ant sting was more than enough), I gazed at the spectacle unfolding before me. The army of ants surged forward. It was a tremendous feat of chemical coordination and communication among tens of thousands of individuals. Somehow all those ants, working without language or leaders, had managed to organize an effective hunt. ◆

Ernest Just (1883—1941)

In 1908 Ernest Just began his career at the Marine Biology Laboratory in Woods Hole, Massachusetts. For 20 years he studied embryology, or the development of organisms. While researching marine invertebrates, he also served as head of the Department of Physiology at Howard University.

Earnest Just was born in Charleston, South Carolina, in 1883. After high school, he worked his way north and enrolled at Kimball Union Academy in New Hampshire. Just continued his education at Dartmouth College, where he graduated with honors in 1907. In 1916 he received his Ph.D. in zoology from the University of Chicago.

Just spent many years in Europe, teaching and doing research. He wrote several books and numerous articles on cell physiology. His work was respected by eminent biologists of his day, and in 1915, the NAACP awarded Just the first Spingarn Medal for his efforts to improve the quality of medical education at Howard. ◆

Jane Goodall (1934—)

Jane Goodall is best known for her study of the behavior of wild chimpanzees. In June 1960 she set up camp in the Gombe Stream Game Preserve in Tanzania. At first she had trouble just finding chimpanzees. When she did find them, she established a routine for observing them at a distance, using binoculars. After nearly a year, the wild chimpanzees became accustomed to Goodall's presence. She was then able to record firsthand observations of nearly every part of their behavior. She found that chimpanzees make and use simple tools and that they hunt for and kill other animals.

Jane Goodall was born in London, England, in 1934. As a child, she developed a strong interest in animals. She studied them and recorded observations of their behavior in her notebooks. At the age of 18, she took a job at Oxford University to earn money for a trip to Africa. Once there, she met paleontologist Louis S. B. Leakey and became his assistant at the Coryndon Museum of Natural History in Nairobi, Kenya.

Although she began her studies of wild chimpanzees without formal education, she later earned her doctorate from Cambridge University. Today Goodall continues her research. One of her concerns is the increasing pressure on wild chimpanzee communities from the interference of humans. ◆

Peter Aromando, SCIENCE TEACHER

After placing the last plant on the table, Peter Aromando steps back and makes a mental note. There are slides, forceps, a scalpel, a microscope, and a plant at each lab station. Everything is set up. The next day's laboratory is ready. He then looks at the clock and heads out the door for a conference. Aromando teaches science at Deland Middle School in Deland, Florida.

Before becoming a teacher, Aromando was a medical technologist. He worked at a blood bank and in hospital laboratories. "What has helped me a great deal in my career as a science teacher," says Aromando, "is my practical work experience in the science field. This work experience helps me bring real-life lab skills into the classroom. I try to teach students safety in the lab, accuracy in experiments and in gathering data, and the proper care of equipment."

Aromando is the science teacher on the seventh-grade teaching team. He and the other teachers on the team meet daily to plan activities and field trips and to review their students' progress. Using the team concept, Aromando is able to combine the skills he teaches in his science class. Aromando feels the science classroom is an excellent place to demonstrate the use of all of a student's skills. "I believe the science classroom should be an active classroom with stimulating lab activities. Students should view the science classroom as a dynamic place where they can learn about the environment," says Aromando.

In the science lab, students get to learn how science works. Using tools such as the microscope, balance, and models, Aromando tries to take science from the textbook and bring it to life for students. "Science is everywhere," explains Aromando. He adds, "Students learn best by hands-on experiences. By using their senses, students can learn about the world around them."

Aromando believes that science also helps all students develop analytic and critical-thinking skills. They can learn how to analyze facts, gather data, and solve problems. He feels science is an excellent foundation for many careers. All of the health fields—medicine, nursing, lab technology, physical therapy, nuclear medicine, and so on—require a basic science education. Other science-related career opportunities are environmental studies, research, space exploration—and, of course, teaching. ◆

Discover More
For more information about careers in teaching science, write to the
National Science Teachers Association
Attn: Public Information Office
1742 Connecticut Avenue, N.W.
Washington, DC 20009

Saving Endangered Wildlife

The population of the giant pandas of China is dwindling as human settlements destroy their natural territory. The okapi, a giraffelike animal that lives in the rain forests of Zaire, is losing its home to farming and lumbering. As their habitats disappear, countless species are threatened with extinction.

The Role of Zoos

Over the last few decades, zoos have become increasingly involved in the breeding of endangered species. Unfortunately, saving animals from extinction is more complex than simply supplying

Embryo transfer allowed this eland to give birth to a bongo calf.

This horse was a surrogate mother for the baby zebra.

them with mates. Animals in captivity often do not get along. The stresses that go with living in small areas often keep animals from reproducing. In addition, zoos cannot presently house the number and variety of individual animals needed for long-term survival of a species.

To counter these problems, researchers have created a computer database of animals in captivity. It is called the International Species Inventory System (ISIS). ISIS helps identify likely candidates for breeding programs. A British database, the *National Online Animal History,* offers information that indicates how closely animals may be related to each other. This informa-

tion is needed because breeding close relatives limits the genes available for the next generation.

Technology to the Rescue

The most commonly used breeding method is *artificial insemination.* It involves the collection of sperm from a male and its placement into the reproductive tract of a female. Another technique, *in vitro fertilization,* involves placing an ovum and sperm together in a glass dish. After fertilization, the resulting embryo is transferred to a selected female.

Embryo transfer, another breeding method, consists

of placing the embryo of one female into the reproductive tract of another female, called the *surrogate mother.* Scientists have even done cross-species embryo transfers between species that are closely related. For example, biologists placed the embryo of a bongo antelope into a surrogate eland antelope, which later gave birth to the bongo calf.

Because the reproductive systems of exotic animals differ from one species to another, researchers have successfully adapted artificial breeding methods to only about 20 species. Perhaps advances in reproductive technology will make artificial breeding possible with a greater variety of endangered animals. ◆

THE HUMAN BODY

Mmmmm. Faced with all this food, what would you choose? Every day you make choices about the food you eat. Do you make the best choices?

Scientists keep learning about how the body works and what is necessary to maintain fitness. The type and amount of food you eat can have an impact on the health of your body. More than ever the daily choices you make can shape the future of your body's fitness.

20 *Reproduction and Development* 542

You began as new life and developed into who you are today.

Science

P A R A D E

SUPPORT AND MOVEMENT

Tanya took a deep breath and tried to relax her muscles and focus on her opening move. Years and years of training and preparation had brought her to this point. Her last routine in the state gymnastic finals was about to begin.

The balance beam had always been one of Tanya's most difficult events. The very thought of trying to perform a perfect routine of exercises along a narrow wooden beam sent chills up her spine. But Tanya was confident that the years of practice would pay off. And if she received a score of 9.6 on this routine, she would win the overall championship.

A good performance would require perfect balance and coordination. Tanya mounted the bar cleanly and her muscles became rigid. Her first move was to stand on the toes of her right foot while her left leg and both arms were extended to balance the framework of her body. The twenty-six small bones of her foot supported the entire weight of her body. Tanya's next move was a cartwheel followed immediately by a handstand. Her confidence increased as the bones and muscles of her body worked together to help her deliver an almost flawless performance.

*T*ying your shoes, peeling an orange, writing a letter, throwing a ball, and walking upright are only a few of the many activities that your bones and muscles help you perform. These actions are possible because human beings have advanced systems of bones and muscles that provide support and permit movement. However, if you have ever seen a gymnast perform a bar routine, you realize that some people have trained their bones and muscles to help them do some truly amazing things!

After several more moves, all that remained was the ever-important dismount. Tanya's arms swept forward over her head and down toward the beam. Her back and leg muscles flexed to propel her strong body off the beam. Flipping through the air, she locked her knee joints, and her leg muscles became rigid. As her feet hit the mat, shock waves rippled upward through her bones and muscles. Her leg muscles immediately tight-ened, leaving her feet securely planted on the mat exactly where they had first touched. With her arms stretched out, Tanya stood tall and erect, knowing that she had given the performance of her life.

For Your Journal

- What purpose do bones and muscles serve?
- What are bones and muscles made of?
- How do muscles work? Illustrate your answer.

The Skeletal System

Objectives

Summarize *the functions and structure of bones.*

Distinguish *among the different parts of the skeleton.*

Compare *the types of joints and* ***contrast*** *their functions.*

As a gymnast, Tanya needs to know about bones and how they work because such knowledge is important to her sport. But bones and their functions are important to you, too. Beneath your skin lies a complex system of 206 bones known as your **skeletal system.** These bones are much like the steel beams that make up the framework of a skyscraper or the wood that forms the frame of a house. As your body's framework, bones give you shape and help you move. Take a moment to feel the parts of your skeletal system.

DISCOVER BY *Doing*

Use your thumb and index finger to feel the small bones of your hands and feet. Probe along your spine at the base of your neck and below your waistline. Feel the large bones of your legs and arms. Examine your ribs and shoulders. Check out the bones of your jaw and skull. Feel around your joints. Your skeletal system extends from the top of your head to the tips of your toes and fingers. As you can feel, the bones that make up this system are different shapes and sizes. ✎

Figure 16–1. If you didn't have bones, your shape would be quite different.

What Bones Do

When you think about what your bones do, you probably think about support. That seems to be the most obvious thing that bones do. But they do more than just provide support.

Bones perform five major functions. First, bones do support other parts of the body. In doing so, they also give shape to your body. Without a skeletal system, you might look like the character in this cartoon!

The second function of bones is to help parts of your body move. Look at your hand while you move your fingers back and forth. Feel the bones move as you wiggle your fingers. These movements are accomplished not only by bones. The bones must team up with muscles to perform the movements.

The third job of bones is protection. Sturdy bones protect many of the delicate organs in your body. Remember when you felt your skull and ribs? Your skull is like a helmet surrounding perhaps the most important organ in your body—the brain. Your ribs protect your heart, lungs, kidneys, liver, and other important organs. Why do you think it is called a rib cage?

Other functions performed by the bones are less obvious than the first three. Their fourth function is to provide an important source of minerals for emergency use by your body. For example, calcium is necessary for muscle contraction and blood clotting. When your diet does not contain enough calcium, your body uses its emergency supply stored in the bones.

Figure 16–2. Bone marrow, found in the center of many large bones, performs many vital life processes.

The fifth and final function is performed inside some of the larger bones of your body. In the center of these bones is a soft substance called **marrow** (MAR oh), which makes blood cells. Every minute, marrow produces millions of blood cells. These new cells move out of the bones and into the bloodstream and spread throughout your body. Since blood cells live only about 120 days, and since you need billions of them, the marrow is busy making new blood cells 24 hours a day.

 ASK YOURSELF

What are the five major functions that bones perform?

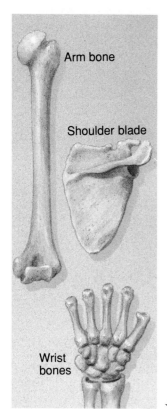

Figure 16-3. Long bones (top), flat bones (middle), and short bones (bottom) are found in different parts of the body. However, the primary functions of all bones are support, movement, and protection.

Types of Bones

If you have ever looked closely at the lumber used to build a house, you may have noticed that it comes in many different shapes and sizes. Long, thick pieces are used to support the floor and roof. Long, thinner pieces form the framework of the walls. The framework is then covered with flat plywood boards.

Like the lumber used to build a house, the bones of your skeletal system are different sizes and shapes. The size and shape of the bones in each group are tied to the bones' functions. For example, the long bones of your arms and legs make up one group of bones. Because of their shape and size, these bones can help support your weight and assist with movement. The smaller long bones of your fingers and toes are also part of this group.

Some bones are flat to help protect organs. The ribs, skull bones, and hip bones are examples of flat bones. Flat bones also contain marrow. Yet another group is the short bones. Most of the short bones are in the wrists and ankles. The shape of these bones helps them to slide easily over one another, giving these areas maximum movement.

 ASK YOURSELF

How are the three types of bones different?

The Composition and Structure of Bones

Bones are amazingly strong. In fact, they are often compared to concrete, granite, or cast iron. Some bones can withstand pressure of over 220 kg per square cm. While bones are as strong as concrete and iron, they are much more flexible and weigh much less. Unlike concrete and iron, bones are living tissue capable of changing and growing.

About 45 percent of bone is made of nonliving minerals. These minerals are mostly calcium and phosphorus, which give bones their strength. Living bone cells and blood make up 30 percent of bone tissue, and the remaining 25 percent is water.

Look at the cutaway drawing of a long bone on the next page. You can see that the bone has three layers. The first layer is a very thin covering of cells that forms a tough, protective membrane around the bone. This membrane produces the new bone cells needed for the growth and repair of bone tissue. If you break a bone, this membrane is responsible for mending the damaged area.

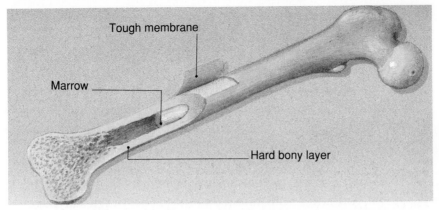

Tough membrane

Marrow

Hard bony layer

Figure 16–4. Many bones consist of three distinct layers. What would happen if your bones could not make any new bone cells?

Beneath the protective membrane is a layer of hard, white, rocklike minerals. This layer is what most people picture when they think of bones. Tiny chambers spread through the mineral layer like caves in a mountain. Each microscopic chamber contains a single living bone cell. This hard layer of minerals and bone cells surrounds a soft core of marrow. Passageways spread through the hard layer of the bone like a maze of interconnecting tunnels. These tiny tunnels connect the bone cell chambers and marrow with the surface of the bone. The tunnels are filled with blood vessels. Blood carries the nutrients and oxygen needed to keep the cells alive.

The center or core of the long bone is filled with marrow. Marrow is lighter than the other material making up bones and, therefore, helps reduce the weight of long bones. Why is it important for the long bones of the arms and legs to weigh less than other bones?

All bones do not have the same three-layer structure. Bone structure varies depending on the type of bone and its function. For example, the short bones and many of the flat bones are mostly spongy bone surrounded by a thin layer of hard bone. The following activity focuses on the composition of a bone.

DISCOVER BY *Observing*

Get a clean leg bone from an uncooked chicken and place it in vinegar. Let it set for five to seven days. Then remove the bone. Feel it and bend it. Write your observations in your journal. How did the bone change? Why did this happen?

▼ ASK YOURSELF

Describe the three layers of a long bone.

Figure 16–5. Human bone cells (top) develop from cells of cartilage (bottom).

Growth and Formation of Bones

The bones of your body did not begin to form until about seven months before your birth. At that time, your skeleton was made of a softer and more flexible substance called **cartilage**. Your nose and ears are made of cartilage. If you hold the tip of your nose or ear and wiggle it, you will know what cartilage feels like.

Slowly, minerals are deposited around the cells, and the cartilage gradually hardens and turns to bone. It is not until you enter your late teens that most of the cartilage is replaced by bone. At the time you were born, your body contained about 300 individual bones. After your birth, some of these bones joined when the cartilage that separated them hardened. So instead of 300 bones, an adult has only 206 bones.

Once your bones are formed, they continue to grow until you are in your early twenties. This growth takes place in an area toward the ends of the growing bones. During your growing years, cartilage is constantly being made in this area. The cartilage then hardens into bone, resulting in longer bones. In time the whole bone hardens, leaving no cartilage at the ends of the bones. This signals the end of growth for bones.

Even after your body stops growing, you continue to produce new bone tissue. The new bone tissue is formed when existing bone cells divide. This new tissue is needed to repair and replace damaged or worn-out cells.

▼ **ASK YOURSELF**

How do your bones change from before you are born to the time you stop growing?

Figure 16–6. As a child grows and develops into an adult, the long bones gradually grow in length. This growth takes place in growth areas made of cartilage.

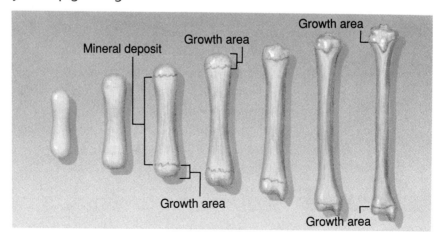

Mineral deposit

Growth area

Growth area

Growth area

Growth area

Growth area

The Skeleton

The word *skeleton* comes from a Greek word that means "dried up." The skeleton does look like a completely dried-up human being! In fact, the skeleton alone can give us a good idea of what an organism looked like. For example, although no one has ever seen a dinosaur, scientists have determined what dinosaurs looked like by putting together the remains of their skeletons.

In some ways your skeleton is like a tree. A tree has two main parts—the trunk and the branches. Your skeleton can also be divided into two parts. One part, the *central skeleton,* is made up of the bones that form the trunk of your body. These include the skull, spine, and rib cage. The other part is like the branches of a tree and is called the *appendages.* The appendages include the bones of your arms and legs and the bones of your hips and shoulders to which your arms and legs are attached.

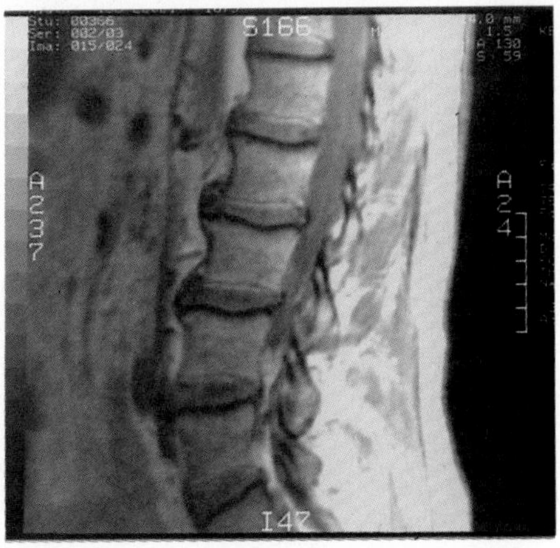

The Central Skeleton   The spine is the core of the central skeleton. The spine consists of 24 small, irregularly shaped bones, called *vertebrae,* which are stacked one on top of the other. The spine is the only support for the upper body. Without your spine, you would not be able to hold yourself upright. The bones of the spine also protect the soft nerve tissue that carries messages between your brain and other parts of your body. The illustration on the following page shows some of the major bones of the human skeleton.

The vertebrae are separated from each other by disks of cartilage. These disks act like cushions and prevent the vertebrae from grinding against each other. Sometimes a disk can rupture or slip out of place. This can be a very painful condition because of the many sensitive nerves that run through your spine. Use the following activity to find out another interesting fact about the spine.

Figure 16–7. Disks of cartilage between the vertebrae can be seen in this photograph. The disks absorb shock when you run or jump. In addition, the whole spinal column is shaped like an *S.* This shape acts like a spring to absorb shock.

DISCOVER BY *Calculating*

Your height changes during the day. You can check to see how much it changes by measuring your height when you first get up in the morning. Later in the day, measure your height again. How much did you shrink? Check with your classmates. Do tall people shrink more than short people? What is the average shrinkage for all the people you checked? ✎

The Skeletal System

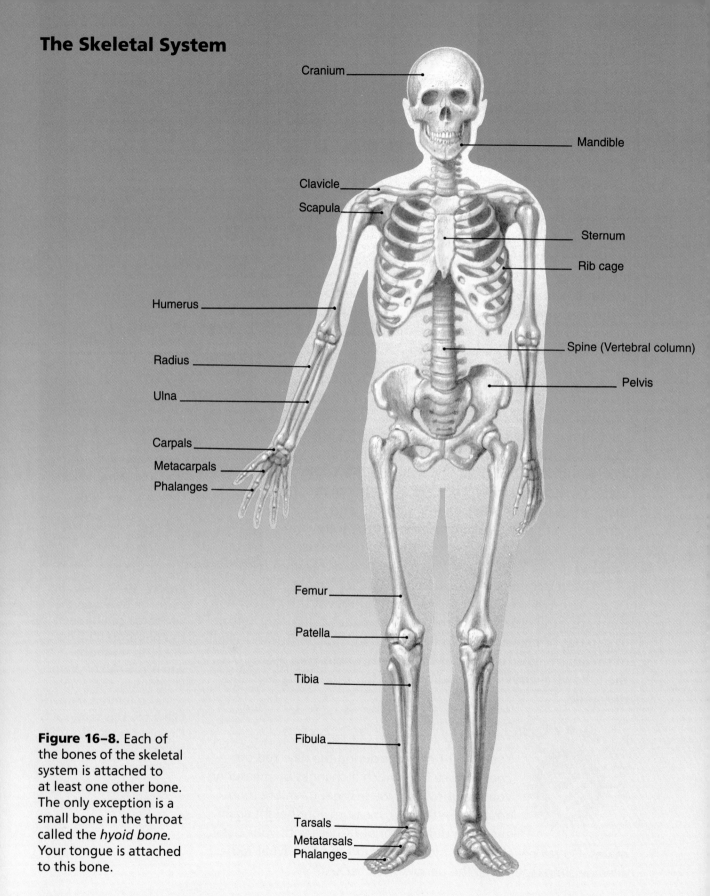

Cranium

Mandible

Clavicle

Scapula

Sternum

Rib cage

Humerus

Spine (Vertebral column)

Radius

Pelvis

Ulna

Carpals

Metacarpals

Phalanges

Femur

Patella

Tibia

Fibula

Tarsals

Metatarsals

Phalanges

Figure 16–8. Each of the bones of the skeletal system is attached to at least one other bone. The only exception is a small bone in the throat called the *hyoid bone.* Your tongue is attached to this bone.

Connected to the spine are 12 pairs of ribs. They combine to form the rib cage. Each pair of ribs is attached to a different vertebra of the spine. The top 10 pairs of ribs are also attached to the breastbone, forming your chest. The lowest two pairs of ribs attach to vertebrae, but their front tips are not connected to any other bone. Why do you think these last two pairs of ribs are often called the floating ribs?

Attached loosely by muscle to the top of the spine is the skull. The skull consists of 22 bones. All but one of these bones are locked in place and cannot move. The one bone that can move is the lower jaw, which is used in chewing food.

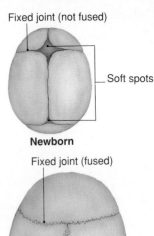

Fixed joint (not fused)

Soft spots

Newborn

Fixed joint (fused)

Adult

Figure 16–9. The soft areas of a baby's skull (top) are gradually filled with bone as the child grows and develops. In the adult skull (bottom), these spaces are completely closed.

ᴰᴵˢᶜᴼᵛᴱᴿ ᴮʸ *Observing* _____

Place your thumb against your upper teeth and perform a chewing motion. Now place your thumb against your lower teeth and do the same chewing motion. What did you observe? ✐

In a newborn, the bones of the upper skull are not joined. This is important because the head is the largest part of a newborn baby, and the head has to be a little flexible to get through the birth canal. The gaps in the skull slowly fill in during a child's first two years of life.

The Appendages   Hanging from the central skeleton are the bones of the shoulders, hips, arms, and legs. These bones form the appendages of the human body. The arms are attached to the central skeleton by two sets of shoulder bones. The arm is divided into three parts—the upper arm, the lower arm, and the hand. These bones are the upper appendages.

The largest bone in the body, the upper leg bone, is attached directly to the hip bone. The hips are fused to the lower spine forming one solid mass of bone. The lower appendages continue below the knee with two long bones that make up the lower leg. The feet and ankles complete the bones of the lower appendages.

▼ ASK YOURSELF

Trace an imaginary line from your skull to your foot. Describe the bones along this line.

Joints

Try to walk without bending your knees. Try to write without bending your fingers, wrist, or arm. Bones are tough, rigid structures that do not bend. Yet in order to move and perform tasks, the parts of our body must bend, twist, and turn. Fortunately, where bones join, we have structures called **joints.** Joints make different kinds of movement possible.

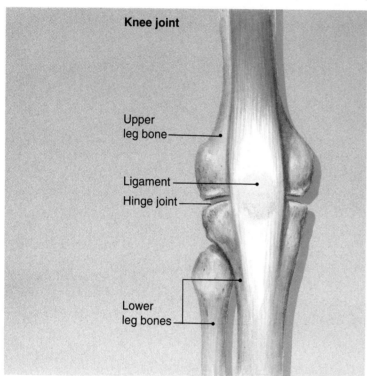

Knee joint

Upper leg bone

Ligament

Hinge joint

Lower leg bones

Figure 16–10. The bones of the upper and lower leg are held together at the knee by a ligament.

When two adjoining bones move against each other, they create friction. To help reduce friction, the ends of bones are covered with a smooth layer of cartilage. The cartilage lets bones slide rather than grind against each other. In addition to cartilage, the moving joints produce a thick fluid that acts like oil to lubricate the joints.

How are bones that meet at a joint held in place? Muscles alone cannot do the job. It takes strands of a very strong, tough tissue, called **ligaments,** to hold the bones in place. The ligaments are attached to each of the bones at a joint.

There are 65 joints that join the bones in your body. Not all of these joints allow movement. One type of joint, called a *fixed joint,* locks the joining bones in place. The bones of your skull, except for your lower jaw, are fixed joints. Most of the remaining joints can be divided into four kinds of movable joints. Each type of joint allows a particular kind of movement. Study the illustrations on the next page to better understand how the joints work.

DISCOVER BY *Observing*

Look around your home, neighborhood, and school for things with moving parts that are joined together. Make a list of at least five things in your journal. Describe how their parts are joined together. Next to each item, write the type of skeletal joint that the connection most resembles. ✎

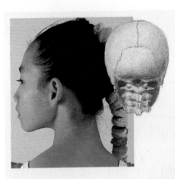

A *pivot joint* allows a rotating or rolling motion. The two top vertebrae form this type of joint, which allows you to roll and rotate your head.

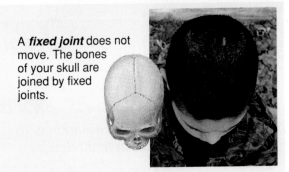

A *fixed joint* does not move. The bones of your skull are joined by fixed joints.

A *gliding joint* allows the bones to slide forward, backward, and sideways. The bones of your ankles and wrists are joined by gliding joints.

A *hinge joint* allows only a back-and-forth movement. It is much like the hinge on a door. Your elbows and knees are hinge joints.

A *ball-and-socket joint* allows movement in nearly all directions. It gets its name because the ball-like end of one bone fits into the socketlike end of another bone. Your hips and shoulders are ball-and-socket joints.

▼ ASK YOURSELF

Describe each type of joint that you find along an imaginary line from your fingertip to the top of your head.

SECTION 1 *REVIEW AND APPLICATION*

Reading Critically
1. What is the function of bone marrow?
2. How does cartilage differ from bone?
3. How is a hinge joint different from a ball-and-socket joint?

Thinking Critically
4. What problems might be experienced by a person whose bones are weakened by disease?
5. Why do you think broken bones heal faster for people under the age of 27 than for people over the age of 27?
6. How are the joints of the skull related to its major function?

SKILL

Organizing Information

1. One way to organize information is to put the list of topics in alphabetical order.

2. Another way to organize information is to make an outline of the topics. For example, you might begin outlining bones in the following manner:

I. Central framework
 A. Skull
  1. mandible

3. Clustering is a grouping technique used to organize information. For example, make a diagram like the one below that groups the bones of the body in circles by location. The distance between circles indicates the closeness of the relationship.

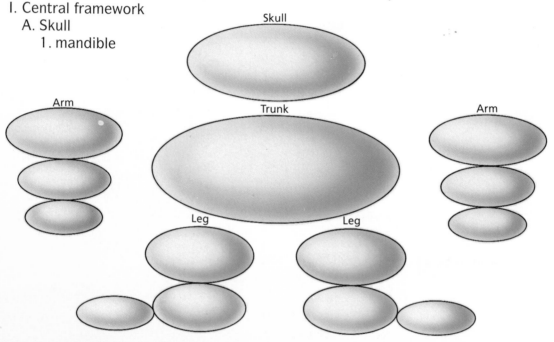

▶ **APPLICATION**

1. Use Figure 16-8 on page 440 and a model human skeleton, if available, to find the location of each of the following bones:

mandible	pelvis	cranium	ulna
phalanges	fibula	sternum	tibia
vertebrae	patella	scapula	rib
humerus	radius	clavicle	femur

2. Use one of the three methods of organizing information described above. Organize the same information using each of the other methods.

3. Select the method of organization that seems to work best for you. Explain why you like this method.

4. Study the information and have a partner test your memory when you feel ready.

✳ *Using What You Have Learned*
Use the diagram of the muscular system on page 446, and organize the muscles as you did the bones.

The Muscular System

As amazing as your skeletal system is, it does not and cannot move itself. For every motion, from a heartbeat to the wink of an eye to a handstand on a balance beam, your body depends on its muscular system. Tanya could never have competed in or won the state gymnastics meet had she not trained her muscles to work smoothly with each other and with her skeletal system to perform the moves of her routines.

Your body has about 650 separate muscles. These muscles, working together to coordinate the movements of your body, form your **muscular system.** Not all the muscles in your muscular system are the same. Different muscles are needed to do the many different jobs required of them. But what is incredible is that muscles themselves, no matter what type they are, can perform only one motion. They can only *contract*, or make themselves shorter. Yet muscles work to help you do all the things you do in a typical day.

Types of Muscles

Muscles move your bones, churn the food in your stomach, cause your lungs to fill with fresh air and push the old air out, and pump blood through your blood vessels. Because muscles do a number of different kinds of work, there are different kinds of muscles. Scientists have identified three basic kinds of muscle—*skeletal, smooth,* and *cardiac.*

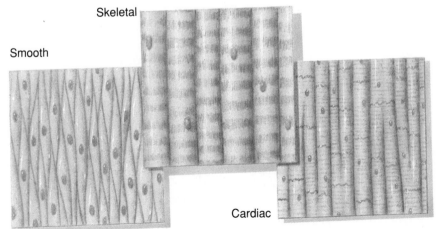

Skeletal

Smooth

Cardiac

Figure 16–11. The three types of muscle in your body are shown here—smooth muscle (left), skeletal muscle (center), and cardiac muscle (right).

The Muscular System

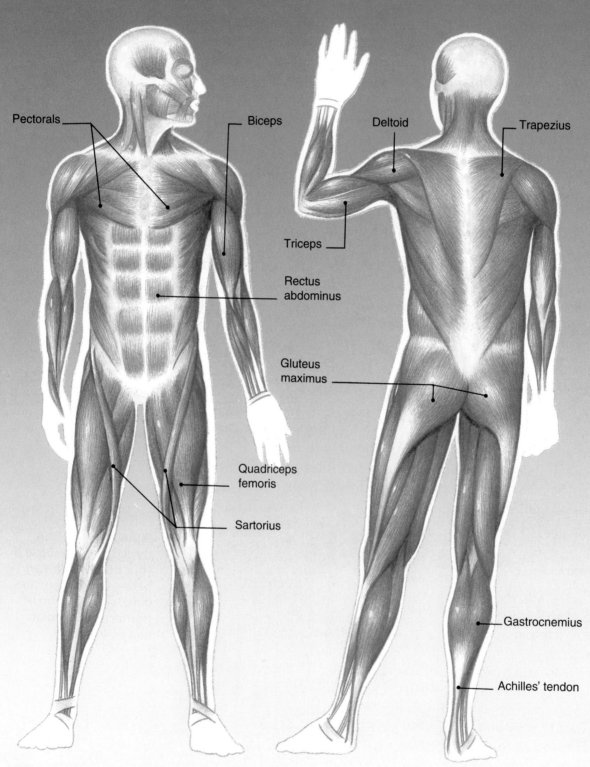

Pectorals

Biceps

Deltoid

Trapezius

Triceps

Rectus abdominus

Gluteus maximus

Quadriceps femoris

Sartorius

Gastrocnemius

Achilles' tendon

Skeletal Muscle For the most part, skeletal muscles are muscles that are attached to bones. But skeletal muscles are not directly attached to bones. Instead, tough bands of connective tissue, called **tendons,** attach the muscles to bones. Tendons are very strong and do not stretch. In one instance, scientists testing the strength of tendons found that a two-cm thick tendon was able to support over 8000 kg! In the next activity you can observe the tendons in your ankle.

Figure 16–13. The skeletal muscles move when you want them to. You can control and coordinate their action.

DISCOVER BY *Doing*

Hold your ankle and move your foot. Do you feel the tendons? Put your fingers on the front of your ankle and your thumb on the back. Point your toes up, then down. You should feel two strong cords of tissue moving. What do those cords do? The tendon behind your ankle connects a muscle in the back of your leg to the bones of your toes. When this muscle shortens, it pulls the toes down. A muscle in the front of your leg pulls on another tendon, raising the toes. ✐

Tendons are also present in your fingers. In fact, there are no muscles at all in your fingers. Instead, skeletal muscles in your forearm pull on the tendons to control the movements of your fingers.

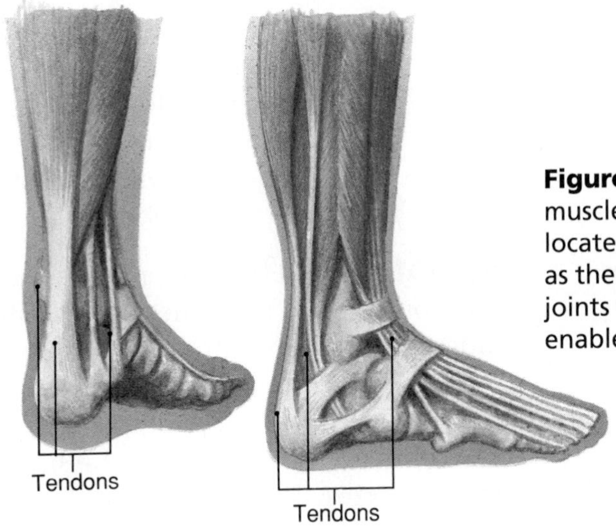

Skeletal muscles are also known as **voluntary muscles** because you have control over their movement. When you want to wiggle your toes, throw a ball, smile, or frown, you can start and stop the action by sending signals from your brain to skeletal muscles.

Tendons

Tendons

Figure 16–14. The muscles and tendons located near joints such as the ankle hold the joints together and enable movement.

Muscles cannot push; they can only pull. Therefore, skeletal muscles must work in pairs. One muscle contracts to pull a bone in one direction, and the other contracts to pull it in the opposite direction. In order for a body part to move, the muscles in a pair must relax and contract at alternate times.

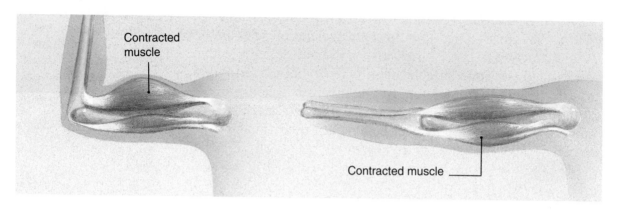

Contracted muscle

Contracted muscle

Figure 16–15. As the biceps muscle contracts to lift the arm, the triceps muscle relaxes. In order to straighten the arm, the biceps must relax and the triceps must contract.

Look at the illustration of the muscles in the upper arm. The muscle on top, the *biceps,* pulls the lower arm up by contracting. At the same time, the muscle on the bottom, the *triceps,* is relaxing. When you want to lower your arm, the triceps contracts while the biceps relaxes. This kind of action controls the movement of all appendages in your body.

Remember, each muscle is attached to bones by tendons. One end of the muscle is attached to a movable bone while the other end is attached to a bone that does not move. It is a bit like a spring attached to a screen door. One end is connected to the movable door, and the other end is connected to the immovable door frame. When the spring contracts, like a muscle, it pulls the door closed.

The cells of skeletal muscle are long and thin. They form long fibers that can be as long as 50 cm. The fibers lie side by side in long, narrow, untwisted strips. If you look at these fibers under a microscope, you will see that they have light and dark bands that make them look striped.

Smooth Muscle   The muscles of internal organs such as the stomach and intestines are smooth muscle. This type of muscle moves food through your digestive tract. Smooth muscles also control the size of blood vessel openings and change the size of the pupils in your eyes.

Unlike skeletal muscles, the smooth muscles are **involuntary muscles.** This means that they work without signals from your brain. Usually you are not aware of the movements of smooth muscles and couldn't stop them even if you wanted to.

Smooth muscles also differ from skeletal muscles in that their cells are smaller and not as long as the cells of skeletal muscles. In addition, there are no light and dark bands across the fibers of smooth muscle. Instead of being bundled together like skeletal muscles, smooth muscle fibers are arranged in compact layers.

Figure 16–16. You have no control over smooth muscle.

Cardiac Muscle The third type of muscle, cardiac muscle, is found only in the heart. Cardiac muscle has characteristics of both of the other types of muscle. Like skeletal muscle, it is striped and arranged in bundles. Like smooth muscle, its cells are shorter than those in skeletal muscle, and they work involuntarily. The fibers are joined in a continuous network that makes cardiac muscle very strong. Why is it important that cardiac muscle be strong?

Figure 16–17. Like smooth muscle, cardiac muscle is an involuntary muscle.

DISCOVER BY *Calculating*

In most people, the heart contracts about 70 times a minute during normal activity. It continues contracting at this rate for a lifetime. How many times does it contract in an hour? A day? A year?

▶ **ASK YOURSELF**

Describe the three types of muscles.

How Muscles Work

Even though your body has different kinds of muscles, they all work in the same way, by contracting and relaxing. In fact, contracting is the only activity a muscle can do. If it is not contracting, it is doing nothing, just relaxing. Yet this one simple activity enables you to do an amazing amount of work.

Muscle fibers can contract many times in a single second. All of this contracting takes a lot of energy. This is why the muscle tissue receives a rich supply of blood. The blood is continually bringing nutrients and oxygen to the cells, while at the same time, the blood removes waste products from the cells.

When muscle fibers contract, they contract completely. After contracting, the fibers must relax. In order to contract again, a fresh supply of nutrients and oxygen must enter, and waste products must leave. These cycles of contraction and relaxation happen very quickly. In fact, they can occur as rapidly as 10 per second.

Figure 16–18. Coordination between muscles and groups of muscles is necessary for any physical activity. Athletics requires coordination, strength, and endurance.

DISCOVER BY *Doing*

Slowly draw your fingers into a loose fist. Now squeeze your fist more tightly and hold the squeeze for 30 seconds. You control the speed and force of contraction. Can you continue squeezing even as the muscles tire? Why?

ACTIVITY

How do the muscles, bones, tendons, and ligaments in your hand work?

MATERIALS
blank white paper, pencil, colored pencils, crumpled paper, tape, coin

PROCEDURE
1. Place your nonwriting hand palm down on a sheet of white paper. Spread your fingers and thumb as far apart as you can and carefully trace around your hand and wrist with a pencil.
2. Examine the bones, muscles, joints, and tendons of your nonwriting hand and wrist. On your drawing, sketch in the bones of the hand and wrist. Try to determine the length, thickness, and shape of the bones.
3. Using a colored pencil, put a capital *L* on your sketch wherever you think a ligament is located.
4. Move your fingers, thumb, and wrist in all directions and squeeze a crumpled piece of paper in your hand repeatedly so you can see the tendons. Use a different colored pencil to draw lines where you think these tendons are located.
5. Tape the thumb of your writing hand securely to the side of the palm and try either to pick up a coin or to button a shirt. Using your thumb-restricted hand, write your initials on your sketch.

APPLICATION
1. How many bones do you think you have in your fingers, thumb, palm, and wrist?
2. What types of joints do you have in your fingers, thumb, and wrist?
3. Where are the muscles that pull the tendons of the hand and wrist when you clench your fist?
4. What purpose does your thumb serve? How would your life be different without your thumbs?

All this contracting of muscle fibers contributes to the heat your body needs to stay warm. The more active muscles are, the more fuel they burn and, therefore, the more heat they produce. This explains why you feel much warmer when you exercise.

Before a muscle can contract, it needs a signal from the nervous system. The skeletal muscles get their signals from the brain. During your waking hours, these muscles are being bombarded with messages at the rate of about 50 per second. The smooth and cardiac muscles get their messages more directly from the spinal cord, and the messages arrive at a slower rate.

Most skeletal muscles remain partially contracted at all times by alternately contracting individual muscle fibers. For this reason, a healthy muscle still feels firm even though it is relaxed. We say that such muscles have good *muscle tone*. Regular exercise develops good muscle tone. Without regular exercise, muscle loses its tone. For instance, when a cast is removed from a leg that was broken, the muscles are smaller. The fibers have decreased in both size and number. Exercise will make the muscles stronger by increasing the size of the fibers.

While a muscle fiber must totally contract, a complete muscle does not. For example, when you pick up your textbook, that movement requires fewer fibers of your biceps to contract than when you pick up a large pail of water. But in both situations, those fibers in the muscle that do contract must fully contract. You can observe muscle contractions by doing the following activity.

DISCOVER BY *Observing*

Bend a 9-cm piece of wire into a U-shape. Balance the wire on a dinner knife. Hold the knife with your arm extended and place the ends of the wire so that they barely touch your desktop. Keep your arm steady, but don't brace it against your body or on your desk. Can you see the tiny contractions of your arm muscles? Now try holding a heavy book with your arm extended until your arm becomes very tired. Then repeat the balanced-wire exercise. What difference did you observe?

ASK YOURSELF

What are the two benefits of muscle contraction?

SECTION 2 *REVIEW AND APPLICATION*

Reading Critically
1. How do muscles work together to produce movement of parts of the body?
2. Why are tendons needed?
3. Describe the difference between voluntary and involuntary muscles.

Thinking Critically
4. Mitochondria are organelles that release energy in body cells. How do you think the number of mitochondria in muscle cells compares to the number in other kinds of cells? Why?
5. What would happen if a muscle constantly contracted and never relaxed?
6. Why is it important that cardiac muscles are not voluntary muscles?

INVESTIGATION

*O*bserving Chicken Wing Muscles and Bones

▶ MATERIALS

● chicken wing, raw ● dissecting pan ● scissors ● forceps ● paper towels

---

▼ **PROCEDURE**

1. **CAUTION: Sharp instruments may cause injury. Use them with care and follow the directions of your teacher.**
2. Rinse the chicken wing and dry it with a paper towel. Place the wing in the dissecting pan.
3. With the forceps, carefully lift the skin at the cut end of the wing and insert the scissors.

4. Cut the skin to the tip of the wing and carefully peel the skin away from the muscle.
5. Observe the muscle structure of the wing. Note where the muscles attach to the various bones.
6. Cut away the muscles.
7. Observe how bones fit together. Move the bones and observe how the joints work.

8. **CAUTION: Wash your hands carefully with soap and water after handling the chicken wing. Clean your work area with a disinfectant.**
9. Make a drawing of the muscles, tendons, and bones that you were able to observe.

▶ **ANALYSES AND CONCLUSIONS**
1. Describe the shapes of the muscles.
2. How do tendons and muscles differ in appearance?
3. How does the structure of a tendon relate to its function?
4. To how many bones is each muscle attached? Why?
5. On your drawing, indicate two muscles that work as a pair. Describe the movement that they would cause.

▶ **APPLICATION**
How would knowing about the way muscles work help a body builder develop his or her muscles?

✳ *Discover More*
Reexamine the joints of the chicken wing. What types of joints are found in the wing? How do they compare to the types of joints in human appendages?

The Big Idea

Our bones and muscles form systems that interact with each other to perform important functions for us; in particular, these systems provide support and movement. But the skeletal and muscular systems do more than that. The bones protect important organs, produce blood cells, and serve as an emergency mineral reserve. The muscles are an important source of heat needed to keep the body warm, and they are part of the processes of digestion, respiration, and circulation.

All muscles can perform only one action: they contract. But this one action allows muscles to perform an amazing amount of work. There are three types of muscles, each with its own particular structure and function. Some permit voluntary movement, while others function involuntarily. Skeletal, smooth, and cardiac muscles make up your muscular system.

For Your Journal

Do you think you have learned a lot about bones and muscles from this chapter? Look back at what you wrote in your journal before you began the chapter. How has your understanding of these important body systems changed since then? How would you revise your ideas on the purpose, structure, and function of bones and muscles?

Connecting Ideas

Copy the unfinished concept map into your journal and fill in the blank boxes.

Two body systems are

- Skeletal
  - helps with
    - ( )
    - ( )
    - protection
    - blood cell production
- ( )
  - helps with
    - ( )
    - circulation
    - ( )

REVIEW

Understanding Vocabulary

Explain the connection between the terms in each pair.

1. skeletal system (434), muscular system (445)
2. cartilage (438), joints (442)
3. marrow (435), bone
4. tendons (447), ligaments (442)
5. involuntary muscles (448), voluntary muscles (447)

Understanding Concepts

MULTIPLE CHOICE

6. Which of the following is *not* a function of the skeleton?
   a) carrying nerve messages
   b) giving support
   c) providing a mineral reserve
   d) producing blood cells

7. Which of the following words most accurately describes what muscles can do?
   a) expand   b) rotate
   c) push   d) contract

8. Which of the following is *not* part of the central skeleton?
   a) skull   b) spinal cord
   c) ribs   d) vertebrae

9. Two functions of muscles are movement and
   a) waste removal.
   b) blood supply.
   c) heat production.
   d) protection of body parts.

10. Cartilage is important to the skeletal system because it
    a) joins muscle to bone.
    b) joins bone to bone.
    c) reduces friction.
    d) strengthens bone tissue.

SHORT ANSWER

11. Why do bones have different shapes and compositions?

12. Describe the differences in form and function among the three types of muscles.

Interpreting Graphics

13. Describe the action of muscles A and B when the bone in the following diagram moves from position 1 to position 2.

14. Describe the action of muscles A and B when the bone in the following diagram moves from position 1 to position 2.

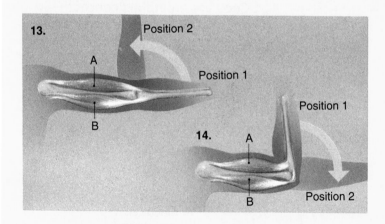

Reviewing Themes

15. *Systems and Structures*
Describe bones and muscles in terms of systems and explain the interaction between the two systems.

16. *Systems and Structures*
Describe the structure of a long bone and the structure of skeletal muscles.

Thinking Critically

17. Unlike human bones, many bird bones are hollow. What advantage do hollow bones have for birds?

18. How would your life be different if your spinal column were made of a single, hollow bone rather than being made of separate vertebrae?

19. Why do muscle cells get a richer supply of blood than many other types of cells?

20. Skeletal muscle cells are long and skinny. Bone cells are surrounded by hard mineral deposits. How does the structure of these cells help each cell perform its job?

21. Determine the effect of temperature on muscle contraction by doing the following experiment: Write your name 20 times on a piece of paper. Soak your hand and forearm in a pan of ice water for 30 seconds and then write your name 20 more times. Compare the two sets of signatures. What is your conclusion?

22. Compare the bone structure of the human leg to that of the human skull. How are the differences related to the functions of these bones?

23. Study the diagrams of the bone structures of the human arm, a bird's wing, a whale's flipper, and a dog's front leg. How are these structures similar and how are they different? How have the bones been adapted for different purposes?

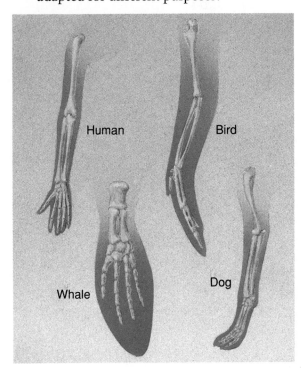

Discovery Through Reading

Parker, Steve. *The Skeleton and Movement.* Watts, 1989. Presents human anatomy and the skeletal system in an up-to-date, easy-to-read format.

CHAPTER 17

DIGESTION AND CIRCULATION

A *fantastic voyage in which you are made smaller and travel through a person's circulatory system is not fact, but fiction. However, in the future, it may become fact. In the meantime, you can share some of the sights and sounds of the digestive and circulatory systems by taking your own voyage through these pages.*

...Has any man ever been miniaturized?"

"I'm afraid," said Michaels, "that we have the honor of being the first."

"How thrilling. Let me ask another question. How far down has any living creature—any living creature at all—been miniaturized?"

"Fifty," said Michaels, briefly.

"What?"

"Fifty. I mean the reduction is such that the linear dimensions are one-fiftieth normal."

"Like reducing me to a height of nearly one and a half inches."

"Yes."

"But we're going far past that point."

"Yes. To nearly a million, I think. Owens can give you the exact figure."

"The exact figure does not matter. The point is it's much more intense a miniaturization than has ever been tried before."

..."Now, Grant, you can understand what we must do: save Benes. Why we must do it: for the information he has. And how we must do it: by miniaturization."

"Why by miniaturization?"

"Because the brain clot cannot be reached from outside. I told you that. So we will miniaturize a submarine, inject it into an artery, and with Captain Owens at the controls and with myself as pilot, journey to the clot. There, Duval and his assistant, Miss Peterson, will operate."

Grant's eyes opened wide. "And I?"

"You will be along as a member of the crew. General supervision, apparently."

Grant said, violently, "Not I. I am not volunteering for any such thing. Not for a minute."

...the loudspeaker sounded almost at once: ATTENTION, PROTEUS. ATTENTION, PROTEUS. THIS IS THE LAST VOICE MESSAGE YOU WILL RECEIVE UNTIL MISSION IS COMPLETED. YOU HAVE SIXTY MINUTES OBJECTIVE TIME. ONCE MINIATURIZATION IS COMPLETE, THE SHIP'S TIME-RECORDER WILL GIVE THE SIXTY READING. YOU ARE AT ALL TIMES TO BE AWARE OF THAT READING WHICH WILL BE REDUCED ONE UNIT AT A TIME, EACH MINUTE. DO NOT—REPEAT, DO NOT—TRUST YOUR SUBJECTIVE FEELINGS AS TO TIME PASSAGE. YOU MUST BE OUT OF BENES' BODY BEFORE THE READING REACHES ZERO. IF YOU ARE NOT, YOU WILL KILL BENES REGARDLESS OF THE SUCCESS OF THE SURGERY. GOOD LUCK!"...

from Fantastic Voyage
by Isaac Asimov

For Your Journal

- How do you think a miniaturized ship would reach the brain if injected into an artery?

- How are the digestive system and the circulatory system related?

- How does eating help supply your body with nutrients?

Digestion and Nutrition

Compare and contrast *mechanical digestion and chemical digestion.*

Diagram *the events that occur during the digestion of a typical meal.*

Evaluate *the nutritional content of foods.*

Has your body ever told you it was hungry by making your stomach rumble and your mouth water? These are ways your body reminds you to replenish your energy supply by eating. People tend to treat a plate full of delicious food in an interesting way. Even though great effort was taken to prepare it and care was taken to arrange it on a plate in an appealing way, what do people do? They respond by chewing, grinding, tearing, shredding, churning, pummeling, and mashing this mouth-watering meal into a thin paste!

Let's imagine taking a field trip through the digestion system. Since a big yellow school bus could not take you on this trip, you must travel inside an inner-space capsule that is about the size of a tiny pea. Climb in and let's follow a meal as it moves through the digestive system. All Aboard!

Figure 17–1. The organs of the digestive system

The Digestive System

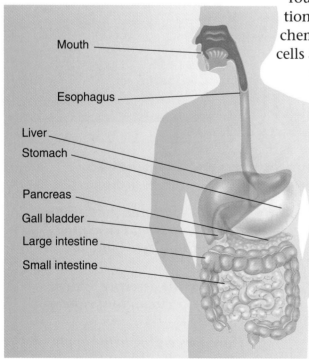

Mouth
Esophagus
Liver
Stomach
Pancreas
Gall bladder
Large intestine
Small intestine

Your trip begins with some background information about the digestive system. A **nutrient** is a chemical substance that your body needs to build cells and to keep those cells alive. Foods contain nutrients, and eating balanced meals is the best way that you can supply your body with the nutrients you need for good health. Nutrients are obtained from the foods you eat by the process of digestion. **Digestion** is the process that changes food into a form that the body can use.

Your digestive system changes large, complex substances, like many different foods on a plate, into nutrients that can pass through the cell membranes of your digestive system and into your bloodstream. The digestive system is a group of organs that work together to digest food for use by the body.

Me and My Big Mouth The digestive process begins with *ingestion,* or eating. The mouth prepares food for travel down a nine-meter tube called the *digestive tract.* The digestive tract winds its way through the human body like a long tunnel. Each part of the tube performs a different function that helps digest food.

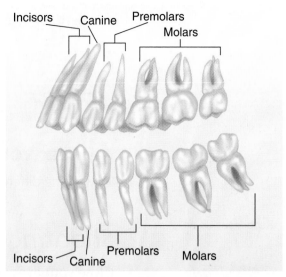

Figure 17–2. Notice the shapes of the teeth. Each shape is related to the tooth's function.

The first step in the digestive process is called mechanical digestion. *Mechanical digestion* is the physical breaking of large pieces of food into smaller pieces in the mouth. Hold on tightly as the tongue and cheeks move your capsule and food around the mouth. As you watch, giant, gleaming white teeth cut, tear, and grind food. Run your tongue over your own teeth. How many different kinds of teeth can you feel? The following activity will help you identify the different types of teeth in your mouth.

DISCOVER BY *Observing*

Using a hand mirror, examine both your upper and lower teeth. How many teeth do you have? Draw a diagram that illustrates the shape and number of your teeth. Identify the purpose of each different type of tooth. Use encyclopedias or other reference books if necessary. ✎

From the activity you learned that front teeth have a narrow, knifelike edge. This makes them well suited for slicing and cutting. On each side of the mouth, next to the front teeth, is a large, pointed canine tooth. In other animals, canine teeth are used for capturing and holding live prey. Why do you think canine teeth are found in a human's jaw? The teeth at the rear of the mouth are broad and flat. They grind and mash food.

Chemical digestion, which breaks large molecules into smaller ones, also begins in the mouth. Glands release *saliva,* a mixture of mucus, water, and enzymes. About 500 mL of saliva flow through the mouth each day. Saliva moistens food, making it easier to swallow. The enzymes in saliva begin breaking down starch into smaller and simpler molecules. The following activity will help you better understand chemical digestion.

Put an unsalted soda cracker in your mouth. Without swallowing, chew the cracker until you notice a change in taste. Describe what happens to the cracker and why the change occurs. ✏

ASK YOURSELF

What is the difference between mechanical and chemical digestion?

Shooting the Food Tube

Close your eyes and concentrate on what happens when you swallow. Can you describe what happens? Hold on tight, because suddenly the tongue presses against the roof of the mouth, forcing food and your capsule down the throat into a narrow tube called the *esophagus* (ih ꜱᴀʜꜰ uh guhs). This action sends you on a 25-cm journey to the stomach.

Your windpipe moves upward as you swallow. This movement closes the entrance to the windpipe, preventing food from entering. Sometimes the opening to the windpipe does not close fast enough. What happens when you try to eat and talk or laugh at the same time?

Your capsule and food do not fall down the esophagus. A series of muscular contractions called *peristalsis* (pehr uh ꜱᴛᴀᴡʟ sihs) pushes material through the tube. The illustration shows how food moves through the esophagus. Contractions squeeze part of the tube shut to force the food farther along. You might feel something like toothpaste being squeezed through a tube.

Fill It Up   Your trip down the esophagus to the stomach takes about five seconds. When you first enter the empty stomach, it looks like a cylinder. As it fills with food, it stretches into a J-shaped sac slightly larger than your closed fist. When completely expanded, the stomach holds about 1.5 L of food.

Bottoms Up   If you thought being squeezed through the esophagus resulted in a rough ride, hold on because the stomach provides an even rougher ride. Three layers of muscle form the outer walls of the stomach. Each layer moves the stomach in a different way. Thus, the stomach acts like a mixer, throwing you first in one direction and then in another. Waves of contractions occur about every 20 seconds. As you tumble over and over, you hear the sounds of food and liquids gurgling

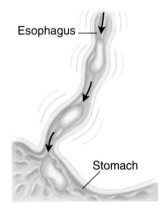

Figure 17–3. Muscular contractions called peristalsis move food down your esophagus.

Esophagus

Stomach

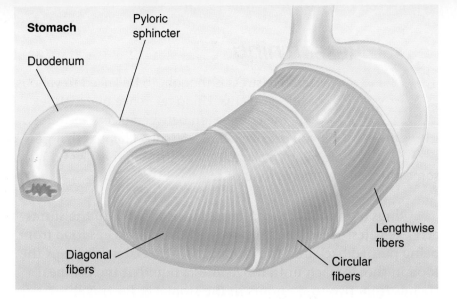

Stomach
Pyloric sphincter
Duodenum
Lengthwise fibers
Diagonal fibers
Circular fibers

Figure 17–4. The stomach has three layers of smooth muscle, each running in a different direction. These muscles squeeze and churn the food to help break it apart.

and sloshing as the stomach churns its contents and your capsule.

During a pause in the mixing, you shine a light on the inner walls of the stomach. *Gastric juice,* a clear, colorless liquid containing enzymes and acid, drips from 35 million tiny pits in the walls. These gastric pits produce about 3 L of gastric juice each day. Thick, slimy mucus glistens on the stomach wall, protecting it from damage by the acidic gastric juice. Fortunately, your capsule protects you from exposure to this harsh environment, although gastric juices and mucus have smeared your capsule.

Even with a protective coating of mucus, the surface layer of stomach cells is attacked by the harsh chemicals inside the stomach. In a healthy person, this layer of cells is replaced every three days. Sometimes, gastric juice or irritating medicines, such as aspirin, eat away at the cells faster than they can be replaced. When this happens, gastric juice burns the stomach wall and causes a painful sore, called an *ulcer.*

Figure 17–5. This electron micrograph shows the pits that line the inner walls of the stomach. Glands within these pits release digestive enzymes and acid.

Excessive stress is thought to be linked to various health problems. Using reference materials, explore the connection between ulcers and stress. If you discover a link, describe changes that people might make to reduce stress in their lives. List some effective treatments for ulcers. ✎

Rings of muscles guard the entrance and exit of the stomach like gates. The entrance gate is a valve that keeps food from being pushed back into the esophagus. The exit valve keeps the food in the stomach until it becomes a liquid. A typical meal spends from two to six hours in the stomach. Eventually, the exit valve opens, and your capsule moves through along with some liquid food.

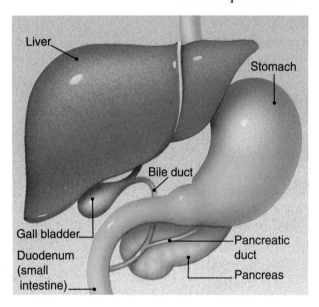

Figure 17–6. Bile is produced by the liver but concentrated and stored in the gall bladder.

An Absorbing Story with a Lot of Twists Although the small intestine is more than 6 m long, most digestion occurs within the first meter. In the small intestine, peristalsis pushes food and your capsule along at a much slower rate than in the esophagus. After your turbulent time in the stomach, the small intestine seems almost still and quiet.

Glands in the walls of the small intestine release additional digestive juices into the liquid food, continuing the breakdown of nutrients into smaller molecules. The liver and pancreas also share a relationship with the small intestine. The liver produces bile, and the pancreas produces enzymes, both of which are used in the small intestine. The following activity will help you understand how bile aids digestion.

Put a spoonful of vegetable oil in a dish. Pour some liquid dish detergent on top of the oil. Stir the mixture and then allow it to stand for 15 to 30 minutes. What happens to the oil? Why do you think this happens? If bile works the same way on fatty substances in the small intestine, what happens to the fatty substances? ✎

Interior of small intestine

Villi

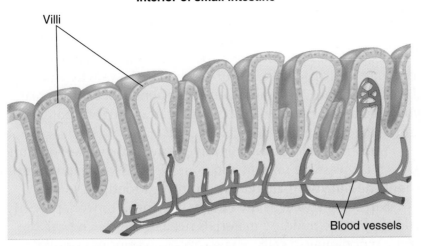

Blood vessels

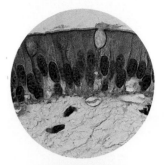

Figure 17–7. Villi increase the surface area of the small intestine and improve the body's ability to absorb nutrients into the bloodstream.

Bile works in much the same way as the detergent worked on the oil in the dish—breaking large clumps of fat into tiny droplets. Enzymes from the pancreas help to neutralize stomach acid so that it does not damage the intestine.

The rippled walls of the small intestine look much different than the walls of the stomach. As you look closely, you see thousands of fingerlike projections sticking out of the walls. These projections, or *villi,* contain many blood vessels and greatly increase the working surface area of the small intestine. As you observe the action, you activate the shrinking mechanism in your capsule. Your capsule becomes smaller, and it and nutrient particles are absorbed by the blood vessels in the villi. **Absorption** is the movement of nutrient molecules into these blood vessels.

Large Intestine: A Wasteful Organ

As your capsule is absorbed by the villi, material that the body cannot digest moves into the large intestine. The large intestine is named for its diameter, not its length—it is only about 1.5 m long. The walls of the large intestine absorb extra water from the undigested material. The entire process of digestion releases about 7 L of fluid into the digestive tract each day. All but about 500 mL of this fluid return to the bloodstream before the waste leaves the body.

Bacteria in the large intestine break down much of the remaining waste material. Then the large intestine forms the waste into a soft, solid mass called *feces.*

 ASK YOURSELF

Describe the digestive process.

Nutrition

Your body needs about 45 different nutrients to remain healthy. Although you may not have known this, you probably know or have a good idea about other aspects of food, such as which foods are healthy, which are not so healthy, and why a balanced diet is important. The more you know about foods, the greater your likelihood of growing to your greatest physical potential and living a long, healthy life.

Your trip through the digestive system has made you hungry—it is mealtime for you and your crew. As you eat, you think about the nutrition you receive from your food.

Nutrients give cells the energy they need to work properly and supply the building materials needed to grow new tissue and repair damaged cells. The six major groups of essential nutrients include carbohydrates, lipids, proteins, vitamins, minerals, and water. Nutritionists call these the essential nutrients because they cannot be made by the body and therefore must be supplied by the foods we eat.

Fuel for Energy   Sugars, starches, and cellulose are examples of *carbohydrates*. Most animals cannot produce carbohydrates. Animals must rely on plants to produce carbohydrates. Sugars are the simplest group of carbohydrates. The cells of your body rapidly break down simple sugars to release energy. The simple sugar fructose gives apples, oranges, and other fruits their sweet taste.

Figure 17–8. Honey is a sugar that is an animal product. Most sugars, however, come from plants such as sugar cane.

Sometimes, two simple sugar molecules join to make a double sugar. More time is needed to break down a double sugar and release its energy than is needed for simple sugars because double sugars are larger than simple sugars. Table sugar is an example of a double sugar.

The Chain Gang Long chains of simple sugars form very large, complex carbohydrate molecules. Pasta, breads, rice, whole-grain cereals, and potatoes contain large amounts of one type of complex carbohydrate—starch. Some athletes, especially marathon runners, often eat these foods the evening before an event.

While gram for gram, all types of sugars provide the same amount of energy, complex carbohydrates take longer to digest than double sugars. Thus starches release their energy in a more balanced and prolonged fashion. The following activity will give you a hands-on opportunity to explore complex carbohydrates.

Figure 17–9. These foods contain starch. The process of digestion changes starches into sugars.

ACTIVITY

How is starch digested?

MATERIALS
soda crackers (2), paper towels, dropping bottle containing amylase solution, dropping bottle containing iodine solution, white potato, cooked egg white, carrot slice

PROCEDURE
1. Lay two soda crackers on a paper towel. Put five drops of amylase solution on one cracker and wait several minutes.
2. Test both crackers for the presence of starch by dropping two drops of iodine on each cracker. When testing the cracker that contains the drops of amylase, be sure to put the iodine on the same spot.
3. If starch is present, the iodine will turn the starch dark purple to black. Record your results.
4. Test all other foods in the same way you tested the crackers.

APPLICATION
1. Does amylase digest starch? Explain.
2. Which foods contained starch?
3. Where do you think amylase is found in your body?

The starches you discovered were large, complex carbohydrate molecules. Another large carbohydrate molecule is cellulose, or fiber, which is produced by plants. Humans cannot digest cellulose, yet cellulose is an important part of a healthy diet because it adds bulk to food. Bulk helps your body move waste through the digestive system. Grains and raw vegetables give your body the cellulose it needs.

Energy Warehouses   Fats and oils are examples of *lipids*. Lipids store large amounts of energy. A molecule of lipid stores more than twice as much energy as a molecule of most sugars.

You may already be familiar with the lipid products shown in the photograph. Lard, butter, margarine, red meats, and corn oil are examples of foods that contain lipids. Fats and oils carry important vitamins, so they should be part of your diet but only about 30 percent of your daily intake. The following activity will help you identify a food that contains lipids.

Figure 17–10. A healthy diet includes small amounts of lipids, such as these.

ᴅɪꜱᴄᴏᴠᴇʀ ʙʏ *Doing*

Cut three squares of brown paper from a bag. Rub shortening on one square, egg white on the second square, and a slice of raw potato on the third square. Let the pieces of paper dry. Hold each square up to a bright light. Fat allows light to pass through the paper. Which of the foods that you tested contained lipids? Which did not? Test some additional foods for lipids. ✎

Figure 17–11. These foods are high in protein.

Chains of Amino Acids *Proteins* are very large molecules made of chains of many smaller molecules called *amino acids.* Common sources of proteins include eggs, meat, fish, cheese, milk, and peanut butter. Your hair, your fingernails, and your toenails are made of proteins. Muscles and skin also contain large amounts of protein.

Your body arranges various combinations of 20 amino acids to create thousands of different proteins. How can only 20 kinds of amino acids build so many proteins? They do so in much the same way as the 26 letters of the alphabet are used to make up tens of thousands of words. The words are made up by arranging different letters in different quantities and orders. Proteins are made up by different arrangements of the various amino acids. By changing the number, kind, and order of amino acids in the chain, the body makes thousands of different proteins.

Figure 17–12. This computer-generated image shows the complexity of a protein molecule.

DISCOVER BY *Calculating*

Twenty different amino acids are used by plants and animals to create proteins. Although not all combinations of amino acids create proteins, how many different combinations of 20 amino acids would be possible? ✐

Protein makes up about three-quarters of the solid material in your body. Growth, repair, and replacement of cells require a continuous supply of new protein molecules.

Some protein molecules are *enzymes*. Enzymes are molecules that speed up certain chemical reactions. Without enzymes, most chemical reactions would not occur fast enough to keep the cell alive.

A, B, C, and More   *Vitamins* are molecules that work with certain enzymes to speed up chemical reactions. Without vitamins, many enzymes cannot work properly. Vitamins are necessary for proper growth and repair of your body. Because your body cannot make the vitamins it needs for good health, you must get the vitamins from the food you eat. Vegetables, fruits, and dairy products are good sources of vitamins.

Vitamins are divided into two groups. The vitamins in one group dissolve in water and are called water soluble. Your body does not store these vitamins, so they should be in your diet each day. Any extra water-soluble vitamins eaten in your food leave the body in your urine. Some of the vitamins in this group are B_1, B_{12}, and C.

The vitamins in the second group dissolve in fat but not in water. Fat-soluble vitamins that are not immediately used can be stored by the body. When your diet does not include needed fat-soluble vitamins, your body uses those you have stored in your fatty tissues. Because your body stores extra fat-soluble vitamins, they can build up in your tissues. Excess fat-soluble vitamins can collect in your liver and harm it. Vitamins A, D, E, and K are fat soluble. Through the following activity you will gain more information about vitamins.

Figure 17–13. Fruits and vegetables are good sources of vitamins.

DISCOVER BY Researching

Choose a vitamin—A, B_1, B_2, B_6, B_{12}, pantothenic acid, biotin, C, D, E, folic acid, K, or niacin—to research. Gather information about sources of the vitamin, what the vitamin does for the body, and problems that may result from a deficiency of that vitamin. Share your findings with your classmates, and together compile a chart that includes information for all the vitamins. Illustrate your chart with photographs or drawings. ✐

Figure 17–14. Water can be obtained from milk, fruit juices, and other liquids.

Cool, Clear Water

Water makes up about two-thirds of your body. Water is not a source of energy, but it is a necessary component of all tissue. Water is also important because it dissolves most other nutrients. Solutions of water and other nutrients move throughout your body and surround every cell.

Each day, you lose about 2 L of water through your skin and lungs and during waste removal. Unless you replace this water, you will become dehydrated. You could probably live without food for several weeks surviving on the nutrients stored in your body. However, you could live without water for only a few days. Even a 10 percent loss of water from your body would cause dehydration. A 20 percent loss could cause death.

Mineral Lode

Minerals form parts of your body and help perform some body processes. You need most minerals in only small amounts. Calcium, phosphorus, and sodium are typical minerals your body needs to help blood clot, to allow nerves to function properly, and to build strong bones and teeth. Green vegetables and milk are good sources of minerals.

Figure 17–15. Food pyramid

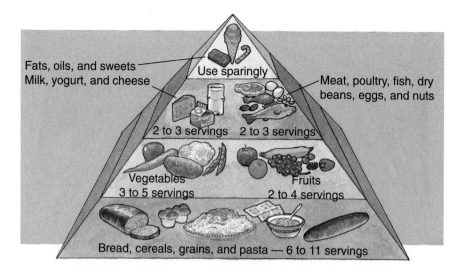

Fats, oils, and sweets
Milk, yogurt, and cheese

Use sparingly

Meat, poultry, fish, dry beans, eggs, and nuts

2 to 3 servings   2 to 3 servings

Vegetables
3 to 5 servings

Fruits
2 to 4 servings

Bread, cereals, grains, and pasta — 6 to 11 servings

What's for Dinner?   Your daily diet should include foods that provide all six types of nutrients. In the early 1990s, the Department of Agriculture of the United States revised its guidelines for nutrition. The food pyramid shows the recommended number of daily servings for each type of food. Use the pyramid and the information about nutrients to help you complete the following activity.

ᴅɪꜱᴄᴏᴠᴇʀ ʙʏ *Problem Solving*

Using information from the food pyramid, plan a balanced diet for one week. Be sure you include foods that provide all the nutrients your body needs to be healthy. ✎

 **ASK YOURSELF**

Give three reasons why a well-balanced diet is important.

SECTION 1   *REVIEW AND APPLICATION*

Reading Critically
1. Trace the path of food through the digestive tract.
2. Compare and contrast mechanical digestion and chemical digestion.
3. State the functions of the six basic nutrients needed for good health.

Thinking Critically
4. Would it be easier to survive without your stomach or without your small intestine? Explain.
5. Can a diet be balanced if it does not contain meat? Explain.
6. Why do you think the Department of Agriculture recommends guidelines for nutrition?

SKILL Classifying Nutritional Information

▶ MATERIALS
- food products (2 or 3) ● paper ● pencil

▼ PROCEDURE

1. Choose two or three food products to study. Make a table like the one shown.

TABLE 1: NUTRITION INFORMATION			
Food Product	Vitamin	Percentage of RDA	Mineral

2. List the name of each product on separate lines down the side of your paper. List the vitamins contained in each food product.

3. Use the "Nutrition Information" on the label of each product to study the product's vitamin content. On your table, record the percentage of the recommended daily allowance (RDA) for each vitamin listed on the label.

4. Study the food products for mineral content in the same way. Add mineral information to your table.

• RICH IN FIBER AND IRON

NUTRITION INFORMATION
SERVING SIZE:1.4 OZ. (1 OZ. BRAN FLAKES WITH 0.4 OZ. RAISINS; 39.4g, ABOUT ¾ CUP)
SERVINGS PER PACKAGE: 14

	CEREAL & RAISINS	WITH ½ CUP VITAMINS A & D SKIM MILK
CALORIES	120	160*
PROTEIN, g	3	7
CARBOHYDRATE, g	31	37
FAT, TOTAL, g	1	1*
UNSATURATED, g	1	
SATURATED, g	0	
CHOLESTEROL, mg	0	0*
SODIUM, mg	210	270
POTASSIUM, mg	240	440

PERCENTAGE OF U.S. RECOMMENDED DAILY ALLOWANCES (U.S. RDA)

PROTEIN	4	15
VITAMIN A	15	20
VITAMIN C	**	2
THIAMIN	25	30
RIBOFLAVIN	25	35
NIACIN	25	25
CALCIUM	**	15
IRON	100	100
VITAMIN D	10	25
VITAMIN B_6	25	25
FOLIC ACID	25	25
VITAMIN B_{12}	25	35
PHOSPHORUS	15	25
MAGNESIUM	15	20
ZINC	25	30
COPPER	10	10

*2% MILK SUPPLIES AN ADDITIONAL 20 CALORIES, 2 g FAT, AND 10 mg CHOLESTEROL.
**CONTAINS LESS THAN 2% OF THE U.S. RDA OF THIS NUTRIENT.

INGREDIENTS: WHEAT BRAN WITH OTHER PARTS OF WHEAT, RAISINS, SUGAR, CORN SYRUP, SALT, MALT FLAVORING.

VITAMINS AND MINERALS: IRON, NIACINAMIDE, ZINC (OXIDE), VITAMIN B_6 (PYRIDOXINE HYDROCHLORIDE), VITAMIN B_2 (RIBOFLAVIN), VITAMIN A (PALMITATE), VITAMIN B_1 (THIAMIN HYDROCHLORIDE), FOLIC ACID, VITAMIN B_{12}, AND VITAMIN D.

▶ APPLICATION

1. Which of the food products contains many different kinds of vitamins?

2. For each vitamin, which food product supplies the largest percentage of the RDA?

3. Which of the food products contains many different kinds of minerals?

4. For each mineral, which food product supplies the largest percentage of the RDA?

✳ Using What You Have Learned

Choose five additional food products to study. Read the labels carefully, and record their vitamin and mineral content as outlined in the Procedure. Expand your table to include substances such as preservatives and other additives that are not considered nutrients.

Blood

Objectives

Summarize *the functions of blood.*

Compare and contrast *red blood cells, white blood cells, and platelets.*

Describe *ways in which the human body develops immunity to disease.*

As your meal concludes, you remember your occasionally bumpy journey through the digestive system, and you wonder if your upcoming trip through the blood and circulatory system will be similar or different. You are sure of one thing—if you decided to travel all of the paths through the circulatory system, you would travel a journey of more than 96 000 km! So instead of traveling the entire system, you decide to take a trip that will show you the highlights of the circulatory system.

As your journey begins, you remember shrinking your capsule before you were absorbed by blood vessels in the villi. To be able to travel through many of the tubes that carry blood, the capsule must be much smaller than it was in the digestive system.

You will be moving in a red, living river. A muscular organ that weighs about as much as a large orange keeps this river moving. This organ propels the river through the tubes about 2.5 billion times in one lifetime. Hour after hour, day after day, year after year, the river keeps flowing.

The Functions of Blood

Blood connects all the other tissues of the body. It is the river of life that helps keep all other cells alive. As you watch from the window of your capsule, you see that blood transports nutrients and oxygen throughout the body. Blood delivers this important cargo to the cells. After the delivery, blood picks up a load of waste materials. It carries these waste materials to the lungs and kidneys for removal from the body.

Blood also helps control your body temperature. When you are hot, vessels open, or dilate, allowing more blood to flow through the skin and heat to escape. When you feel cold, the reverse happens. Vessels close, or constrict, limiting the flow of blood to the skin. Your skin loses less heat and your body stays warmer.

Figure 17–16. Blood samples are often tested in laboratories.

Blood also helps you to fight disease. You see specialized cells and chemicals patrolling the bloodstream, attacking disease- or allergy-causing substances 24 hours a day.

ASK YOURSELF

Blood has many functions. Which is most important? Why?

Components of Blood

Although blood flows like a river, it is part liquid and part solid. Water and other molecules make up the **plasma,** which is the liquid part of blood. You feel surprised when you notice that plasma is not red but a clear, yellowish liquid. Water makes up about 90 percent of plasma. Dissolved substances, such as proteins, form the remaining 10 percent and give plasma its color. Plasma carries dissolved nutrient molecules, such as amino acids and sugars, to all the cells of the body. It also carries carbon dioxide and other wastes away from the cells.

Much of the plasma can pass through the walls of the blood vessels. Outside of the blood vessels, this liquid is called *lymph.* You see this colorless lymph when a blister breaks open. Lymph surrounds every cell bringing the dissolved nutrient molecules closer to the cell membrane for easier transfer.

Blood Is Thicker Than Water Soon you realize that this river of life is much "thicker" than water. Plasma makes up only about 55 percent of the blood. Your capsule bounces from side to side as you travel through the blood.

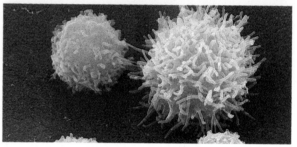

Figure 17–17. Red blood cells (left), platelets (top right), and white blood cells (bottom right) make up the solid part of blood.

In the Red Red blood cells make up about 44 percent of your blood. These cells are disk shaped, with the center of each side pushed in. They look like doughnuts with the holes not fully formed. Red blood cells act like barges loaded down with oxygen. Each cell carries millions of oxygen molecules from the lungs to other parts of the body. The unusual shape of the cell allows oxygen to move into and out of it quickly.

As you look more closely at red blood cells, you discover that they lack a nucleus. Without a nucleus, red blood cells cannot reproduce. A normal red blood cell lives only about four months. Don't worry! You won't run out of red blood cells because your bone marrow and your spleen produce new red blood cells at a rate of 2.5 million per second. A tremendous number of these cells are produced in your lifetime.

DISCOVER BY *Calculating*

Given that your body produces new red blood cells at a rate of 2.5 million per second, determine how many new red blood cells will be produced by your body today.

Red blood cells get their color from an oxygen-carrying chemical called *hemoglobin* (HEE muh gloh buhn), which contains iron. Hemoglobin molecules take the oxygen you obtain by inhaling and transport it throughout the body. Each hemoglobin molecule can carry more than one oxygen molecule. When an iron-containing hemoglobin molecule attaches itself to one or more oxygen molecules, the blood becomes bright red in color. The same thing happens when you receive a small cut—the blood you lose turns bright red when it contacts the oxygen in the atmosphere. Which cells in your body have the greatest need for oxygen at this moment?

People with sickle-cell disease have abnormally shaped red blood cells. As a result, these cells move through the blood vessels with much more difficulty. Often they stick to one another or to the walls of the vessels. A person with sickle-cell disease often has pain because much of the needed oxygen does not reach all the cells throughout the body quickly enough.

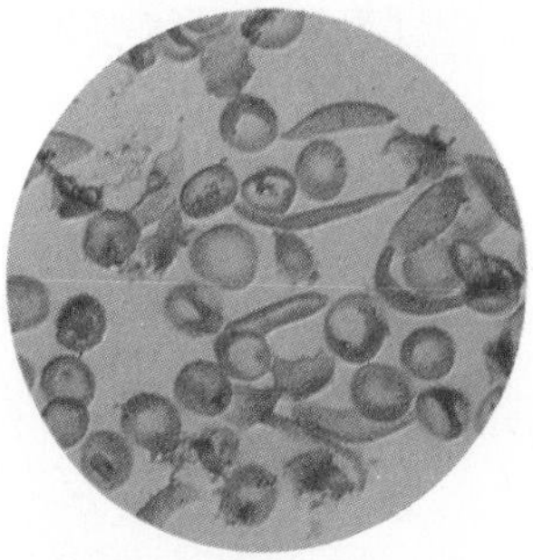

Figure 17–18. Normal red blood cells carry oxygen. Sickled red blood cells often block blood vessels.

Time to Close   Soon the red river carries you and your capsule to a small cut in the skin. You watch as tiny packets of cytoplasm from the blood burst open and begin to clot and seal the damaged area. These small, colorless packets of cytoplasm, called *platelets,* are another solid part of the blood. Platelets break apart in a wound and cause the blood to clot. The clot plugs the injured blood vessel, and the bleeding stops.

To the Rescue   Although only a small amount of blood escaped through the cut, you notice that many bacteria entered the body. While there are millions of red blood cells in a drop of blood, they seem helpless when it comes to defending the body. Then a white giant—a white blood cell—appears. Although larger than red blood cells, white blood cells are much fewer in number. Only one white blood cell exists for every 700 red blood cells. Unlike the red cells, the white cells have a nucleus. As you watch, this giant white blood cell surrounds a bacterium. Slowly, the white cell kills and consumes the invader. Then it goes looking for more.

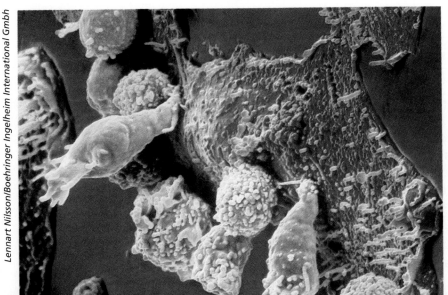

Lennart Nilsson/Boehringer Ingelheim International Gmbh

Figure 17–19. White blood cells also attack abnormal cells. This cancer cell is being attacked by white blood cells.

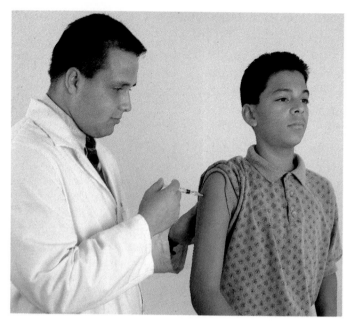

Some white blood cells make proteins, known as *antibodies,* that attack and kill invading organisms. These antibodies make you immune to a disease. Have you ever had chicken pox? If you have, your body has built up antibodies that now make you immune to the disease. You can never again contract chicken pox. Vaccinations are now available for many diseases. The vaccinations help your body develop antibodies that prevent you from getting the diseases. Some vaccinations develop antibodies that last a lifetime; others develop antibodies that eventually disintegrate. You have to be vaccinated more than once to be protected from diseases for which only temporary vaccinations are available.

Figure 17–20. Vaccination is an important method of providing individuals with immunity to many diseases.

Observe a prepared slide of human blood, using the microscope's low and high power. Try to find red blood cells, white blood cells, and platelets. Describe each type of cell. In your journal, draw and label what you see. ✏

 ASK YOURSELF

Describe the composition of blood.

SECTION 2 *REVIEW AND APPLICATION*

Reading Critically

1. What are four functions of blood?
2. How do red blood cells differ from white blood cells?
3. How are platelets different from red and white blood cells?

Thinking Critically

4. Is the function of red blood cells more like the function of white blood cells or the function of plasma? Explain your answer.
5. If you have received a vaccination for mumps, will you get the mumps? Why or why not?

INVESTIGATION

*O*bserving Hemoglobin

▶ MATERIALS
- beakers, 250 mL (2) ● water (dechlorinated) ● *Daphnia*
- medicine dropper ● yeast suspension ● plastic food wrap

▼ PROCEDURE

1. Obtain two beakers from your teacher. Label one beaker *A* and the other beaker *B*.
2. Fill each beaker with 5 cm of dechlorinated water or with water that has been standing for at least 24 hours.
3. Add a culture of *Daphnia* to each container. Arrange the containers so that they are not in direct light and the temperature of the water remains fairly constant.
4. Using a medicine dropper, feed the *Daphnia* a few drops of a yeast suspension. Be careful not to add so much yeast that the water becomes cloudy and foul smelling. If you add too much, replace most of the clouded water with fresh, dechlorinated water. Feed the *Daphnia* again in two or three days.
5. Cover beaker *A* with plastic food wrap to reduce the amount of oxygen entering the container. If you need to add food or water to beaker *A*, do so by lifting only a small area of the covering. Replace the covering quickly. Leave beaker *B* open to the air.
6. Care for your *Daphnia* and observe them daily. Record your observations daily. After five or six days, look for a color change. As you look at the containers, place a white sheet of paper behind them so you can see a color change more easily.

▶ ANALYSES AND CONCLUSIONS
1. Was there a color change in either one of the containers? If so, explain why.
2. Use what you know about hemoglobin's reaction to oxygen to help you explain the color change in the *Daphnia*.

▶ APPLICATION
If there is a lack of oxygen in the environment, would you expect an organism to increase or decrease its hemoglobin content? Why?

✳ *Discover More*
Use a hand lens to observe the *Daphnia* in both containers. Watch the heartbeat of the *Daphnia*. Count how many times the *Daphnia's* heart pulses in one minute. Are the heartbeats of the *Daphnia* in both beakers the same? If not, make a hypothesis to explain the difference. How would you test your hypothesis?

Circulation

Describe the structure of the heart and **trace** the flow of blood through the heart and lungs.

Explain what causes a heartbeat and pulse.

Compare and contrast the three types of blood vessels found in the circulatory system.

Your movement through the circulatory system is not smooth and constant—your capsule seems to move in spurts. It almost feels like a giant wave picks you up and hurls you forward. Soon you hear sounds like Lub-Dup! Lub-Dup! The sounds grow louder and louder until your capsule races out of a blood vessel into the heart.

DISCOVER BY Calculating

Use a watch or a clock that reads in seconds to determine how many times per minute your heart is beating. Determine how many times your heart will beat in a day. There are 1440 minutes in a day. Find out how many days there are until your next birthday. Then estimate how many times your heart will have beaten by your next birthday. Calculate how many times your heart has beaten in your life so far. Record your results in your journal. ✎

A Story with Heart

Your heart is a muscle located between your lungs, almost in the center of your chest. Protected by your breastbone, your heart produces muscular contractions that push blood throughout your body. During each minute, almost all of the blood in your body moves through your heart. During an average lifetime, a heart will beat about 2.5 billion times.

The Heart of the Matter Figure 17–21 shows how the right side of your heart receives blood from the body and pumps it to your lungs. In the lungs, blood picks up oxygen and releases its load of waste carbon dioxide. The left side of your heart receives blood from the lungs and pumps it to the rest of the body. This blood is rich with oxygen and very red. A thick, muscular wall separates these two sides of the heart.

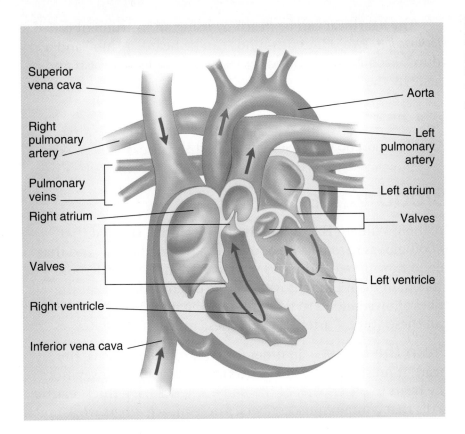

Superior vena cava

Right pulmonary artery

Pulmonary veins

Right atrium

Valves

Right ventricle

Inferior vena cava

Aorta

Left pulmonary artery

Left atrium

Valves

Left ventricle

Figure 17–21. The heart is a two-sided pump. One side of the heart pumps blood to the lungs. The other side of the heart pumps blood to the rest of the body.

Each side of your heart contains two chambers. An *atrium* (AY tree uhm) forms the upper chamber of each side. Each lower chamber is a *ventricle* (VEHN trih kuhl). Blood flows through these chambers in only one direction. Valves between the chambers act like one-way swinging doors. The valves open to allow blood to flow in one direction, then close to prevent blood from flowing backward.

The Beat Goes On   Your heart beats many times in a day without you thinking about it. While you rest, your heart usually beats about 60 to 70 times each minute. When you exercise, your heart can beat as many as 200 times each minute. The number of times your heart beats each minute is called your *heartbeat rate*.

DISCOVER BY Researching

Have you ever wondered if your heartbeat rate changes when you sleep? Use reference materials to find out the heartbeat rate for an average person during sleep. Do you think your heartbeat rate increases or decreases when you dream? 🖉

The heartbeat can be heard by using an instrument called a *stethoscope*. Two different sounds can be heard, a "lub" and a "dup." Each lub-dup sound represents a single heartbeat and is made by the closing of valves within the heart.

 **ASK YOURSELF**

What function do the valves in the heart serve?

Blood Vessels

In 1868, William Harvey, an English scientist, showed that blood moves through the body in a series of "tubes," and that blood flows in a continuous, closed system. Harvey demonstrated, in front of the Royal College of Physicians, that the tubes allowed blood to flow in only one direction. Today, we call these "tubes" arteries, capillaries, and veins.

From the Heart   In order for your capsule to leave the heart, it must enter an artery. **Arteries** carry blood away from the heart to other parts of the body. Arteries have thick, elastic walls that contain muscle fibers. These muscle fibers work much like rubber bands. As blood spurts from the heart, an artery stretches in diameter to make room for it. When the artery returns to its normal size, the blood is pushed or squeezed farther along the vessel.

The farther you travel from the heart, the narrower you notice the artery becoming. Now the speed your capsule is traveling decreases. The red river of blood is flowing more slowly. The artery is beginning to divide into smaller and smaller vessels, or **capillaries.** In fact, capillaries are so narrow that your capsule and red blood cells must move through them in single file. As you observe capillary walls, you seem to be able to see right

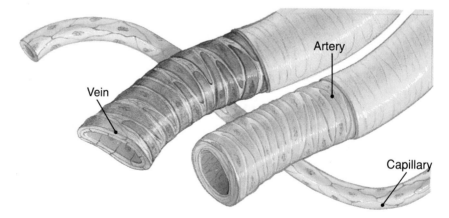

Figure 17–22. How do these three types of blood vessels differ from each other?

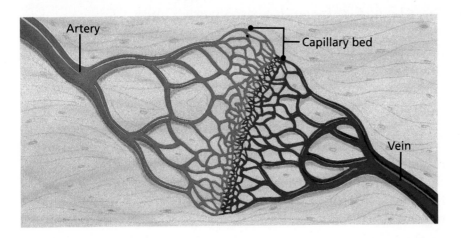

Figure 17–23. Capillaries connect arteries to veins.

through them, because the tiny capillaries have walls only one cell thick. The movement of nutrients, oxygen, and waste molecules between the blood and the tissues takes place through these thin capillary walls. You observe nutrients and oxygen moving from the blood into the fluid spaces and on into body cells. At the same time, waste products, including carbon dioxide, move from the body cells into the blood.

A Journey Made in "Vein"

Your capsule flows through the capillary into a small vein. Small veins join to form larger veins. **Veins** carry the blood back to the heart. Veins have thinner walls than arteries. Unlike arteries, veins contain one-way valves. These valves prevent the blood from falling backward as it is pushed toward the heart.

Skeletal muscles surround many veins. As a person moves, these muscles squeeze the veins and push blood through the blood vessel. This squeezing helps the blood return to the heart. People who must stand in one place for a long time learn to flex the muscles of their legs. This helps to keep the blood moving. Otherwise, the blood may pool in the legs, and the flow of blood to the brain may decrease. Eventually, the person might faint from inadequate oxygen in the brain. What kinds of jobs require people to stand in one place for a long period of time? Astronauts in the space shuttle do not have this problem. Because of the lack of gravity, their circulatory systems do not have the tendency to pool blood in the lowest portions of their bodies. Blood can, however, pool in the chest, the abdomen, and the head of an astronaut in space.

 **ASK YOURSELF**

How are the types of blood vessels different?

Path of Blood

Blood travels through the body in a continuous path. This continuous path is divided into two parts—the pulmonary system and the systemic system. Blood in the pulmonary system moves from the heart to the lungs through pulmonary arteries. In the lungs, the blood obtains oxygen and releases waste carbon dioxide. Then the blood moves to the heart again through pulmonary veins. The oxygen-rich blood then is pumped throughout the body through the systemic system. The systemic system supplies oxygen to all parts of your body, including the heart.

Clogged Pipes   Proper movement of blood through the blood vessels of the heart is necessary to keep the heart tissue healthy. If arteries or veins in the heart become blocked, the supply of nutrients and oxygen that the heart receives will be reduced. Such blockage can damage the heart muscle and is called a "heart attack."

Figure 17–24.
Pulmonary and systemic circulation are shown in this illustration.

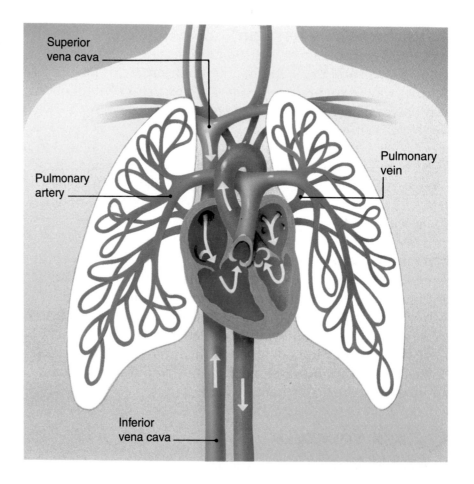

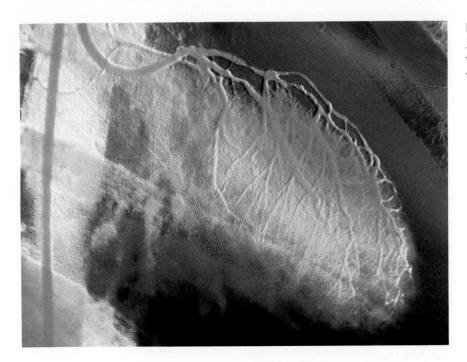

Figure 17–25. This X-ray picture shows the blood vessels of the heart.

As shown in Figure 17–26, arteries can become clogged with deposits of cholesterol and other substances. *Arteriosclerosis* is a condition characterized by abnormal thickening and clogging of artery walls. As you grow older, your arteries will gradually begin to thicken naturally—it's a part of the aging process. However, you can increase the rate at which your arteries become clogged through a diet heavy in fats and cholesterol. Cholesterol is a fatty substance that can stick to the walls of arteries. When this happens, the flow of blood through that artery is restricted to a smaller area. This restriction causes blood pressure to increase and can lead to blockages that cause heart attacks. These blockages result from deposits breaking off from the artery wall and blocking a smaller vessel elsewhere in the body.

Figure 17–26. Arteries can become clogged with deposits of cholesterol.

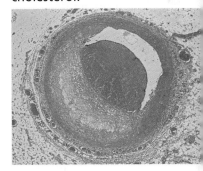

Under Pressure   A physician is sometimes interested in the force at which the blood moves through the arteries. This force is known as the *blood pressure*. Blood pressure increases as the ventricles push the blood through the arteries. Blood pressure then decreases while the ventricles refill with blood. Having a normal blood pressure is an important health concern. Blood pressure that is too high makes the kidneys work harder and can damage valves and muscle tissue of the heart. Blood pressure that is too low can cause poor circulation. In a cold climate, poor circulation is especially dangerous because a person might not realize when extremities of the body are becoming frostbitten.

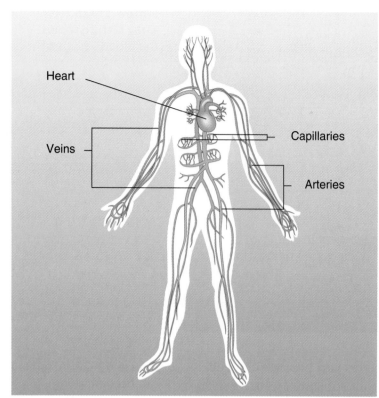

Heart

Veins

Capillaries

Arteries

Figure 17–27. The human circulatory system is a complex, continuous pathway.

Your journey through the complex, continuous circulatory system reaches a conclusion with a note about your pulse. Your pulse can be detected at various locations throughout your body. A *pulse* is felt as an artery expands, then relaxes after blood surges through the artery. Your pulse can be found at points where an artery is near the surface of your body. Common pulse points are found on the insides of the wrists, ankles, and thighs. The pulse can also be felt on the temples and the sides of the neck. How do you think your pulse rate compares to the pulse rate of someone else?

DISCOVER BY Calculating

Survey the pulse rate for some of your friends or family members. In your journal, record each person's age, sex, and pulse rate. Graph your results. Compare age to pulse rates. Do all of your family members have the same pulse rate? What factors might explain some of the differences? ✎

▶ ASK YOURSELF

Why must blood circulate through the body?

SECTION 3 REVIEW AND APPLICATION

Reading Critically

1. Explain the function of a valve.
2. What causes the "lub-dup" sound in the heart?
3. How are arteries different from veins?

Thinking Critically

4. What kind of damage might high blood pressure do to valves of the heart?
5. Which system—pulmonary or systemic—contains more kilometers of blood vessels? Why?
6. Is a bruise more likely to be the result of damage to an artery, a vein, or a capillary? Explain.

*H*IGHLIGHTS

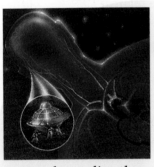

The Big Idea

All parts of the human body require a consistent source of energy, which is provided by nutrients. Nutrients are processed by the digestive system. The circulatory system absorbs the nutrients and oxygen and supplies them to the entire body. Powered by the heart and flowing through capillaries, veins, and arteries, blood transports nutrients and oxygen to the body and carries waste products in return. The normal operation of the systems and their interactions enable the body to function properly. As a result, people are able to lead active and healthy lives.

At the beginning of the chapter, you wrote your ideas about traveling to the brain through arteries, the relationship of the digestive and circulatory systems, and the role of nutrients in maintaining health. Reread your journal entry. Revise it to reflect changes in your understanding. Be sure to include a list of things you could do to take better care of your digestive and circulatory systems.

Connecting Ideas

Copy this unfinished concept map into your journal. Complete the concept map by writing the correct term in each blank.

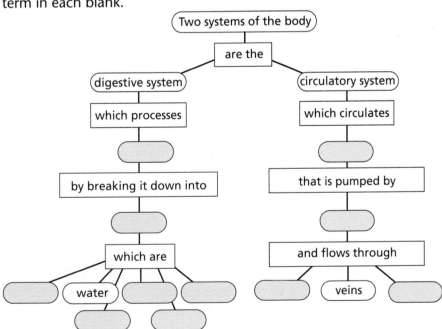

Two systems of the body

are the

digestive system — circulatory system

which processes — which circulates

by breaking it down into — that is pumped by

which are — and flows through

water — veins

Understanding Vocabulary

For each set of terms, explain how they are related.

1. digestion (458), ingestion (459)
2. chemical digestion (459), mechanical digestion (459)
3. nutrients (458), absorption (463)
4. plasma (473), red blood cells (474)
5. white blood cells (475), platelets (475)
6. arteries (480), veins (481), capillaries (480)
7. pulmonary system (482), systemic system (482)

Understanding Concepts

MULTIPLE CHOICE

8. A small bruise is most likely to show damage to which type of blood vessel?
   a) artery
   b) vein
   c) capillary
   d) Each vessel is equally likely to show damage.

9. Why do people sometimes take a drink of something while chewing food?
   a) to aid mechanical digestion
   b) to aid chemical digestion
   c) to aid the absorption of nutrients
   d) to speed the digestive process

10. A diet that is high in cholesterol should be a concern for which group of people?
    a) infants
    b) youths
    c) adults
    d) people of all ages

11. A well-balanced diet provides
    a) cell energy.
    b) cell growth.
    c) cell repair.
    d) all of these.

SHORT ANSWER

12. In what way is the structure of the small intestine well suited for the absorption of nutrients?

13. How are heart contraction, pulse, and blood pressure related?

14. What function does each of the six nutrient groups provide for the human body?

15. Why is it often suggested that food be chewed slowly and completely?

Interpreting Graphics

16. The illustration shows how a network of fibers grows over a wound to stop the flow of blood. Explain how these fibers stop blood flow.

17. Some people must take medication that thins their blood. Do you think such people have more or less difficulty than usual stopping the flow of blood from a wound? Why?

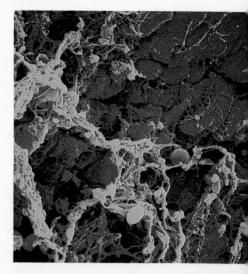

Reviewing Themes

18. *Systems and Structures*
Explain how the digestive and circulatory systems work together to supply nutrients and oxygen to all parts of the body.

19. *Energy*
Explain why a balanced diet is necessary for the proper functioning of the digestive and circulatory systems.

Thinking Critically

20. Will your digestive system function effectively if you eat a meal that is not well balanced? Explain.

21. A person may vomit when ill. Why does the mixture of food and gastric juice produce a burning sensation in the throat but not in the stomach?

22. Arteries are usually found farther from the surface of the body than veins. Why do you think this is true?

23. How is it possible for a person to eat large amounts of food and still suffer malnutrition?

24. Arteriosclerosis causes the walls of the arteries to become less elastic. How would this disease affect the heart?

25. An intestinal virus often prevents the large intestine from absorbing water. What effect might this have on a person if the condition lasted for several days or longer?

26. Explain why a wound that cuts an artery is more dangerous than a wound that cuts a vein.

27. This photograph shows red blood cells infected with the protozoans that cause malaria. Using your knowledge of what red blood cells do, explain why their breaking open is harmful.

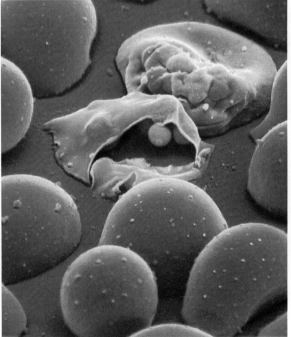

Lennart Nilsson/Boehringer Ingelheim International Gmbh

Discovery Through Reading

Lambourne, Mike. *Down the Hatch: Find Out about Your Food.* Millbrook Press, 1992. This book about food and digestion is illustrated with colored drawings and diagrams and includes investigative activities.

CHAPTER 40

RESPIRATION AND EXCRETION

A variety of sights, sounds, and odors greets the spectators at any race. At a swim meet, they smell the chlorine and hear the splashing of water. At a wheelchair race, they hear the shouts of supporters and watch the powerful movements of the racers. At a car race, they smell exhaust fumes and hear the roar of high-performance engines.

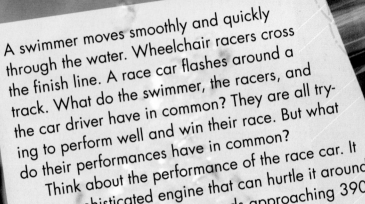

A swimmer moves smoothly and quickly through the water. Wheelchair racers cross the finish line. A race car flashes around a track. What do the swimmer, the racers, and the car driver have in common? They are all trying to perform well and win their race. But what do their performances have in common?

Think about the performance of the race car. It has a sophisticated engine that can hurtle it around the track at straightaway speeds approaching 390 km/hr! It takes about 40 seconds to travel once around a four-km track! Race car engines always contain all of the latest automotive technology and innovations. Designers of these engines pay special attention to the waste products produced by the engines. The removal of these waste products in an efficient manner ensures that the engines perform at peak efficiency—a crucial factor in winning a race.

In many ways, the human body is like a race car engine, only much more sophisticated. The body too must dispose of the waste products it generates. Proper disposal of waste products by the respiratory and excretory systems ensures that the body operates at peak efficiency—a crucial factor in winning a race for people as well as cars.

For Your Journal

✏️ In what other ways is your body like an engine?

✏️ List as many of the structures of the respiratory and excretory systems as you can think of.

✏️ What might happen if waste products were not removed from an engine or from your body?

Respiration

List *each structure of the respiratory system and **describe** its function.*

***Relate** the structure of lungs to their function.*

***Summarize** the process of breathing.*

Imagine that you are standing next to the engine shown in the photograph. Although you don't look like the engine, you and the engine are alike in many ways. Both of you are machines. Both of you burn fuel to release energy. Both of you take in oxygen and give off waste products. If the waste products produced by each machine are not released, neither machine will run efficiently. In fact, if the waste products from the machines are allowed to build up, the machines will stop functioning.

Figure 18–1. How is your body like this engine?

The Respiratory System

Your digestive system changes food into nutrient molecules that you use for fuel. Your circulatory system then transports oxygen and these nutrients to cells in all parts of your body. Inside each cell, the powerhouses of the cell—mitochondria—use the nutrients and oxygen in a process called *cellular respiration.* This process releases energy and produces carbon dioxide and water as waste.

Breathe In, Breathe Out   Your respiratory system brings oxygen into your body and removes carbon dioxide. Your nose, your lungs, and a number of tubes that connect your nose and lungs make up the major components of your respiratory system. However, to bring air into your body, your respiratory system needs help from your skeletal and muscular systems. The systems work together to fill your lungs with air.

Air enters your body through either your nose or your mouth. Even though both ways bring oxygen into your body, breathing through your nose is healthier than breathing through your mouth. When air enters through your nose, your nasal passages warm, moisten, and clean the air. A layer of wet, sticky mucus coats the inner walls of your nose. Most of the dust and dirt in the air sticks to the mucus. Your nose acts like the air filter in an engine; it cleans particles and debris from the air that don't belong in the system. The tissues in your mouth cannot perform these functions.

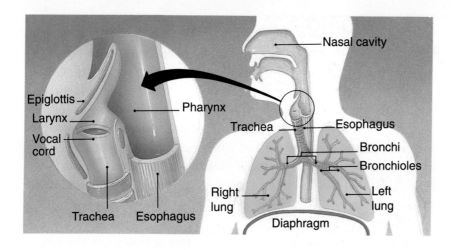

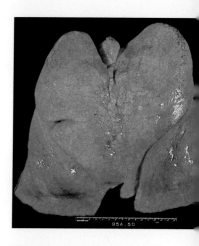

Figure 18–2. The healthy lung (right) is an essential component of the human respiratory system.

Notice in the illustration that the passages of the nose and the mouth come together at the back of your throat. The place where your nasal passages and mouth meet is called the *pharynx* (FAR inks). As you swallow, food pushes up the cartilage at the back of your mouth. This cartilage covers the opening that leads to your nose and keeps food from entering your nasal passages.

Two tubes lead deeper into the body from the pharynx. One tube, the esophagus, carries food down to the stomach. Air moves to the lungs through the other tube, called the windpipe, or **trachea** (TRAY kee uh). You can feel the trachea beneath your skin by gently rubbing the front of your neck. You can feel the rings of cartilage that hold the trachea open like a stiff but flexible garden hose. These cartilage rings hold the trachea open so that air can move through it more easily. A flap of tissue called the **epiglottis** (ep uh GLAH tis) swings down over the opening to the trachea when you swallow. This helps keep food from blocking airflow.

Say "Ahhh"

The *larynx* (LAR inks), or voicebox, forms the upper part of the trachea. Thin ligaments stretch across the opening of the larynx, forming your vocal cords. As air passes through the space between the vocal cords, they vibrate and make sound. This is the sound of your voice. You change the pitch of the sound by changing the length and thickness of your vocal cords. Rest your fingers on your throat while you make an "Ahhh" sound. Change the pitch of the sound, first higher and then lower.

The action of the vocal cords works in a way similar to the tightening and loosening of a rubber band or guitar string. A model of the vocal cords can be made in the next activity.

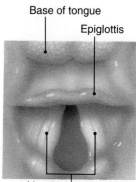

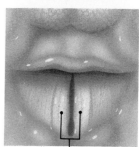

Figure 18–3. To create sounds, vocal cords open and close to allow air to pass through them.

Obtain three different-sized boxes with lids and three elastic bands that are the same size. Cut a square out of each box lid. Stretch an elastic band around the length of each box so that the elastic band passes over the square hole in the lid. Pluck each elastic band and listen to the sound. Describe its pitch. Why is there a difference in pitch? Which elastic band has the highest pitch? Which has the lowest pitch? Compare the model to your vocal cords. ✐

The sounds made by the boxes in the activity are simple. The sounds you make are much more sophisticated. Your tongue, cheeks, and lips help shape the sounds into words. The volume of the sound depends on the force of the air as it flows over the vocal cords.

Beneath the larynx, the trachea widens and branches into two smaller tubes called **bronchi** (BRAHN kee). Each tube leads to a lung where it branches into smaller tubes called **bronchioles** (BRAHN kee ohlz). As the bronchioles go deeper into the lungs, they branch into still smaller and smaller tubes like the limbs of a mature tree. For this reason, the tubes are often called the bronchial tree. Each of these tiny tubes ends in a group of tiny air sacs called **alveoli** (al VEE uh ly). These tiny air sacs make up the tissue of the lungs. About 150 million alveoli are found in each of your lungs.

The trachea, bronchi, and bronchioles are lined with mucus that traps dust, dirt, and other foreign material brought in with the air. Cilia, which are tiny, hairlike projections, continually sweep the mucus upward and out of the body.

Figure 18–4. People produce sound as air is forced past the vocal cords.

▼ **ASK YOURSELF**

How do your vocal cords produce sounds of different pitch?

Breathing

When you breathe, your chest first expands, then contracts. It might seem as if your lungs pull air in and then push air out when you breathe, but this is not the case. Your lungs are not made of muscle tissue, so they cannot pull air in, any more than a balloon can pull air in. Your respiratory system works together with your muscles and your bones to move air into and out of your body.

Take a Deep Breath When the size of your chest cavity changes, it causes you to breathe. However, you need muscles to change the size of your chest cavity. One of these muscles, the **diaphragm** (DY uh fram), is a thick sheet of muscle that forms the floor of your chest cavity. When relaxed, your diaphragm is dome-shaped, but it flattens when it contracts. Contraction moves the diaphragm down, making the chest cavity larger. When your diaphragm relaxes, it rises to its original position and your chest cavity gets smaller.

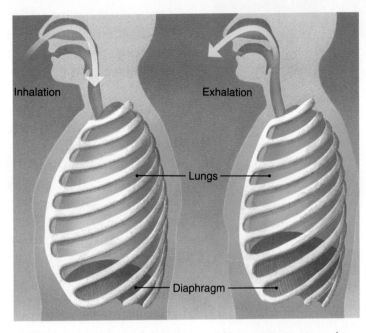

Inhalation

Exhalation

Lungs

Diaphragm

As your diaphragm moves down, other muscles pull your rib cage up and away from your backbone. This action further increases the size of your chest cavity. Your chest cavity is a flexible container that is sealed from the outside environment. The lungs hang inside this container like two balloons.

Figure 18–5. How do the positions of the rib cage and the diaphragm change during breathing? How do these changes affect the pressure in the chest cavity?

DISCOVER BY Doing

Using the description given in the text, design and create a model of the human lungs. What materials did you use? How does your model demonstrate the breathing process? Share your model with your classmates. ✎

All Bottled Up What happens when you squeeze a container such as a plastic detergent bottle? Squeezing the container increases the pressure on its walls. This reduces the volume of the container and increases the pressure inside. With less volume inside, there is less space for the contents. The increased pressure forces some of the soap out through the open top.

What happens when the top is closed and the bottle is squeezed? Since the contents cannot escape from the container, the pressure inside increases. The bottle pushes back on your hand, but the soap cannot escape. If you could increase the size of the container by pulling the walls outward, the reverse would happen. The pressure inside the container would decrease and, if the bottle were open, air would rush in to fill the extra space.

Figure 18–6. How is the process of breathing like squeezing this liquid detergent bottle?

Your chest cavity acts in somewhat the same way as the plastic detergent bottle. Your diaphragm pulls downward and your rib cage rises. This increases the volume of your chest cavity and decreases the pressure. Since the chest cavity is sealed, air cannot rush in to fill the extra space. However, the lungs, which are open to the environment, hang inside the chest cavity. The pressure on the lungs inside your chest cavity is lower than the air pressure outside your body. Thus, air outside the body forces its way through your nose and down the trachea into the lungs. The higher air pressure outside the body fills the lungs with air. The lungs expand and fill the extra space in the chest cavity.

You breathe out by relaxing your diaphragm and lowering your rib cage. The volume of the chest cavity decreases and the pressure inside the chest cavity increases. This increased pressure squeezes the air from your lungs. In other words, as you inhale, air is pushed into your body from the outside. As you exhale, your body pushes the air back out.

 ASK YOURSELF

What causes you to breathe in?

Exchange of Gases

The pie graph on page 495 shows that air is a mixture of gases. Nitrogen gas makes up the largest component of air. Oxygen makes up only about 21 percent, and carbon dioxide about 0.03 percent of air. All of the air that you breathe in does not reach your bloodstream. Your blood can carry only oxygen and carbon dioxide in large quantities.

When you breathe, fresh air moves into your lungs. Oxygen moves from the air in the alveoli into the surrounding capillaries. Red blood cells absorb the oxygen and carry it through

Figure 18–7. The X-ray on the left shows the lungs after air has been inhaled. The X-ray on the right shows the lungs after air has been exhaled.

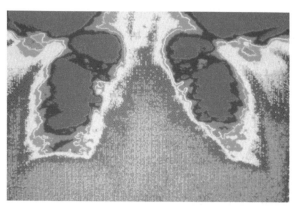
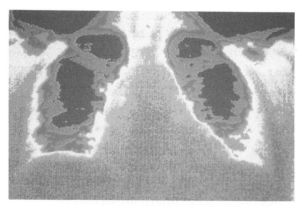

your body. At the same time, carbon dioxide moves from the blood's plasma into the alveoli. The carbon dioxide moves out of your body when you breathe out. This process of taking in oxygen and giving off carbon dioxide is called *gas exchange.*

Gas exchange also takes place through the walls of capillaries in your body tissues. Oxygen in the blood diffuses into the cells. At the same time, carbon dioxide diffuses from the cells into the blood.

Gas exchange occurs because there is a difference in the concentration of the gases on each side of a membrane. In the lungs, the concentration of oxygen is greater in the air inside the alveoli than it is in the blood. The opposite is true for carbon dioxide; the concentration of carbon dioxide is greater in the blood than in the air inside the alveoli. The gases move from areas of high concentration to areas of low concentration. When blood reaches the body tissues, the difference in concentrations of these gases is reversed. As a result, oxygen moves into the cells while carbon dioxide moves into the blood.

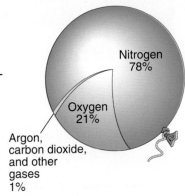

Figure 18–8. This pie graph shows the percentages of different gases in the air you breathe.

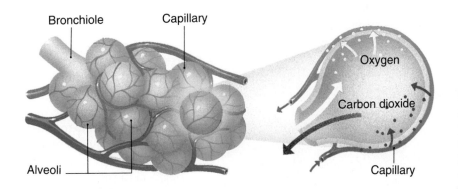

Figure 18–9. In gas exchange, oxygen moves from the alveoli to the blood, and carbon dioxide moves from the blood to the alveoli.

 ASK YOURSELF

Why does gas exchange occur?

SECTION 1 *REVIEW AND APPLICATION*

Reading Critically

1. Why does your body need oxygen?
2. How is the air cleaned as it travels to the lungs?
3. Through what structures does air pass as it moves from the nose to the lungs?

Thinking Critically

4. Why does the health of the respiratory system depend, in part, on the health of the muscular system?
5. People sometimes take several deep breaths before swimming underwater. Why might this practice be dangerous?
6. Why is the image of human lungs as a large balloon not accurate?

INVESTIGATION

▶ Measuring Lung Capacity During Exercise

▶ MATERIALS

- mouthpiece holder ● lung volume bag ● rubber band
- mouthpiece (disposable) ● paper towel ● pencil ● paper

▼ PROCEDURE

1. **CAUTION: Advise your teacher of any medical condition that prevents your participation in this activity. Also be sure to use only your own mouthpiece and then throw it away.**

2. Working with a partner, insert the mouthpiece holder into the end of the lung volume bag. Secure it by wrapping a rubber band around it tightly. Insert the disposable mouthpiece into the mouthpiece holder.

3. Remove as much air from the bag as possible by flattening it.

4. Take a deep breath and blow all of your breath into

the bag. Immediately trap the air by rolling the mouthpiece end of the bag in your hand until the bag becomes stiff with your trapped air.

5. Read the numbers on the bag to find your lung capacity in liters. On a sheet of paper, make a chart and record the number of liters.

6. Repeat steps 2-4 three times and calculate your average lung capacity.

7. Remove your mouthpiece and save it by placing it on a clean paper towel.

8. Rest while your partner is measuring his or her lung capacity.

9. Run in place for three minutes. Immediately following this exercise, again measure your average lung capacity and record it on your chart. **CAUTION: Be sure to use your own mouthpiece and then throw it away.**

▶ ANALYSES AND CONCLUSIONS

1. Before exercising, how did your average lung capacity compare to your partner's? Explain any similarities or differences.

2. After exercising, was your lung capacity larger, smaller, or the same? Why?

3. Your lung capacity can change. Name several things you could do to increase your lung capacity.

▶ APPLICATION

How does exercise affect your lung capacity? If you exercised every day, how do you think your lung capacity would change?

✳ Discover More

Design an experiment that measures the lung capacities of volunteer members of one of your school's athletic teams. Record their capacities before the season begins, on several occasions during the season, and after the season ends. Graph and display your results. Do your results support the hypothesis you made for the Application question?

Excretion

Race cars, like all vehicles, burn fuel and give off gases. Have you ever seen an automobile, truck, bus, or train engine giving off clouds of exhaust fumes? Have you ever traveled down a street or road that was littered with empty bottles, cans, and paper? An environment polluted with trash and exhaust fumes is not only ugly, but unhealthy. To help make the environment a cleaner place, the government requires cars to meet emission standards, and many workers and volunteers regularly pick up trash from streets, highways, and lots. Just like a city or an engine, your body also produces wastes that need to be removed.

Objectives

Name the waste products typically eliminated by the excretory system.

Describe the way in which the excretory system works.

Relate the importance of the excretory system to the overall health of the human body.

Waste Removal

The chemical activities that your body constantly performs are called your *metabolism,* and the waste products of these activities are called *metabolic wastes.* As your body performs the chemical activities that keep you alive, waste products such as carbon dioxide, nitrogen, water, and heat are produced. In order for you to remain healthy, your body must remove the metabolic wastes that you generate. **Excretion** (ihks KREE shuhn) is the process by which these wastes are removed from your body.

A number of different organs and systems work together to remove wastes from your body. Most of the waste your body produces is removed by your lungs and skin. Your lungs remove carbon dioxide and water vapor. When you breathe on a mirror or cool window glass, you can see the water vapor that an exhaled breath contains as the mirror or glass becomes cloudy. Your skin removes the waste product nitrogen, as well as water, and helps control your body temperature by removing extra heat. The following table shows the approximate amount of water gained and lost each day by an adult.

Figure 18–10. How are city streets and the human body alike?

Table 18-1	**Daily Adult Water Gain and Loss**			
	mL			**mL**
Input:		**Output:**		
Produced by metabolism	400	Feces		100
Eating	900	Skin and lungs		1000
Drinking	1300	Urine		1500
Total: 2600			**Total:** 2600	

The table shows that the excretory system removes most of the water that is lost by the human body.

▼ **ASK YOURSELF**

Name the waste products removed by your lungs and skin.

The Excretory System

The **kidneys** are the main organs of the excretory system. Your two kidneys are located above your waist on either side of your spine. To find them, place your hands on your back just below the ribs. Your kidneys are located just under your hands.

Finicky Filters   When cells of your body break down certain compounds and proteins, a nitrogen waste called *urea* (yoo REE uh) is produced. Your kidneys work constantly to filter your blood and produce *urine* (YOOR ihn). Urine is a yellow liquid that contains urea and excess water, minerals, and salt filtered out of your blood. The following activity will help you discover the serious problems that might occur if a person puts an additional strain on his or her kidneys and excretory system.

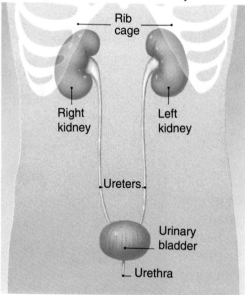

Figure 18–11. The excretory system is responsible for removing some of the waste products produced by the human body.

When some people feel like they are getting a cold, they take very large doses of vitamin C in hopes of warding off the cold. Other people take massive doses of various vitamins on a regular basis. Use reference materials to discover what serious effects these megadoses of vitamins can have on the excretory system. Record your findings in your journal.

Your kidneys work in a way that is similar to the way an air filter works in a car engine or a coffee filter works in a coffeepot. Each filter traps unwanted materials. Your kidneys filter your blood, helping to keep it clean. In an average person, about 1600 L of blood pass through the kidneys each day, creating about 1.0 to 1.5 L of urine.

Each kidney contains more than a million tiny filtering units called *nephrons* (NEF rahnz). The illustration shows that a nephron consists of a long, thin, folded tubule surrounded by a group of capillaries. Liquid from the blood flows through the capillary wall into the nephron tubule. The liquid contains urea as well as water, sugars, and important minerals. As the liquid moves through the tubule, sugars and needed minerals and water move back into the capillaries, leaving the urine behind. Ninety-nine percent of the liquid that leaves the capillaries returns to the bloodstream.

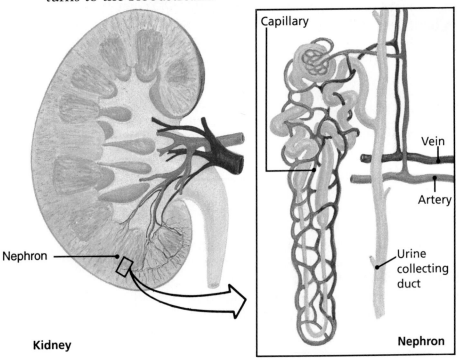

Figure 18–12. Each kidney contains more than a million nephrons, which filter the blood.

Nephron

Kidney

Capillary

Vein

Artery

Urine collecting duct

Nephron

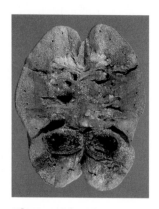

Figure 18–13. A diseased kidney (left) cannot filter wastes as efficiently as a healthy kidney (right).

Moving On All the nephron tubules of the kidney connect to a larger tubule called the *ureter* (yoo REET uhr). The ureter from each kidney carries the urine to a large sac, the *urinary bladder*. The bladder stores urine until it can be removed from the body. Once the bladder is filled, its muscular walls contract, forcing the urine down the *urethra* (yoo REE thruh) and out of the body.

The composition of urine is often checked during a physical examination. This process, known as a urinalysis, tests the urine for signs of kidney malfunction or diseases of other body organs. It can also detect the presence of drugs.

Sometimes, due to disease or injury, a person's kidneys do not function as they should. A person with kidneys that do not function will die because waste products are not being removed from the bloodstream. Too many waste products in the bloodstream will poison the human body. Fortunately, something can be done to help these people, and the following activity will give you the chance to explore what can be done to help.

DISCOVER BY *Researching*

Kidney diseases can be very serious illnesses. Prepare a brief oral report on kidney dialysis. In your report, explain what a dialysis machine is, how it works, who uses dialysis, and how often dialysis must be done. ✎

▶ ASK YOURSELF

Describe how your kidneys remove wastes from your blood.

SECTION 2   *REVIEW AND APPLICATION*

Reading Critically

1. Why is excretion important to the body?
2. What structures make up the excretory system?
3. How do the nephrons produce urine?

Thinking Critically

4. Undigested material from the digestive tract is "eliminated" rather than "excreted" from the body. How are these two processes different?
5. Are the kidneys more like a swimming-pool filter or a strainer used to drain noodles? Explain your answer.

▼ PROCEDURE

A flow chart is a type of diagram that shows the steps in a process or problem. Arrows lead the reader from one step to the next. Flow charts can be used to make sequencing much easier. A flow chart of pulmonary circulation is shown here. Make flow charts that show the sequence involved in each of the following processes:

- the movement of oxygen from the environment to the lungs and the movement of carbon dioxide back to the environment
- the circulation of blood
- the excretion of urea from the blood and its removal from the body as urine

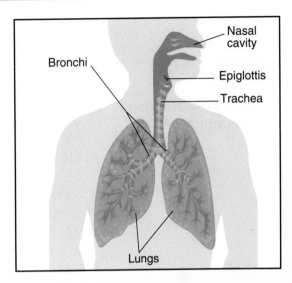

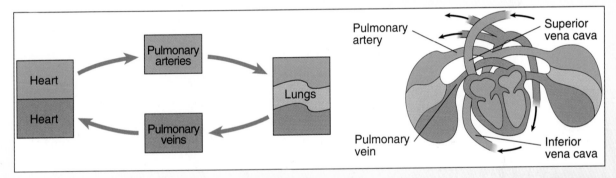

▶ APPLICATION

Combine the information from your three flow charts and the flow chart of pulmonary circulation into one flow chart. This flow chart should show the structures involved in:

- the movement of oxygen from the environment to the cells
- the movement of carbon dioxide from the cells to the environment
- the movement of urea from the cells to the environment

✳ Using What You Have Learned

1. How is a flow chart different from a written passage in which you might describe a process?
2. Is a flow chart a good substitute for a written description of a process?
3. Which method of presenting information allows the reader to visualize the details of the process more easily?
4. Can a flow chart act as an aid in remembering detailed information?
5. Which method of presenting information allows the reader to combine the information more easily?

The Skin

If someone asked you to write a list of the excretory organs of your body, how long would your list be? If your list didn't include skin, you would not have listed all of the excretory organs of your body! It's easier to think of excretory organs such as the kidneys and bladder. But your skin is also an organ that performs excretory tasks. Just as the exhaust system of a race car engine helps eliminate waste products, your skin helps eliminate waste products from your body.

The Structure of the Skin

Your skin performs many important tasks that help keep you healthy and alive. Many kinds of cells and tissues are needed to perform the jobs of the skin. These different cells and tissues are arranged in two layers. The thinner outer layer of skin is called the **epidermis** (ep uh DUHR mis), and the thicker inner layer is called the **dermis** (DUHR mis).

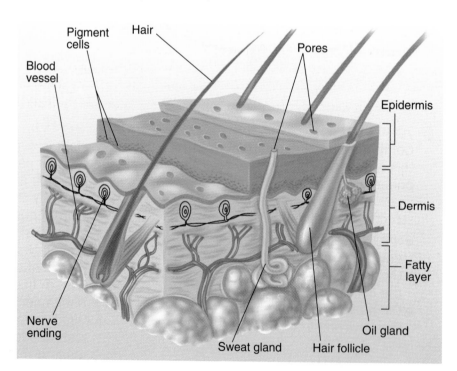

Pigment cells
Hair
Pores
Blood vessel
Epidermis
Dermis
Fatty layer
Nerve ending
Sweat gland
Hair follicle
Oil gland

Figure 18–14. How many layers of tissue make up the skin?

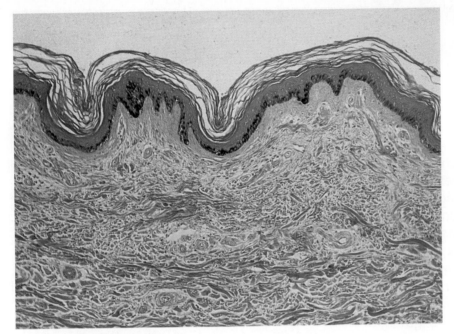

Figure 18–15. The pattern of ridges and valleys on the skin of the fingertips makes each person's fingerprints unique.

Epidermis   The epidermis has an outer layer of dead cells over an inner layer of living cells. Cells of the outer epidermis die because the supply of nutrients and oxygen is poor in that area. Capillaries do not reach into the epidermis. The epidermis gets its nutrients from the fluid that surrounds the dermis. Therefore, only the cells closest to the dermis can survive. As those cells reproduce, the new cells push the older cells farther away from the dermis, where they die. These dry, dead cells form a waterproof barrier between the living cells and the environment outside the body.

Dead cells are constantly being shed from the body. Each time you scratch your skin or rub something against it, these dead cells break free. The epidermis sheds over a million dead cells each hour! Cells from below replace these lost cells. This loss of cells keeps the body's surface fresh and clean. How long do you think it takes the skin of your body to completely replace itself?

Dermis   The dermis is thicker, tougher, more elastic, and more complex than the epidermis. The dermis gets its food from the blood in the capillaries. Within the dermis are many glands, nerve endings, and hair follicles.

The dermis of your fingers, palms, toes, and soles of your feet is thicker than the dermis on other areas of your body. Why do you think this is so? In this thicker dermis, the upper layer forms ridges and valleys. In the next activity, you can find out more about the ridges and valleys on the skin of your fingers.

What are some characteristics of skin?

MATERIALS
plain white paper, pen or pencil, ruler, ink pad, hand lens

PROCEDURE
1. Trace around each of your hands on a plain white sheet of paper. Then examine your hands and make a list of the locations and causes of any blisters or calluses you have.
2. Using a hand lens, examine the skin creases or folds around the joints of your hands. Find the longest line on your palm and draw it on your sketch. Record its length in centimeters on the line.
3. Using the following procedure, add your fingerprints to the bottom of your drawing.
   a. Roll your thumb onto the ink pad from left to right until the area from joint to tip is covered with an even layer of ink.
   b. Placing the left side down first, lay the inked thumb on your paper and roll the thumb until it is resting on its right side.
   c. Lift the thumb straight up off the paper to prevent smudging.
   d. Repeat this procedure for your other fingers.

4. Use a hand lens to compare your fingerprints to those of your classmates. Do you see any similarities or differences?

APPLICATION
1. Why do you think your hands have so many creases on them?
2. What do fingerprints have in common with tire treads?
3. Explain how the calluses on a person's hands can be used to guess the occupation of the person.

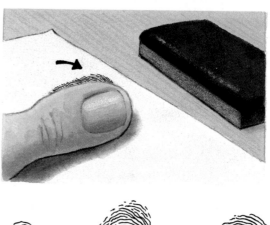

Your fingernails and toenails are produced by certain cells in your skin. Your hair is made of dead cells in the epidermis. Tiny pits in the epidermis push deep into the dermis and form tubelike hair follicles. Dead skin cells are pushed up through the follicle and form a strand of hair. Muscles fastened to the hair follicle can contract, pulling the hair upright. This causes the "gooseflesh," or "goose bumps," you get when you are frightened or cold. In the next activity, you can examine a hair more closely.

Mount a hair from your head or eyebrow on a clean microscope slide. Examine the hair under the low and high powers of a microscope. In your journal, make a diagram of what you see. A typical hair has a hair shaft and a root. Label these on your diagram. ✎

The bottom of the hair shaft you looked at in the activity was pulled out of an oil gland. Oil glands are found around the bottom of each of your hair follicles. These glands produce an oily mixture that keeps your skin and your hair soft. A thin coat of oil on the skin also reduces the amount of water given off through perspiration. Acne occurs when these oils mix with dead skin cells, plug up the hair follicles, and allow bacteria to be trapped and to multiply.

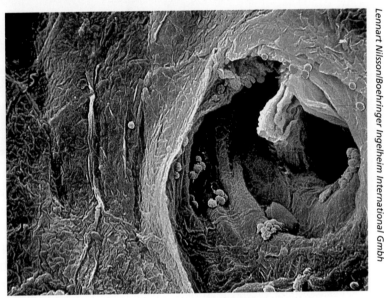

Lennart Nilsson/Boehringer Ingelheim International Gmbh

Figure 18–16. Sweat glands, such as this one, give off excess water and salts from the body.

Sweat glands are coiled tubes connected to the surface of your skin by pores. Capillaries surround these tubes. Water, salts, and even small amounts of urea move from the blood in your capillaries into the sweat glands and then out through your pores.

ASK YOURSELF

Compare the epidermis and the dermis.

The Functions of the Skin

Your skin performs several important functions for you. Skin protects other tissues in your body from harmful bacteria and other organisms. Skin keeps your body from drying out and protects it from the harmful effects of the sun. Skin helps to control your body temperature, senses your external environment, and manufactures vitamin D. In addition to all of these, your skin also functions as an excretory organ. It removes excess water, salt, urea, and heat from your body.

Figure 18–17. The amount of melanin manufactured in the epidermis determines the shade of your skin.

The outer layer of cells forms your body's first line of defense. Dead cells, salty sweat, and oil form a protective outer shield that keeps bacteria, other microbes, and many harmful chemicals away from your living cells. Most microscopic organisms die when they come in contact with this outer surface.

Your skin also keeps out harmful radiation. The chemical responsible for your skin color is called *melanin*. Melanin helps protect your body by blocking the ultraviolet radiation from the sun. More melanin is made when the skin is exposed to sunlight. This results in a "tan" or darkened skin color. However, scientific studies have shown that repeated exposure to ultraviolet radiation can cause skin cancer and may cause premature wrinkles.

 ASK YOURSELF

Name several important functions of the skin.

SECTION 3   *REVIEW AND APPLICATION*

Reading Critically
1. Which layer of the skin is partially composed of dead cells?
2. Why do the cells of the outer skin layer die?
3. What causes "gooseflesh," or "goose bumps"?

Thinking Critically
4. Why is the skin considered to be an organ?
5. Why doesn't it hurt when you get a haircut?
6. Explain under what conditions, if any, a person could live without skin.

The Big Idea

The respiratory system is responsible for bringing oxygen into your body and removing carbon dioxide. But to fulfill this vital function, the respiratory system must interact with the skeletal system and the muscular system. Lungs are not muscle tissue and, therefore, cannot expand and contract on their own to bring air into the body. In the breathing process, the lungs are helped by muscles, such as the diaphragm, and by bones, such as the rib cage.

Cells eventually use the oxygen supplied by the lungs to perform chemical activities and create energy. The creation of energy, however, results in the creation of waste products, such as carbon dioxide, nitrogen, water, and heat. Both the respiratory and the excretory systems work to eliminate these waste products. Through their combined efforts, your body can function efficiently and remain healthy.

For Your Journal

Look back at the ideas you wrote in your journal at the beginning of the chapter. Revise your journal to include more ways in which your body is like an engine. Also include things that could be done to keep the skin and respiratory system of your body in peak operating condition.

Connecting Ideas

Copy the unfinished flow chart into your journal. Complete the flow chart by writing the correct term in each blank. Make a similar flow chart for the respiratory system.

Parts of excretory system

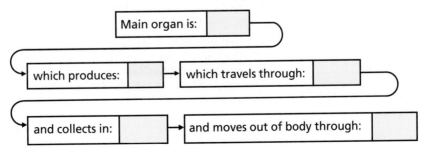

Main organ is:

which produces: → which travels through:

and collects in: → and moves out of body through:

REVIEW

Understanding Vocabulary

For each set of terms, explain the relationships in function.

1. trachea (491), epiglottis (491)
2. metabolism (497), excretion (497)
3. kidney (498), urinary bladder (500)
4. alveoli (492), diaphragm (493)
5. epidermis (502), dermis (502)

Understanding Concepts

MULTIPLE CHOICE

6. Clogged alveoli will result in
   a) shortness of breath.
   b) increased excretory rate.
   c) decreased heart rate.
   d) nothing; alveoli cannot clog.

7. When the water intake of a person increases,
   a) urinary output increases.
   b) urinary output remains constant.
   c) urinary output decreases.
   d) urinary output is not affected because water has nothing to do with urinary output.

8. The dead cell layer of the epidermis protects people from
   a) water.
   b) germs.
   c) harmful rays of the sun.
   d) all of these.

9. The epiglottis functions in a way similar to which of these?
   a) a screen door
   b) a revolving door
   c) a hinged door
   d) a sliding door

SHORT ANSWER

10. Why is breathing through your nose healthier than breathing through your mouth?

11. How is food prevented from entering your lungs?

12. What are the main excretory organs of the body? Which metabolic wastes are removed by each of these organs?

13. How can pores in the skin become clogged?

Interpreting Graphics

14. The illustrations show a piece of artificial skin that has been used on the hand of a burn patient. Why is not replacing burned epidermis and dermis tissue dangerous?

15. How do you think the recovery of the patient with artificial skin compares to the recovery of a patient using natural skin? Explain.

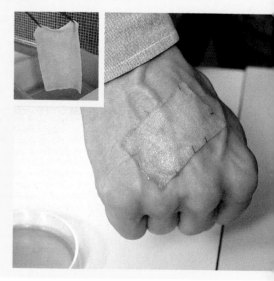

Reviewing Themes

16. *Systems and Structures*
Explain how the muscular system and the skeletal system work together in the process of respiration.

17. *Energy*
Describe how breathing provides fuel for the cells of your body.

Thinking Critically

18. Bronchitis is characterized by inflammation of one or more bronchi. Explain how bronchitis might affect the breathing process.

19. People are often told to drink large amounts of water when they have a fever. Why is this good advice?

20. Hiccups occur when there is a lack of coordination in the respiratory system. Suggest a possible reason for the occurrence of hiccups.

21. Natural skin "breathes" in the sense that it has pores. Describe the complications that might result from covering an area of a burn victim with artificial skin that does not "breathe."

22. Among other things, tobacco smoke contains tar. Describe how a respiratory system coated with tar might function when compared to a clean respiratory system.

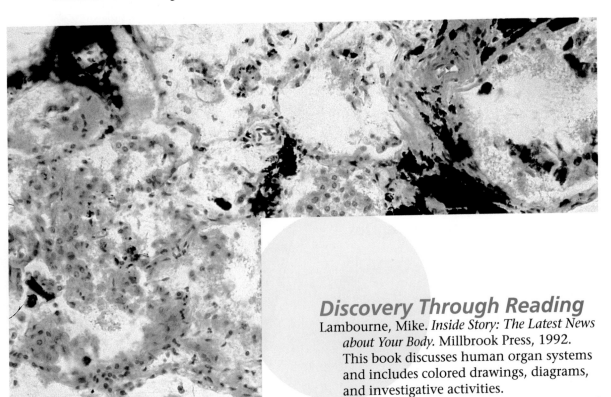

Discovery Through Reading

Lambourne, Mike. *Inside Story: The Latest News about Your Body.* Millbrook Press, 1992. This book discusses human organ systems and includes colored drawings, diagrams, and investigative activities.

CHAPTER 19

Coordination and Control

MUDVILLE

You might say that Casey's nervous system let him down. Twice he let pitches go by, while he took the umpire's call of "strike." And even though he took a mighty swing at the third pitch, he missed completely. Had his nervous system processed the information accurately there might have been joy, instead of sorrow, in Mudville. You will examine in this chapter just how the nervous system operates.

Casey

Ten thousand eyes were on him as he rubbed his hands with dirt;
Five thousand tongues applauded when he wiped them on his shirt.
Then while the writhing pitcher ground the ball into his hip,
Defiance gleamed in Casey's eye, a sneer curled Casey's lip.

And now the leather-covered sphere came hurtling through the air,
And Casey stood a-watching it in haughty grandeur there.
Close by the sturdy batsman the ball unheeded sped—
"That ain't my style," said Casey. "Strike one," the umpire said....

The sneer is gone from Casey's lip, his teeth are clenched in hate;
He pounds with cruel violence his bat upon the plate.
And now the pitcher holds the ball, and now he lets it go,
And now the air is shattered by the force of Casey's blow.

Oh, somewhere in this favored land the sun is shining bright;
The band is playing somewhere, and somewhere hearts are light,
And somewhere men are laughing, and somewhere children shout;
But there is no joy in Mudville—mighty Casey has struck out.

from "Casey At the Bat"
by Ernest Lawrence Thayer

For Your Journal

- *What caused Casey to swing at the ball?*
- *How do you think messages travel to and from the various parts of your body?*
- *Describe how you think the sense organs gather information from your environment.*

The Nervous System

Objectives

Describe the three major parts of the nervous system and their functions.

Compare and contrast the three types of neurons.

Relate the three major areas of the brain to the functions each performs.

Whether or not you hit a ball when you swing a bat depends on how well your nervous system does its job. Your nervous system is both a control system and a communication system. Let's look at the nervous system in terms of the basic jobs it does. Your nerves are responsible for gathering information and delivering it to your brain or spinal cord. The information is then processed, and a response is sent to various body parts telling them how to react. This important job of the nervous system is what enables you to swing a bat.

Another job of your nervous system is controlling the internal organs of your body. Your breathing, heartbeat, and digestion are a few of the functions of internal organs that are controlled by your nervous system.

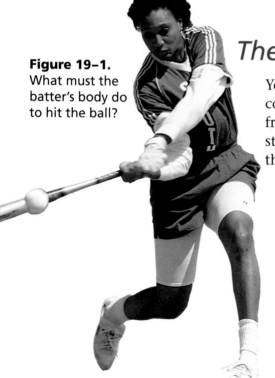

A small, white sphere sails toward the batter at more than 90 km/h. In an instant, her eye tracks the ball and her brain must select an action. Swing! The bat begins to move.

Figure 19–1. What must the batter's body do to hit the ball?

The Parts of the Nervous System

Your nervous system functions much like a video computer game. A video game receives information from the player. By using the keyboard or control stick, the player sends information through wires to the processor, which is the computer's "brain." The processor then interprets this information and selects a response. More wires carry the processor's response to an output device, such as a computer monitor or a TV screen. The screen produces a picture, which represents the processor's response.

Like the wires and processor of the computer, your nervous system has sense organs to collect information, a brain to interpret this information, and nerves to deliver responses to other body parts. Muscles and glands are the body's output devices that respond to directions from the nervous system.

Bring the tips of your thumb and middle finger together so that they just touch. A signal travels up your fingers and arm to your brain, and you feel a slight sensation on your skin. Specialized cells in your nervous system act like the computer's keyboard or control stick. These cells, called **receptors,** receive information from their surroundings. Some receptors are individual cells, such as those in your skin that produce the sense of touch. Other receptors are part of other sense organs, such as your eyes and ears.

Each kind of receptor is affected by a certain kind of stimulus. A *stimulus* is something that causes the nervous system to react. Stimuli such as light, heat, sound, chemicals, and pressure affect different kinds of receptors, causing them to send messages to the brain.

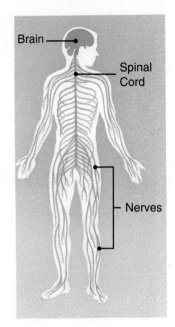

Figure 19–2. The human nervous system enables the body to respond to stimuli.

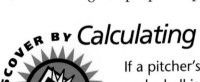 **Two outs, bases loaded, bottom of the eighth, and down by four runs! The batter waits as the pitcher winds up for the pitch. For an instant, the batter sees the tiny, white sphere as it sails toward the plate. Is it in the strike zone?**

The brain and spinal cord collect and interpret information for the receptors. In the case of the batter, light reflects off the ball, enters her eyes, and stimulates receptor cells in the back of them. Information about the size, speed, and position of the ball is gathered by the receptors. This information is then sent to the brain for processing. The brain along with the spinal cord make up the *central nervous system,* or the CNS. The CNS is responsible for interpreting all information gathered by receptors and selecting the proper response.

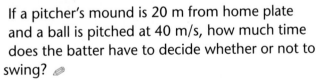 **DISCOVER BY** *Calculating*

 If a pitcher's mound is 20 m from home plate and a ball is pitched at 40 m/s, how much time does the batter have to decide whether or not to swing? 🖋

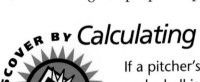 **Should she swing? Can she hit it?**

So how does information travel to and from the brain and spinal cord? The answer lies with chains of specialized cells that form long, thin fibers called **nerves.** Nerves carry messages from receptors to the central nervous system. Nerves also carry responses from the CNS to the muscles, organs, and glands of your body. A stimulus causes a receptor to send messages along a nerve to the CNS. The messages that travel along

Figure 19–3. Your body receives, processes, and acts on information in much the same way as the components of this computer system do.

nerves are in the form of weak electrical signals called *impulses.* The stimuli may be in the form of light, sound waves, or temperature changes gathered by receptors in your eyes, ears, or skin. Whatever the stimulus, nerve impulses are sent racing along nerves to the CNS at speeds of about 90 m/s. While this is a lot slower than electricity, which travels through space at more than 300 000 km/s, it is still fast enough to deliver messages throughout your body in a very short time. In fact, it would take an impulse less than $\frac{1}{49}$ s to move from the toe to the brain of a 183-cm-tall person. Now that is fast!

Let's take another look at the batter. The receptors in her eyes gathered information about the pitch and converted the information into an impulse that was sent to the brain. The brain then determined that the ball was in the strike zone and directed her muscles to go into action and take a swing.

 **Strike One!**

 Doing

Work with a partner in this activity. Hold out your hand with your thumb and forefinger separated. Have your partner hold a ruler vertically above your hand so that the zero mark is between your fingers. Try to catch the ruler as soon as your partner drops it. Record the distance the ruler falls before you catch it. Repeat the procedure several times. What does the distance the ruler falls tell you about your reaction time? What happened to your reaction time as you did more repetitions? Compare the reaction times of several of your classmates. ✐

 ASK YOURSELF

What is the central nervous system?

Nerve Cells

The basic unit of the nervous system is a specialized cell called a nerve cell, or **neuron.** There are billions of neurons in your body. While not all neurons are alike, they do have the same basic structure and perform the same function—carrying impulses to and from different parts of the body.

The Nerve of It All As with all cells in your body, the structure of a neuron is well suited to its job. Look at the drawing of the neuron. You can see that there is a main body to the cell. The main body houses the nucleus and other cell parts that are needed to keep the cell alive. Extending from the cell body are two types of branches. One type of branch, called the *dendrite* (DEN dryt), brings impulses into the cell. Usually, several dendrites branch from the cell body just as limbs branch from a tree trunk.

The other kind of branch is called an *axon* (AK sahn). It is responsible for carrying impulses away from the cell to other neurons. There is usually only one axon extending from a cell body. Bundles of these axons make up what we call nerves. A single nerve can be a fraction of a centimeter long or more than a meter long. The long axons are often wrapped in a protective covering that speeds the nerve impulse.

Going My Way? A neuron acts like a one-way street. Impulses move through the neuron in only one direction. As a result, one set of nerves sends information to the CNS, while other nerves carry impulses away from the CNS back to the body parts. Dendrites receive information from receptor cells or from other neurons. The impulse moves down the dendrites to the cell body. From the cell body, the impulse travels down the axon toward the next neuron in the chain. Neurons are not connected. A small space, or *synapse* (SIN aps), separates one neuron from another. For the impulse to continue down the nerve, it must cross the synapse. When the impulse reaches the end of the axon, it causes the axon to release chemicals. These chemicals spread across the synapse and trigger a new electrical impulse in the dendrite of the next neuron. In this way, information travels from neuron to neuron through the nervous system.

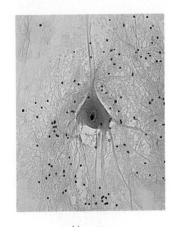

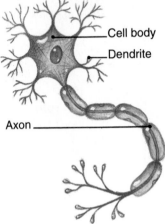

Figure 19–4. The neuron shown here is a single cell. The illustration shows the parts of a neuron.

 ASK YOURSELF

How is the structure of a nerve cell suited to the job it does?

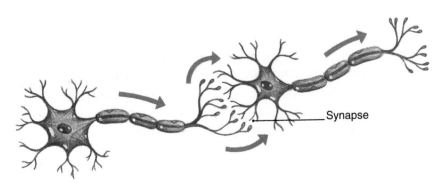

Figure 19–5. Chemicals carry nerve impulses across the gap that separates neurons.

515

Motor neuron

Sensory neuron

Interneuron

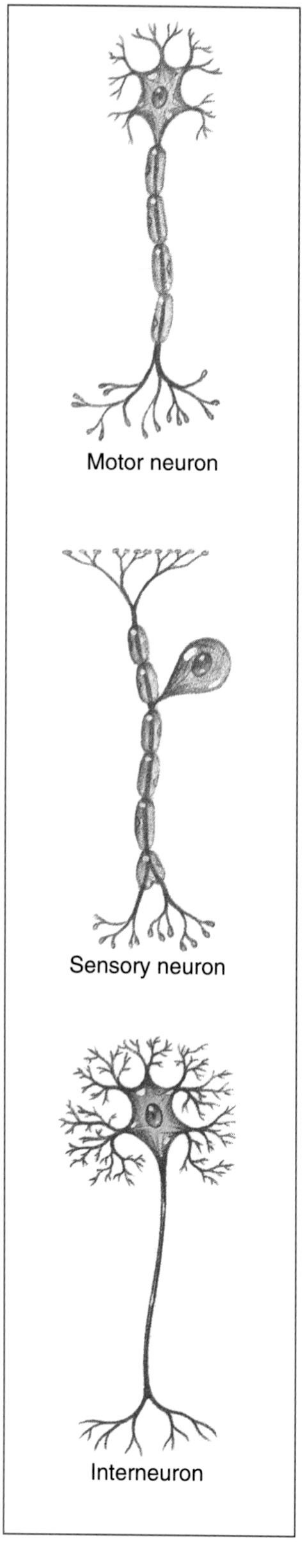

Figure 19–6. How are the three types of nerve cells different? How are they alike?

Types of Neurons

The nervous system contains three types of neurons: motor neurons, sensory neurons, and interneurons. Each kind of neuron has a unique function. Information sent from your brain or spinal cord to other cells in your body moves through *motor neurons*. In order for you to wiggle your big toe, an impulse must travel from your brain and down your spinal cord to motor neurons that lead to your leg. Some of these motor neurons connect to muscles that move your big toe. The impulse from your brain causes the muscles to contract, and your toe moves.

Try pressing a finger against the corner of your textbook cover. Information travels through *sensory neurons* from receptors in your sense organs to your spinal cord and brain. Your brain and spinal cord contain *interneurons* that connect sensory neurons to other neurons. Interneurons in your brain, for example, form connections that allow information to be transferred between different parts of your brain. Your memory is related, in part, to the number of connections between interneurons in the brain. About 97 percent of all the neurons in your body are interneurons.

ASK YOURSELF

Describe the three types of neurons and explain how they differ from each other in function.

Reflex Actions

You are constantly making decisions and choices about all sorts of things. For example, if you are up to bat and the pitcher throws a fast ball high and inside, your eyes send a message to your brain. The brain processes the information. Messages are then sent to your muscles telling them to respond in a manner that will allow you to duck out of the way of the pitch. These kinds of responses are *voluntary responses*. In other words, you think about an action and decide whether or not to respond in a certain way. But not all responses to stimuli are voluntary.

Have you ever grabbed a very hot pan? In an emergency, response time is very important. A fraction of a second can mean the difference between an injury and no injury. In the case of the hot pan, you move your hand away from the pan much faster than you could move it away if you had to think about it. Actually, your hand begins to move even before your brain processes information about the heat and pain. Why?

Your response to the hot pan is very fast because it does not require thought. Rather than relying on your brain to process a response, the impulse is received by the spinal cord, which then passes the impulse directly on to motor neurons. This causes your hand to pull away. A message then goes to the interneurons, which carry the information to your brain. Because your spinal cord, not your brain, sends a signal telling your muscles to release your grip on the pan, you are able to avoid a painful burn. This type of response is called a **reflex.** Reflexes are automatic; they do not require thought.

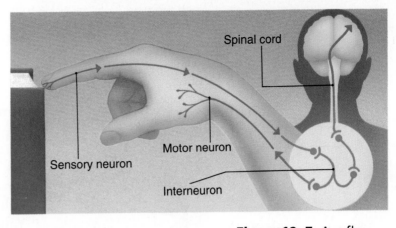

Figure 19–7. A reflex action involves three types of neurons— sensory neurons, interneurons, and motor neurons.

 DISCOVER BY *Doing*

Check your reflexes by sitting on the edge of a table with one leg crossed over the other. Have someone tap your crossed leg just below the kneecap. What happens? Now have someone give a verbal command to kick the lower leg forward. How did the speed of the reaction differ from the tap on the knee? Which response was faster? Why? 🖉

▼ **ASK YOURSELF**

Why are you able to move your foot so quickly if you step on a sharp object?

The Brain

The brain is your body's main control center. Not only does it control the activities of your body, but it also controls your thoughts and feelings. In addition, your brain serves as a storehouse of information known as *memory.* Your brain uses the information it has stored to evaluate new information and select present and future actions.

The average human brain is about the size of a large grapefruit and weighs about 1.3 kg. That's about as much as a head of cabbage weighs. The brain is the most complex organ in your body and contains over 10 billion neurons. Since the brain is so active, it requires a rich supply of blood to bring it fuel and oxygen. In fact, the brain uses one-fifth of the body's energy and

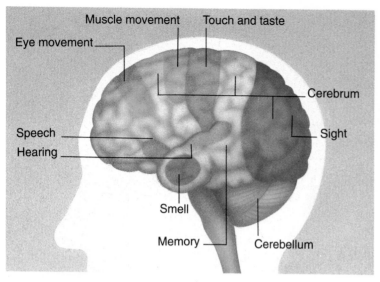

Muscle movement

Touch and taste

Eye movement

Cerebrum

Speech

Sight

Hearing

Smell

Memory

Cerebellum

nearly one-fourth of its oxygen supply. The brain has three major parts: the cerebrum (sehr REE bruhm), the brain stem, and the cerebellum (sehr uh BEHL uhm). You can see these parts in the illustration.

The **cerebrum** fills most of the skull, making up about 80 percent of the brain tissue. It is divided into halves, or hemispheres. Your thoughts, memories, emotions, and decisions begin in the thin outer portion of your cerebrum. This outer layer is often called gray matter because of its color. The gray matter also directs the voluntary movements of the skeletal muscles. In addition, it contains the centers for evaluating information from the sense organs and for speech. A large mass of white tissue lies below the gray matter. This white matter of the cerebrum acts much like a switchboard, controlling the movement of information between the gray matter and other parts of the brain.

Figure 19–8. Each part of the brain (top) has its own job. Sometimes, delicate brain surgery (bottom) is needed to correct a problem. Here, a laser cuts the tissue and stops the bleeding at the same time.

DISCOVER BY *Problem Solving*

Get a sheet of newspaper and crumple the newspaper into a tight ball. Estimate the size of the ball. Then find a smooth ball of approximately the same size. In your journal, compare the surface area and volume of the crumpled-up newspaper ball to the surface area and volume of the smooth ball. Which has the larger surface area? Why? Then compare the crumpled-up newspaper to the wrinkled outer surface of the brain. In relation to surface area and volume, why is the wrinkled surface of the cerebrum an advantage? ✐

Below the cerebrum lies the **brain stem.** The brain stem is the place where the brain and spinal cord meet. The brain stem relays information received from the spinal cord to the rest of the brain. Many of your body's functions are controlled by this part of the brain. Heartbeat rate, breathing rate, coughing, and swallowing are all controlled by the brain stem. The **cerebellum** controls muscle coordination, balance, and muscle tone. It is located below the cerebrum and behind the brain stem.

*W*here are the greatest number of skin receptors located?

MATERIALS

cork, wire, centimeter ruler, blindfold

PROCEDURE

1. Cut two pieces of wire about 4 cm long.
2. Stick one piece of wire into one of the ends of the cork. Measure a distance about 5 mm away from the first wire and stick a second wire into the cork. Check your measurement to make sure the ends of the wire are about 5 mm apart.
3. Blindfold your partner. Tell your partner that you will press one or both tips of the wire against his or her skin. Then press either one or both tips of the wire gently against the skin of the forearm. Ask your partner whether he or she feels one or two points. Do several tests in different regions of the forearm and record your results.
4. Next, test other areas of the body. Try the palm of the hand, fingertips, back of the hand, and the elbow. Record the results of each test.
5. Next, have your partner do the testing while you are blindfolded.

APPLICATION

1. What conclusions were you able to draw about the number of receptors in the different parts of the body?
2. Compare the recorded results of the tests of several classmates and report your findings.
3. How might the results of your testing be useful to you?
4. Why do you think some body parts have more receptors than others?

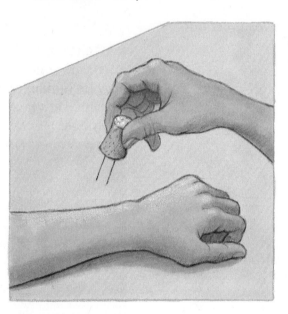

 ASK YOURSELF

List and describe the functions of each part of the brain.

SECTION 1 *REVIEW AND APPLICATION*

Reading Critically

1. What are sensory neurons?
2. Describe how an impulse moves from one neuron to another.
3. What structures of the nervous system are involved in a reflex?

Thinking Critically

4. Describe the nervous system as both a communication system and a control system.
5. Why might a doctor test reflexes such as the knee-jerk reflex during a general examination of your health?

The Senses

Objectives

List and *describe* the five senses.

Identify the major sense organs of the body and *relate* each to the appropriate stimulus.

Explain how each of the major sense organs functions.

Your brain is an amazing organ that performs many functions to keep you alive. Yet, your brain would be useless without receptor cells. Receptor cells provide your brain with information about your internal and external environment.

The buzz of a bee near her ear causes the pitcher to move off the mound for a moment. The batter squeezes the bat tightly and feels every ripple in the wood's grain. The umpire cries, "Let's play ball!" All eyes focus on the pitcher.

Some receptors are scattered throughout your body, while others are concentrated in your sense organs. These receptors provide humans with five senses: touch, taste, smell, sight, and hearing. Let's take a closer look at the sense organs and their functions.

Figure 19–9. When one of the senses is impaired, people devise ways to function without it. The person shown here is visually impaired. The guide dog serves as her eyes.

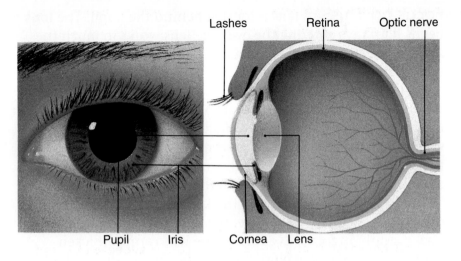

Lashes Retina Optic nerve

Pupil Iris Cornea Lens

Figure 19–10. The eye is a complex sense organ that is stimulated by light.

Vision

A white blur leaves the pitcher's hand and streaks toward the plate. Can the batter's eyes follow this blazing pitch?

Your ability to see things begins with your eyes. Each eye is a slightly flattened ball about the size of a Ping-Pong ball. Three layers of tissue form the hollow eyeball. The center is filled with a clear, jellylike fluid. This fluid gives the eyeball its shape and helps keep the inner surface moist. A tough covering of tissue forms the outer layer of the eye. Only the front portion of this layer is clear; the rest is white. The clear portion, or **cornea,** lets light pass into your eye.

The second layer of tissue absorbs stray light. This layer has many capillaries that bring blood to the eye tissue. The front part of this layer is a colored ring called the *iris*. The *pupil* is the opening in the center of the iris.

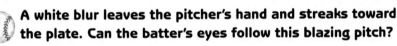

DISCOVER BY *Observing*

Enter a dimly lit room and look at your pupils in a mirror. Now turn on a bright lamp or flashlight. Look at your pupils again. What did you observe? Repeat the test, using a stopwatch to time how long it takes for your pupils to get larger and smaller as you change the lighting.

The opening and closing of the iris changes the size of the pupil. When the light is dim, the pupil gets larger. In bright light, the pupil gets smaller. The changes in the size of the pupil allow for optimal vision.

Focus on This   A lens is located behind the pupil. The **lens** bends light rays entering the eye. The lens works in much the same way as the lens of a camera, enabling you to focus on objects any distance from your eye.

The **retina** (RET ih nuh) is the layer of light-sensitive cells that lines the inside of the eyeball. Millions of these cells make up the surface of the retina. An image is made on the retina in much the same way as an image is made on film. A camera lens moves back and forth to focus the image on the film. The lens in your eye focuses an image onto the retina by changing shape. The lens becomes rounder and thicker for closer vision. As an object moves closer to your eye, the lens thickens to keep the image clearly in focus. Eventually, the lens cannot thicken enough and the image blurs. Extend your right arm and look at your index finger. Slowly move your finger toward one eye until the image is blurred. This is the point at which the lens can no longer thicken.

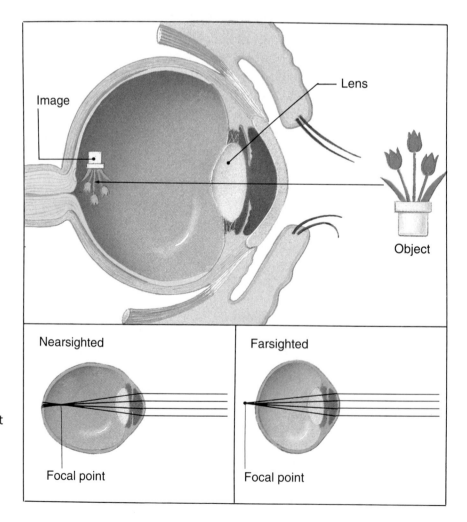

Image

Lens

Object

Nearsighted

Focal point

Farsighted

Focal point

Figure 19–11. Light rays from an object cross after they pass through the lens. The image of the object that forms on the retina is upside-down. What conditions result when the light rays do not form on the retina?

Sometimes the lens is not able to focus an image on your retina. In such cases, the focused image falls in front of or behind the retina. If the image is in front of the retina, the condition is called *nearsightedness*. If it falls behind the retina, the condition is known as *farsightedness*. A nearsighted person sees close objects much better than far objects. It is just the opposite for farsighted people. Fortunately, both conditions can be corrected with eyeglasses or contact lenses.

Seeing Red and Blue

How are you able to see the color of a ball? Two types of receptors make up the outer surface of the retina. Receptors called *rods* are most sensitive to light. Receptors called *cones* are most sensitive to color.

Have you ever noticed that you cannot see colors very well by moonlight? Your inability to see color in dim light is due to the fact that cones work only in bright light. Rods, on the other hand, work well in dim light but produce only black-and-white vision. Night vision depends on rods.

In Sight

Once it is formed on the retina, an image is sent to the brain as a nerve impulse. Sensory neurons attached to each rod and cone come together to form the *optic nerve*. Impulses move along the optic nerve to your brain for processing. Vision is a combination of the image seen by the eye and your brain's interpretation of the information it receives about that image. The brain's interpretation does not always agree with reality. This is why illusions are possible.

Figure 19–12. Even if you have normal vision, your perception may be different from someone else's perception. In this optical illusion, some people see an old woman while others see a young woman. If you look first for one and then for the other, you will probably see both.

 **Observing**

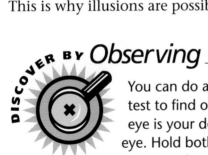 You can do a simple test to find out which eye is your dominant eye. Hold both of your arms out in front of your face. Now use your thumbs and index fingers to form a triangle as shown. With both eyes open, look at a distant object through the triangle. Next, shut one eye and then the other, looking at the same distant object each time. What did you observe? The eye that allowed you to see the same as you did with both eyes is your dominant eye. ✎

 ASK YOURSELF

Name the parts of the eye and describe the function of each part.

Hearing

Figure 19–13. The sense of hearing can be damaged or destroyed by repeated exposure to sounds that are too loud.

Figure 19–14. Sound causes the air to vibrate. The ear translates these vibrations into nerve impulses, which are sent to the brain.

The murmur of the crowd dies away. The crowd can almost hear the whoosh of the ball as it cuts through the air. It's a fly ball. The fielder goes deep to right field and . . . she catches it!

Your ears receive two types of information. You know that your ears enable you to hear. But did you know your ears also help control your sense of balance?

Through the Canal   Your ear is divided into three parts: the *outer ear,* the *middle ear,* and the *inner ear.* The outer ear is what you most often think of as your ear because it is the part you see. The shape of the outer ear helps it collect sound waves and direct them into the ear canal, which leads deeper into the skull. A thin piece of tissue called the *eardrum* stretches across the canal.

Drumbeats   Sound waves travel down the ear canal and strike the surface of the eardrum, causing it to vibrate. The middle ear is found on the other side of the eardrum. Three tiny bones—the hammer, the anvil, and the stirrup—form a bridge across the middle ear. The vibrating eardrum vibrates the hammer. The hammer transfers the vibrations to the anvil, and the anvil, in turn, relays the vibrations to the stirrup. The vibrations then move into the inner ear.

Under Pressure   The middle ear is connected to the throat by the Eustachian (yoo STAY shuhn) tube. The purpose of this tube is to equalize pressure in the middle ear with the pressure of the atmosphere. If you hold your nose and swallow, the tube closes and you can feel the pressure change. You can also feel pressure changes when you dive deep into water or fly in an airplane. If you did not have this tube to equalize pressure on both sides of your eardrum, the eardrum would rupture, causing severe pain.

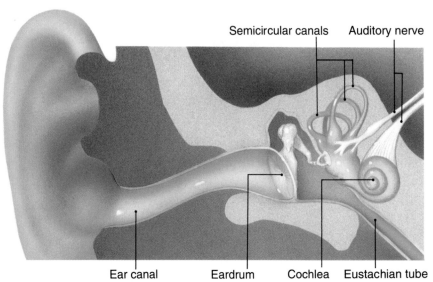

Semicircular canals   Auditory nerve

Ear canal   Eardrum   Cochlea   Eustachian tube

Message Received   A series of curved, fluid-filled tubes makes up the inner ear. The spiral-shaped *cochlea* (KAHK lee uh), which looks somewhat like a snail, forms one part of the inner ear. Receptor cells with tiny hairs on them line the inner walls of the cochlea. Movement of the bones of the middle ear vibrates the membrane of the inner ear, which causes the liquid in the cochlea to vibrate. The vibrating liquid moves the tiny hairs of the receptor cells. The receptors change the vibrations to nerve impulses that travel along the auditory nerve to the brain. The brain processes these signals and interprets them.

Your ears are amazingly sensitive organs. They can distinguish over 300 000 different sounds. Your ears are made of delicate parts that can be easily damaged. Loud noises, especially repeated and prolonged exposure to them, can lead to a loss of hearing. Noise is measured in units called *decibels*. Noise at 120 decibels can be painful. Frequent exposure to noise levels as low as 85 decibels can damage delicate structures in your ears.

TABLE 19-1: MEASUREMENT OF SOUND		
Jet takeoff at close range	140	
	120	Live rock band
Power mower	100	
	85	Subway train
Vacuum cleaner	80	
	70	Telephone bell
Normal talk	60	
	20	Whisper
Normal breathing	10	

Balancing Act   The sensory organs that help control balance are also located in your inner ear. Three fluid-filled tubes, called the *semicircular canals,* are lined with receptors. When you bend over, lie down, spin around, or do a flip, the fluid in these canals moves. The receptors detect this motion and send nerve impulses to your brain. Your brain sends signals telling your muscles to move in ways that help control your balance.

Tip your head to the left. The liquid within the semicircular canals is now pressing on a different group of receptors. Your brain interprets this information and tells you that you have changed position. Sometimes the information from the inner ear does not agree with the information gathered by your eyes.

Have you ever been dizzy or had an upset stomach while you were in a car or an airplane? Information from your eyes indicates that you are not moving. At the same time, information from your inner ear tells your brain that you are moving. This conflict sometimes causes motion sickness.

Figure 19–15. Your sense of balance is located in the inner ear. This acrobat must have an excellent sense of balance.

▶ **ASK YOURSELF**

What happens to sound waves that enter the ear?

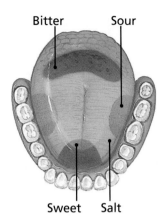

Bitter   Sour

Sweet   Salt

Figure 19–16. Note the location and distribution of taste buds on the tongue.

Taste and Smell

 As you wait for the next batter to come to the plate, you notice her sipping a cool drink. She then chooses a bat, swings it, and stands up at the plate. She's ready for her turn at bat.

The batter can taste the drink because of her senses of smell and taste. These senses depend on chemicals in the food you eat and in the air you breathe. These chemicals affect taste and smell receptors that are found mainly on your tongue and in your nose. The receptors of the tongue are called *taste buds*. There are four kinds of taste buds, and each responds to one of four basic tastes: sweet, sour, salty, and bitter. These four kinds of receptors are grouped in different areas on the tongue. Sweet receptors are found near the tip of the tongue, while bitter receptors are near the back of the tongue and sour receptors are at the sides. Receptors for salty taste are located between the sweet and sour receptors. You can see the location of these receptors in the drawing.

Your sense of smell works closely with your sense of taste. You might wonder how you can taste so many different flavors with only four kinds of taste receptors. The answer lies in your nose, which senses chemicals. A small area, about the size of a postage stamp, at the top of each nostril has many more receptor cells than the tongue. These receptors can sense more than 10 000 different chemicals, giving you your sense of smell. Your sense of smell greatly improves your sense of taste. With the help of your nose, you taste the flavor of strawberries as well as their sweetness. The next activity can show you how much your sense of taste depends on smell.

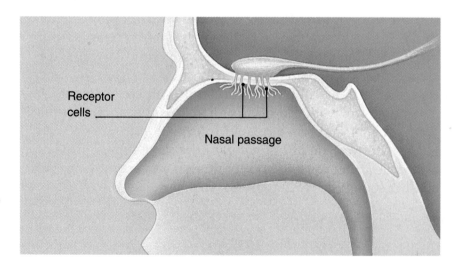

Receptor cells

Nasal passage

Figure 19–17. Receptor cells high in your nose allow you to detect odors.

Discover by Doing

Close your eyes and tightly hold your nose. Have your partner feed you small pieces of apple, potato, and onion. Try to identify each piece without looking. Describe the taste of each piece. Why was it so difficult to identify each food? ✎

As molecules are released from the food you eat, some move into the nasal passages. These molecules dissolve in a sticky substance called *mucus,* which lines the inside of your nose. The dissolved chemicals trigger chemical receptors in the nose. The receptors produce nerve impulses that move along sensory neurons to the brain for processing. Taste results from the brain's evaluation of information received from the nose and tongue. How does food taste when you have a stuffed nose due to a cold? Without a sense of smell, your sense of taste is limited.

▼ ASK YOURSELF

Where are the receptors that detect chemicals located?

Touch

The bat meets the ball with a crack, and a tingle moves up the batter's arms from the bat. The batter doesn't even move as the ball is going, going, gone over the center field fence.

If you were to shut your eyes and close off your ears and nose, you could still learn a lot about your surroundings by using your sense of touch. The sense of touch comes from individual receptors scattered over the entire surface of your body. Along with the sense of touch, the body's surface receptors also detect heat, pressure, and pain. You can feel even the lightest contact on your skin. Touch your finger to an ice cube, and the receptors tell you that it lacks heat. Very hot water can stimulate heat receptors and sometimes pain receptors as well.

Figure 19–18. There are many different types of receptors.

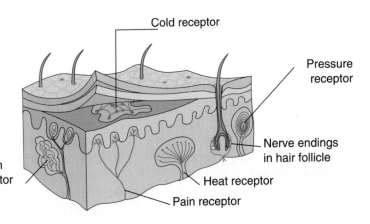

Cold receptor
Pressure receptor
Nerve endings in hair follicle
Touch receptor
Heat receptor
Pain receptor

Figure 19–19. Why is it important that you have pain receptors in most tissues of your body?

Why are your fingertips and lips more sensitive than the backs of your hands? What could cause this difference? Receptors for touch, pressure, and temperature are spread unevenly over the body. A small square area on the front side of your fingertip may have 100 touch receptors, while an equal area on the back side has only ten. Areas of your body that have closely spaced receptors are more sensitive than areas with fewer receptors. Touch receptors lie near the surface of the skin, but pressure receptors lie much deeper. Pain receptors can be found in almost every part of the body. These receptors help protect the body from injury by sending a warning of possible danger to the brain. In the next activity, you can discover an ancient art that is used to control pain.

DISCOVER BY *Researching*

For hundreds of years, the Chinese have practiced a method of pain relief called *acupuncture*. This method involves placing needles into various parts of the body. Go to the library and look up acupuncture. Write a report on this procedure to relieve pain.

◢ ASK YOURSELF

What types of sense receptors are in your skin?

SECTION 2 *REVIEW AND APPLICATION*

Reading Critically
1. Name the sense organs involved with each sense.
2. Why are some areas of the body more sensitive to touch than others?
3. Explain how each sense organ functions.

Thinking Critically
4. What might cause a condition, called night blindness, in which a person has difficulty seeing at night?
5. Why do surgeons wear such thin rubber gloves when doing surgery?
6. Why should workers operating power grinders wear eye and ear protection?

SKILL

Observing Sound Waves

▶ **MATERIALS**

- empty soup can • piece of thin rubber membrane • rubber band • glue
- small piece of mirror • flashlight • can opener

▼ **PROCEDURE**

1. Using the can opener, remove both ends of the soup can. **CAUTION: Be careful of the sharp edges of the can.**
2. Stretch the rubber membrane over one end of the can. Hold the piece of rubber in place with a rubber band.
3. Glue the piece of mirror, slightly off center, to the rubber membrane.

4. Hold the open end of the can near your mouth and shine the light toward the mirror. Position the can so that the light reflecting from the mirror strikes a shaded wall. You should be able to see the mirror's reflection easily.
5. Talk into the open end of the can. Vary the volume and the pitch of your voice.

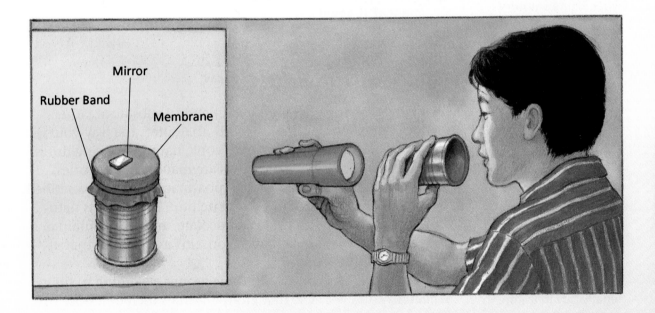

Mirror

Rubber Band

Membrane

▶ **APPLICATION**

1. How do the patterns made by the sound change with volume and pitch?
2. Have two people talk normally into their "sound visualizers" at the same time and with their reflections striking the wall at about the same place. Can you detect any difference in the vibrations?

✳ *Using What You Have Learned*

Describe any changes in the vibrations. How is your "sound visualizer" like an eardrum?

Alcohol, Tobacco, and Other Drugs

Cold medicines can make a person feel better, but they can also affect the person's reaction time. Most people use such medicines only as needed. But some people abuse medicines and other substances. Substance abuse is a problem that touches every part of society. It is a problem of the young, the old, the rich, and the poor. It costs the United States over 50 billion dollars in lost productivity, medical expenses, and crime.

The next batter swings at the ball, and she misses. Could the cold medicine she is taking be affecting her performance? It could be!

Drugs and Medicines

Generally speaking, a **drug** is any chemical substance, other than water, oxygen, and most foods, that alters the way your body works. Drug use is not new. People have been using drugs for thousands of years. Many drugs are made in laboratories, prescribed by doctors, and sold by pharmacies. When prescribed by doctors, these drugs can be very helpful. Many foods naturally contain drugs. Coffee, tea, chocolate, and cola contain a drug called *caffeine*. Tobacco also contains a substance that affects the nervous system.

Drugs can be divided into three groups according to their availability and use. The first group of drugs includes those that cannot legally be bought or sold by anyone in this country for any reason. They are illegal for any use, including medical use, and are extremely dangerous. Examples of drugs in this group are heroin and LSD.

Figure 19–20. Many substances that you can legally buy in a store contain drugs.

Drugs in the second group have medical uses, but they can be just as dangerous as those in the first group if used improperly. These drugs can be used when prescribed by a doctor. Examples of these drugs are tranquilizers, steroids, and morphine. Although these drugs have some legal uses, they are often bought illegally and used without a doctor's supervision.

Drugs in the third group can be bought by anyone and used by following the directions on the package. These drugs are often called over-the-counter drugs. Most headache and cold medicines are examples of over-the-counter drugs. However, even these drugs can be harmful if the directions for their use are not properly followed.

Medicines are drugs that can cure illness, heal the body, and relieve symptoms of disease or injury. Some medicines may poison and kill disease-causing organisms. For example, your doctor may give you *antibiotics* to help your body recover from an infection. They kill harmful organisms that invade your body.

Other medicines contain *psychoactive drugs*. These drugs affect the nervous system and can change a person's behavior. Some psychoactive drugs may cause a person to feel relaxed or sleepy. Others may make a person feel more energetic. If used improperly, psychoactive drugs are extremely dangerous.

Figure 19–21. A pharmacist can give advice on the action and effects of over-the-counter drugs. The label on a prescription drug gives instructions for its use. These instructions should be followed carefully.

 ASK YOURSELF

How are medicines different from other drugs?

Drug Abuse

The incorrect and unsafe use of a drug is called **drug abuse.** Drug abuse includes the taking of drugs for which you have no prescription or no medical reason for using. Taking more of a prescribed drug or taking it more often than recommended also is a form of drug abuse. Almost any drug can be abused. The drug does not have to be illegal or dangerous to be abused. Even over-the-counter drugs that people take for a cold can be abused. Drug abuse can damage body tissues, have a negative effect on behavior, and even lead to death.

Many drugs that are safe in small quantities are *toxic,* or poisonous, in larger amounts. High doses or prolonged use of many drugs can damage the liver, kidneys, lungs, heart, and brain. This damage can increase the risk of disease and death.

If a drug is taken over a long period, the person will gradually get used to the drug, or build up a *tolerance*. This means that a higher dose of the drug is needed to produce the same effect. As a result, the person takes a larger, more dangerous quantity, or *overdose*, of the drug. An overdose may result in serious physical or mental harm or even death.

Some drugs are dangerous because they are habit-forming. A person may use a drug so frequently that using the drug becomes a habit. A drug habit can lead to drug *addiction*, or dependence on a drug. A person with a drug addiction must continue taking the drug to feel good. Such a condition is known as a physical addiction. Psychological addiction exists when a person is emotionally dependent on the drug. A person with this type of addiction depends on a drug to produce a false sense of well-being.

When a person is physically addicted to a drug, his or her body needs the drug. A condition, called *withdrawal*, occurs when an addict does not use the drug. Withdrawal symptoms may include extreme nervousness, uncontrollable shaking, excessive sweating, nausea, and muscle cramps. Extreme feelings of sadness or depression may also occur with drug withdrawal.

DISCOVER BY *Researching*

There are many types of illegal drugs abused by people throughout the world. These include hallucinogens, narcotics, stimulants, and depressants. Select one of these groups of drugs, go to the library, and research the group of drugs. Write about what you learn.

ASK YOURSELF

How is drug abuse different from using drugs properly?

Figure 19–22. People who abuse drugs need help, such as counseling, to overcome their habit.

Tobacco

More than 80 percent of all Americans are aware that smoking can cause cancer. All tobacco products and advertisements carry warnings about the harmful effects of tobacco, yet billions of cigarettes are smoked each year. In the United States alone, between 40 and 50 million people smoke cigarettes. Why do people start or continue smoking when they know it can damage their health?

Figure 19–23. By law in the United States, warnings about the hazards of smoking cigarettes must be printed on the packages.

Many people smoke for social and emotional reasons. Some young people begin because they observe adults smoking or they are influenced by advertisements. Some people think that smoking is a sign of independence or rebellion. Other people smoke because they want to be like their friends. Smoking is a behavior that people try for many different reasons. The next activity can give you some idea as to why some people start to smoke.

DISCOVER BY Researching

Cigarette packages and ads must include several warnings from the Surgeon General of the United States that indicate smoking is harmful to your health. Ask some smokers you know why they started smoking. Write a short report on how to deal differently with these reasons or pressures. Be sure to include how a person could avoid starting to smoke. ✎

People often smoke while doing something else. Soon a person associates one behavior with the other. Once this happens, the person feels the need to smoke whenever the other behavior occurs. Such association makes a smoker psychologically dependent on smoking. Eventually, the smoking habit leads to a physical dependence. Smoking is a costly habit not only in terms of health but also in terms of money. The following activity will help you calculate the monetary costs of smoking.

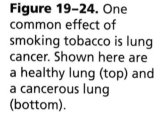

Smoking is an expensive habit. Investigate the cost of a package of cigarettes in three types of stores. Then use an average of these costs to determine how much money you would save in a week, a month, and a year if you never started a one-pack-a-day habit. ✎

Tobacco smoke contains over 300 different chemical substances. At least 15 of these are cancer causing. And one of these

Figure 19–24. One common effect of smoking tobacco is lung cancer. Shown here are a healthy lung (top) and a cancerous lung (bottom).

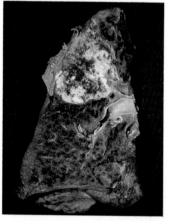

chemicals, **nicotine,** is a psychoactive drug. In small doses, it acts as a stimulant to the nervous system. It speeds the heartbeat rate and raises blood pressure. In large doses, nicotine acts as a depressant. It depresses the nervous system and slows the heartbeat. In doses as small as 70 mg, pure nicotine is poisonous and can cause death. An addicted smoker develops both a psychological and a physical addiction to nicotine. Withdrawal symptoms include nervousness, anxiety, shakiness, and loss of sleep.

Nicotine addiction is only one of the harmful effects of smoking. Tobacco smoke contains gases and tiny particles that damage the body of a smoker as well as those of nonsmokers who breathe the smoke. The respiratory, circulatory, and digestive systems are all harmed by smoke from tobacco products. But smoking does not harm just the body. Careless smoking can result in fires in homes, hotels, and forests. Cigarette butts and wrappers are a major source of litter.

The effects of smoking on the respiratory system are well known. Smoking makes breathing more difficult and increases the chances of respiratory infection and lung cancer. The mucus that lines the air passages traps smoke particles. As these particles cool, they form a sticky substance called *tar.* Tiny, hairlike cilia usually sweep the mucus up the air passages and out of the body so that the air passages are kept clean. However, the cilia have a very difficult time moving mucus that is heavy with tar.

The buildup of tar and other dirt particles damages the cells and cilia that line the air passages. New cells, more resistant to damage but less able to keep the air passages clean, are formed to replace damaged cells.

The dirt and tar that pass through the air passages also collect in the lungs. These sticky substances damage the delicate tissues of the air sacs, or alveoli. Tougher, less flexible cells form scar tissue in the lungs. This new tissue cannot expand as normal tissue does.

Figure 19–25. Many harmful effects are hidden in cigarette smoke. Don't start!

Tobacco smoke also affects the digestive tract. Smoke that is swallowed irritates the lining of the stomach and intestines. Swallowed smoke can cause indigestion and constipation and increases the chance of ulcers forming. Stomach and liver cancer are also more common in smokers than in nonsmokers.

Recently, more attention has been paid to the harmful effect of tobacco smoke on nonsmokers. Breathing smoke from other people's tobacco products is known as *passive smoking*. Studies show that more respiratory infections occur in children of smokers than in children of nonsmokers. Passive smoking may also cause other diseases of the lungs, including cancer, in nonsmokers. Husbands and wives of smokers have 3.4 times the risk of having a heart attack that husbands and wives of nonsmokers have. Concerns about passive smoking have resulted in a ban on smoking in airplanes and in many other public places.

Smoking is not the only dangerous use of tobacco. Thousands of people use smokeless tobacco, such as snuff and chewing tobacco. The National Cancer Society and other agencies have found that the use of such tobacco increases the risk of cancer of the lips and jaws and of other diseases of the mouth. Smokeless tobacco can be as habit-forming and addictive as tobacco that is smoked.

 **ASK YOURSELF**

What is the addictive drug in tobacco?

Alcohol

Like smoking, drinking alcoholic beverages is a learned behavior. People drink for many reasons. Alcoholic drinks are commonly served at social gatherings and business affairs. Some people drink because they feel pressured by others or because they want to feel like one of the crowd. Other people drink because they like the taste or the feeling alcohol provides. As a result, alcohol is the most commonly abused drug in our society.

Figure 19–26. It is estimated that well over one half of all accidents involve drivers who have been drinking alcoholic beverages.

Alcoholic beverages contain water, flavorings, and ethyl alcohol. Ethyl alcohol is a psychoactive drug that acts as a depressant, slowing impulses in the nervous system.

When a person drinks, about 20 percent of the alcohol passes directly into the bloodstream through the stomach wall. The rest enters the bloodstream through the wall of the small intestine. Once in the circulatory system, it is carried throughout the body. Alcohol in the body is slowly broken down by the liver. It takes the liver a little less than one hour to break down the alcohol in two bottles of beer.

Pure alcohol leaves a strong taste and a burning sensation in the mouth and throat. People who drink alcoholic beverages often say that they experience a freer and happier feeling. For these reasons, even though alcohol is a depressant, it is mistakenly thought to be a stimulant.

Alcohol interferes with the function of the brain. Its effects begin with a feeling of relaxation. As more alcohol reaches the brain, drinkers become less aware of their environment and themselves. Most activities that depend on the senses and muscular movement are impaired by the use of alcohol. Alcohol also causes people to lose their self-control and impairs their judgment. Reaction time becomes much slower. Increased consumption usually leads to blurred vision and loss of coordination. It is easy to understand why people who drink alcohol and drive motor vehicles are responsible for thousands of deaths every year.

The effects of alcohol on a person's behavior depend on how much alcohol reaches the brain through the bloodstream. The alcohol level in the blood depends on several factors, including (1) how much and how fast alcohol is consumed, (2) how much the person drinking weighs, and (3) whether or not the stomach is empty. When the percentage of alcohol in the blood reaches 0.10 percent, a person is usually considered to be drunk and should not drive or operate any tool or machine that could cause injury.

As with other drugs, alcohol can create a psychological dependence. Occasionally, it can even lead to physical dependence. People who become dependent on alcohol are called *alcoholics*. They use alcohol as an escape from stress and other situations that they have difficulty facing. Heavy use of alcohol can lead to serious damage of the liver. Alcoholism is the most serious drug problem in the United States, which has over 5 million alcoholics.

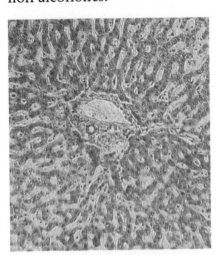 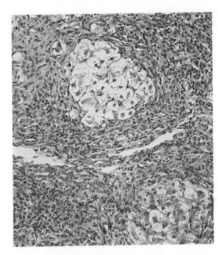

Figure 19–27. Cirrhosis of the liver can result from drinking too much alcohol. The liver on the right was removed from the body of an alcoholic.

 ASK YOURSELF

How does alcohol affect the functions of the brain?

SECTION 3 *REVIEW AND APPLICATION*

Reading Critically

1. Distinguish between drugs and medicines.
2. What are some types of drug abuse?
3. How do tobacco and alcohol affect a person?

Thinking Critically

4. Why is the ability to develop a tolerance to a drug dangerous?
5. What effects do nicotine and caffeine have in common?
6. Why could it be dangerous to work at a construction site with an alcoholic?

INVESTIGATION

Mapping Your Sensory Receptors

► **MATERIALS**

- water-soluble, fine-tipped felt pen • two paper clips • straight pin
- dull pencil • ice cube • hot water • two small cups

▼ **PROCEDURE**

1. Use a fine-tipped pen to draw a 2-cm square on the back of your hand. Divide the square into a grid of 16 equal-sized squares as shown.

2. Make a similar but larger grid on a sheet of paper. This grid will serve as a recording device.

3. Place an ice cube in one cup and hot water in the other. Unbend one section of each paper clip. Put the tip of the straight end of one clip on the ice cube and put the straight end of the other clip in the hot water.

4. You are now ready to locate receptors for pain, pressure, and heat. You will test each part of the grid to determine whether pain, pressure, or hot and cold

sensory receptors are located there. As you test each part, you should invent a symbol for each receptor. You will then use these symbols to record the results on your paper grid.

5. CAUTION: Be careful when using the pin so that

you do not break the surface of the skin. Use a pin for pain, a dull pencil for pressure, the paper clip touched to an ice cube for cold, and the paper clip dipped in hot water for hot. Use each tool in each part of the grid and record your findings.

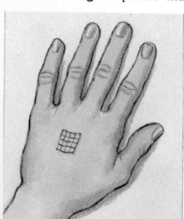

► **ANALYSES AND CONCLUSIONS**

1. What does your skin receptor map look like? Which kind of receptor seems to be most common? Least common?

2. Why do you think you have more of one receptor than another?

3. Compare your receptor map with those of other students. How are they different?

► **APPLICATION**

Suppose you were to go to the doctor for a shot. How might you use a sensor map to

make your visit more pleasant? How could a receptor map help you decide which parts of your body would be most sensitive to cold weather?

✳ *Discover More*

Mark grids on other areas of your body, such as the heel of your foot, the palm of your hand, and your forearm. Make predictions about the types of receptors most common in these areas. Test to find out how accurate your predictions are.

The Big Idea

The nervous system is a control and message system that operates throughout the body. The structure of the system enables the senses to send and receive messages to and from the CNS. The system interprets information in order to control body functions and to assure well-being. The interaction of microscopic receptors and larger sense organs make it possible to see, hear, smell, taste, and touch.

Alcohol, tobacco, and drugs interfere with the proper functioning of the nervous system as well as other systems in the body. This interference can have long-term and short-term effects on individuals who use or abuse these substances. People who use and abuse such substances can cause great harm to themselves and to other people.

For Your Journal

You have probably changed your thinking about how the nervous system works. Review the journal entry you wrote about the nervous system before you began your study of this chapter. How would you explain why Casey struck out? What role might the nervous system have played in his failure to hit the ball?

Connecting Ideas

Copy the unfinished concept map into your journal. Complete the map by filling in the blank boxes.

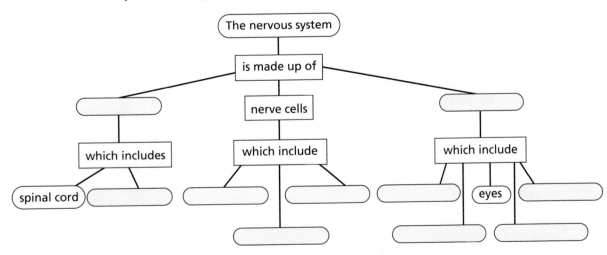

Understanding Vocabulary

Demonstrate your understanding of these terms by using each in a sentence. You may use more than one term in a sentence.

1. receptor (513)   **2.** neuron (514)   **3.** cerebrum (518)
4. cerebellum (518)   **5.** retina (522)   **6.** lens (522)
7. semicircular canals (525)   **8.** drug abuse (531)   **9.** nicotine (534)

Understanding Concepts

MULTIPLE CHOICE

10. Which of the following is *not* a type of neuron?
   a) dendrite   **b)** sensory
   c) motor   **d)** interneuron

11. Imagine being a police officer trying to determine if a driver has had too much to drink. Which of the following symptoms would *not* be helpful in determining if a person has had too much alcohol?
   a) lack of coordination   **b)** age
   c) a slower reaction rate   **d)** slurred speech

12. The organ most likely to be damaged by excessive use of alcohol is the
   a) kidneys. **b)** lungs. **c)** brain. **d)** liver.

13. A distance sense is one that could be used to identify something that is 10 m away. Which of the following would *not* be considered a distance sense?
   a) hearing   **b)** touch
   c) vision   **d)** all of the above

14. How are dendrites and axons similar?
   a) They both are parts of nerve cells.
   b) They both carry messages to another structure.
   c) They both carry messages away from another structure.
   d) All of the above are true.

15. Which part of your brain is responsible for thought processes?
   a) brain stem   **b)** cerebrum
   c) cerebellum   **d)** receptors

16. When you see a car coming at you and you have to get out of the way, which neurons carry the message to your muscles, telling them to act?
   a) sensory neurons   **b)** spinal neurons
   c) interneurons   **d)** motor neurons

SHORT ANSWER

17. Describe the difference between a voluntary action and a reflex action.

18. Explain why it is more difficult to give up drugs after you have become addicted.

19. Trace what happens to a sound wave that enters your outer ear.

Interpreting Graphics

20. Look at the diagram of the eyeball. Describe what happens at each lettered point as the light from the apple enters the eye.

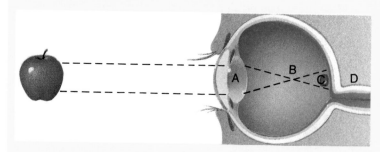

Reviewing Themes

21. *Systems and Structures*
A close look at the nervous system helps you understand that it consists of structures that range in size from microscopic to large enough to be seen by the naked eye. You also learn that the structure of each part is adapted to its function. Give three examples of structures in the nervous system that demonstrate wide differences in scale. Also give three examples in which structure is particularly well suited to function. Explain how the parts are suited to the job they do.

22. *Systems and Structures*
Select two examples within the nervous system of how parts interact to perform a bodily function.

Thinking Critically

23. Neurons damaged by injury cannot be repaired or replaced. Why would serious damage to your brain stem almost certainly result in death?

24. Tiny blood vessels near the skin dilate (widen) when a person consumes a lot of alcoholic beverages. This causes heat to escape from the body through the skin. The escaping heat gives a warm sensation. Some people, therefore, believe that drinking alcohol will keep them warm on a cold day. What is wrong with this reasoning?

25. Why is it so difficult for a long-time smoker to quit?

26. Why are your eyes and ears referred to as the distance receptors? Why is it important to have distance receptors?

27. Why might taking a prescribed amount of antibiotic in half the amount of time as prescribed be less effective than taking the antibiotic in the prescribed amount of time?

28. Most animals that are more active at night are colorblind. Explain why.

29. Cats have the ability to land on their feet. Based on what you have learned in this chapter, explain why this is so.

Discovery Through Reading

Parker, Steve. *The Brain and Nervous System.* Watts, 1990.
Describes the control system of the body: its structure and function, sleep mechanism, reflexes and autonomic nervous system, the two brain hemispheres, and health care.

Reproduction and Development

Twins' convention

*H*ave you ever seen a pair of identical twins? What was your first reaction? Were you amazed or curious? Have you ever wondered what causes identical twins? It's just one part of the story of human reproduction. You may find that the explanation of how two cells can join together and eventually develop into a human being to be just as amazing as the myths and stories about twins.

People have always been fascinated by identical twins.

Twins appear as characters in many myths, legends, and stories. From Shakespeare to Walt Disney, story plots have revolved around identical twins being mistaken for one another. Dozens of movies have focused on the relationship between twins, often characterizing one as "good" and the other as "evil."

In some cultures, twins are regarded as a sign of good fortune. Often twins appear in the legends of these cultures, sometimes as the sun and the moon, or as darkness and light.

Other cultures feared twins, believing that their unlooked-for arrival was a curse and even going so far as to kill one or both twins to protect the community. For some nomadic tribes or societies with limited resources, two babies were a burden for the mother, and one child was killed to ensure the survival of the other. Some cultures associated twins with animals' multiple births. Depending on the culture's attitude toward animals, this association could help or harm twins.

As science has revealed more about the process of human reproduction, superstitions and fears about twins have largely disappeared. In fact, identical twins have been important in scientific research. Because identical twins are born with identical genes, research into the effects of heredity versus environment has often involved pairs of identical twins. Also interesting to scientists are pairs of identical twins who have been raised apart. How similar will two people be if they have the same genetic material but different environmental influences? These studies have provided some intriguing data, proving that twins have not completely lost their power to amaze us!

What is it like to have an identical twin?

For Your Journal

- What myths have you heard about how babies are born? Why do such myths exist?
- What do you think causes identical twins?
- What do you think it would be like to have an identical twin?

The Endocrine System

Objectives

Distinguish *between endocrine glands and other glands in the body.*

Identify *the glands of the endocrine system and* **state** *their functions.*

Explain *what is meant by a feedback control system.*

Your body systems do not act alone; they function together to ensure the health and smooth functioning of your body. For example, the skeletal and muscular systems are both responsible for your body's ability to move. The respiratory and excretory systems both work to remove wastes from your body. Two systems—the nervous system and the endocrine system—help regulate your body's functions. Recall that the nervous system works through electrical impulses. The **endocrine** (EHN duh krihn) **system** controls body functions with chemicals. Your growth rate, the amount of sugar in your blood, and the development of your reproductive organs are three functions controlled by your endocrine system. The nervous system responds to a stimulus in an instant. The endocrine system can take minutes, hours, or even months to respond. The nervous system and the endocrine system work together to control your body systems.

Endocrine Glands

A **gland** is a group of cells that makes chemicals for your body. Salivary glands produce saliva. Gastric glands that line the walls of the stomach produce digestive juices. Sweat glands produce perspiration. Chemicals leave these glands through small tubes, or ducts, and flow directly to where they are needed. The chemicals of endocrine glands, however, go directly into the bloodstream because endocrine glands do not have ducts.

Figure 20–1. The endocrine system works together with the other systems of the body to help you react quickly to fear or excitement.

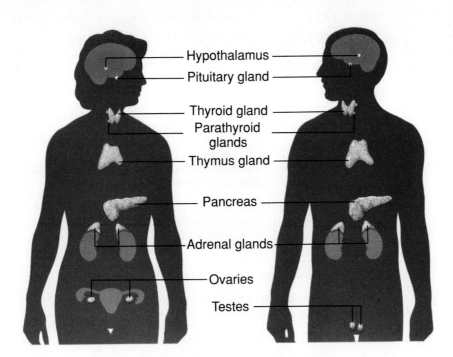

Figure 20–2. The endocrine glands monitor and control many of your body functions.

Labels: Hypothalamus, Pituitary gland, Thyroid gland, Parathyroid glands, Thymus gland, Pancreas, Adrenal glands, Ovaries, Testes

Chemical Messengers

Endocrine glands produce chemicals called *hormones*. **Hormones** are chemical messengers that control other cells and organs. Hormones enter the bloodstream and travel throughout the body. This method of transport allows an endocrine gland to signal cells and organs some distance away. An endocrine gland in the head can control the actions of an organ located elsewhere in your body. By traveling through the bloodstream, a hormone can affect many organs at the same time.

Stressed Out

The endocrine system includes the *adrenal* (uh DREE nuhl) *glands,* which are located on the top of each kidney. They prepare your body to deal with stress. The hormone *adrenalin,* which is released by the adrenal glands, speeds up your heartbeat and breathing rates. Adrenalin also opens your air passages and decreases your rate of digestion. Each of these actions prepares your body either to respond to danger or to run from it. This response is sometimes call the "fight or flight" response. You may have noticed these effects when you were angry or frightened.

There are several other endocrine glands. The following table lists the names and functions of the endocrine glands.

 **ASK YOURSELF**

What are hormones?

Table 20-1

The Endocrine System

Gland	Hormone	Function
Thyroid	Thyroxine	Regulates the release of energy in body; iodine is needed to make thyroxine
Parathyroid	Parathyroid hormone	Controls the use of calcium
Pituitary	Growth hormone and other pituitary hormones	Regulates bone growth; controls the release of hormones from other glands; controls kidney function; regulates blood pressure
Adrenals	Adrenalin	Increases heartbeat rate, blood flow, and the amount of sugar in the blood; activates the nervous system; regulates water and mineral balance in body tissues
Pancreas	Insulin	Allows liver to store sugar; regulates use of sugar
Ovaries (female)	Estrogen (female sex hormone)	Controls the development of reproductive organs and female characteristics
Testes (male)	Testosterone (male sex hormone)	Controls the development of reproductive organs and male characteristics

Feedback Control

The *pituitary gland,* located just below the brain, is often called the body's "master gland." This gland is about the size of a kidney bean, yet it is very important. Hormones given off by the pituitary gland control the actions of the other endocrine glands. The part of the brain called the *hypothalamus* (hy puh THAL uh muhs) controls the pituitary gland. This area of the brain is, in turn, affected by hormones produced by other glands in the endocrine system. For the endocrine glands to work properly, they should produce hormones only when needed. A process called a *feedback control system* controls endocrine glands. Let's look at an example of a feedback system.

Hot Enough for You?   Think about how the temperature inside your home is regulated. Often a home has a furnace with a thermostat. The thermostat can be set at a certain temperature so that it will automatically turn the furnace on and off. A low temperature causes the thermostat to send a message that turns the furnace on. Heat from the furnace then warms the house. Once the house is warm, the thermostat stops sending its message. The furnace turns off and the cycle begins again. The temperature of the house regulates the furnace in this feedback system.

A Personal Furnace   The functioning of the thyroid gland shows how a feedback control system works in your body. When you go outside on a cold day, your cells must produce more heat to keep you warm. A hormone called *thyroxine* (thy RAHK sihn) controls the amount of heat produced by your body's cells.

Large quantities of thyroxine cause the cells to work harder and produce more heat. The amount of thyroxine decreases as the cells work harder. The hypothalamus senses this decrease in thyroxine and sends a message to the pituitary gland. The pituitary, in turn, releases a hormone that increases the activity of the thyroid gland. As a result, the thyroid increases production of thyroxine. Increased thyroxine causes the body's cells to produce more heat.

When the hypothalamus senses the higher level of thyroxine, it stops sending its message to the pituitary gland. If your body temperature falls again, the process begins again. In this feedback control system, the level of thyroxine turns the system on and off. Use the following activity to learn more about the thyroid gland.

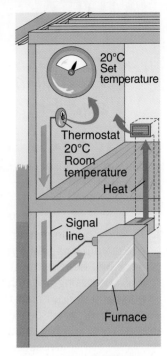

Figure 20-3. A thermostat is an important part of the feedback system that controls the furnace in most homes.

 **ᴰⁱˢᶜᵒᵛᵉʳ ᴮʸ** *Researching* ─────────────

Use reference books to find out more about the function of the thyroid gland. How is the body affected by too much thyroxine? By too little thyroxine? How can these conditions be treated?
Write a short summary of your research in your journal. Illustrate your report with diagrams.

▼ **ASK YOURSELF**

Diagram a feedback control system.

SECTION 1   *REVIEW AND APPLICATION*

Reading Critically

1. How are endocrine glands different from glands in the digestive system?

2. Why is the pituitary gland called the master gland?

Thinking Critically

3. Why would endocrine glands be unable to function as they do if they had ducts?

4. Give another example of a feedback control system.

INVESTIGATION

Predicting Changes in Heartbeat and Breathing Rates

▶ **MATERIALS**
- watch with second hand

▼ **PROCEDURE**

1. To find your heartbeat rate, press your fingers to your neck just below your ear and jaw. Using a watch with a second hand, count the number of beats per minute. You can also count for 10 seconds and then multiply by 6. Repeat the count several times and then find and record your average heartbeat rate while at rest.

2. Now find and record your breathing rate. Count the number of breaths you take in a minute. Again, you can count for 10 seconds and then multiply by 6. Repeat the count several times to find and record your average breathing rate while at rest.

3. CAUTION: Notify your teacher of any medical condition that prevents you from doing this activity. Perform five minutes of aerobic exercise, such as running in place or skipping rope.

4. Now find your heartbeat and breathing rates again as you did in steps 1 and 2. Record your heartbeat and breathing rates after exercise.

TABLE 1: HEARTBEAT AND BREATHING RATES		
	At Rest	**After Exercise**
Heartbeat Rate		
Breathing Rate		

▶ **ANALYSES AND CONCLUSIONS**

1. How did the aerobic exercise affect your heartbeat and breathing rates?

2. What other changes, if any, occurred?

▶ **APPLICATION**

Compare the changes caused by the aerobic exercise with the changes caused by adrenalin when you are excited or frightened.

✳ *Discover More*

Working in a group, list as many stressful situations as you can. Examples might be giving a speech in class, preparing for an important game, or watching a scary movie. Describe the physical changes that occur in the body before, during, and after stress-causing events.

The Reproductive Systems

Compare and contrast the male and female reproductive systems.

Describe the changes that occur during puberty.

Explain the process of menstruation.

A single sex cell from each parent fuses together to form a single cell. This single cell divides again and again. Although the cells start the same, they gradually become different. First, tissues form, then organs, then systems. About 266 days after the formation of the first cell, a new individual made of trillions of cells emerges. A new human being enters the world. How can such an event take place? Although all the body's systems help produce the new individual, the endocrine and reproductive systems are among the most important.

Male Reproductive System

The male reproductive system has two functions in the process of reproduction. It produces sex cells, or *sperm,* and it delivers the sperm to the female. Both the nervous system and endocrine glands control the male reproductive system.

The *testes* (TEHS teez) are endocrine glands that produce hormones and sperm cells. The testes are located outside the body cavity in a sac called the *scrotum* (SKROHT uhm). Sperm are sensitive to high temperatures and survive better at the slightly cooler temperatures found outside the body cavity.

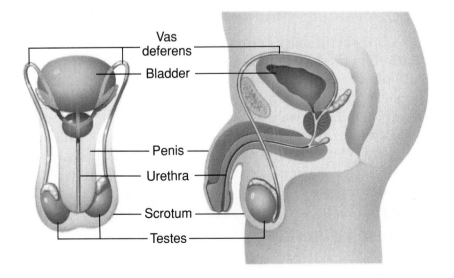

Vas deferens
Bladder
Penis
Urethra
Scrotum
Testes

Figure 20–4. Hormones produced by the endocrine glands affect or control the functions of the organs in the male reproductive system.

Figure 20–5. Muscle development is one change that occurs during puberty in males. The changes that occur during puberty are called *secondary sex characteristics.*

Hormones produced by the pituitary gland and the testes control the male reproductive system. The pituitary hormones control the production of sperm in the testes. A hormone called *testosterone* (tehs TAHS tuh rohn) is produced by the testes. Testosterone causes the changes that occur in the male body during **puberty** (PYOO buhr tee), the stage of life during which a person matures sexually. In males, these changes include the growth of a beard and body hair, as well as the deepening of the voice. During this period, a male's shoulders broaden in proportion to the hips, and the muscles increase in size and strength.

As you can see in the illustration, a sperm cell consists of three regions: the head, middle piece, and tail. The head of the sperm contains DNA, the hereditary material. The middle piece contains many mitochondria, which release the energy needed for movement. The tail section propels the cell forward.

Within the testes are tiny, tightly coiled tubes that produce the sperm. Together, the tubes of the two testes would stretch almost 7 m in length, if they were uncoiled. These sperm-producing tubes join to other sperm ducts, or *vas deferens* (vas DEHF uh rehnz), which lead back into the body cavity. The inner surfaces of the sperm ducts are lined with cilia. The motion of cilia, along with muscular contractions, sweep the sperm forward through the ducts to the urethra.

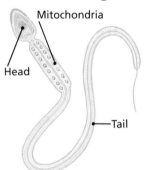
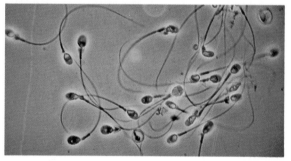

Figure 20–6. The male reproductive system produces sex cells called *sperm.*

As the sperm move through the vas deferens, they mix with a thick fluid produced by glands in the reproductive system. This fluid contains large amounts of sugar and other chemicals. The sugar provides nourishment to the sperm. The other chemicals moisten the sperm ducts and help the sperm move. Together, the sperm and the thick fluid form *semen* (SEE muhn). The urethra carries both semen and urine out of the body through the *penis* but never at the same time. Typically, about 250 million sperm cells are released from the body during each emission of sperm, which is called an *ejaculation.*

ASK YOURSELF

What are two functions of the testes?

Female Reproductive System

The female reproductive system produces hormones and female sex cells, called *ova* [singular, *ovum*], or eggs. It also receives the sperm and provides a place for the development of the offspring. Ova develop near the surface of two *ovaries,* which are located in the abdomen. The ovum is the largest cell in the human body and can be seen without a microscope. Generally, one ovary releases an ovum every 28 days. Once released from the ovary, the ovum enters the *Fallopian* (fah LOH pee uhn) *tube,* which leads from the ovary to the *uterus* (YOO tuh ruhs). Since there are two ovaries, there are also two Fallopian tubes.

Figure 20–7. The female reproductive system produces ova and provides a place for the offspring to develop.

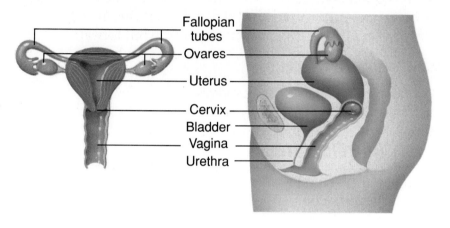

Fallopian tubes
Ovares
Uterus
Cervix
Bladder
Vagina
Urethra

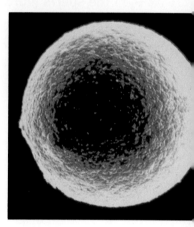

Thick walls of smooth muscle form the pear-shaped uterus. The lower end of the uterus narrows into the *cervix.* The cervix acts as a muscular gate that guards the uterus. The lower end of the cervix opens into the *vagina* (vuh JY nuh).

Hormones play an important role in controlling the female reproductive system. Hormones produced by the pituitary gland, the ovaries, and the uterus control the formation and release of ova. Hormones are also involved in the changes that take place in the female during puberty. These changes include the development of breasts, an increase in body hair, and the widening of the hips.

Sex cells, either male or female, are called *gametes.* In the following activity you can compare male and female gametes.

Figure 20–8. Secondary sex characteristics of females include the widening of the hips and the development of breasts.

ACTIVITY

What do sperm and ova look like?

MATERIALS
compound microscope, prepared slides of human sperm and ova

PROCEDURE
1. Place the slide of the sperm on the microscope stage and focus on low power. Draw what you see.
2. Switch to high power and examine the sperm more closely. Draw what you see.
3. Examine the slide of the ova as you did the slide of the sperm.

APPLICATION
1. Describe the structure of the sperm and the ova. How are they alike? How are they different?
2. Why do you think an ovum is so much larger than a sperm?
3. How does the design of these cells help them carry out their functions?

 ASK YOURSELF

What are the functions of the ovaries?

The Menstrual Cycle

After puberty, several events prepare the body of a female for the possible development of a baby. These events occur during a process called the *menstrual* (MEHN struhl) *cycle,* which begins about every 28 days.

Three events take place within each menstrual cycle. The first two events occur at the same time. First, during the first 14 days of each cycle, an ovum matures within an ovary. Second, the inner lining of the uterus thickens. This lining is made of a spongy tissue that has a rich supply of blood.

At about day 14 of the cycle, the ovum bursts free from the surface of the ovary. The release of a mature ovum is called *ovulation.* The ovum then enters the funnel-shaped opening of the Fallopian tube. As the mature ovum travels down the Fallopian tube, the lining of the uterus continues to thicken. If fertilization occurs, it will take place in the Fallopian tube. However, if fertilization does not take place, the third event of the menstrual cycle occurs. If a sperm does not fertilize the ovum, the thickened lining of the uterus begins to break down. The unfertilized ovum and the tissues of the lining of the uterus leave the body through the vagina. This process is called **menstruation** (mehn STRAY shuhn). Menstruation usually lasts from three to five days. As menstruation begins, hormones cause another ovum to begin maturing in the ovary. The cycle begins again.

DISCOVER BY Calculating

The average female produces about 500 mature eggs in her lifetime, at the rate of about one per month. During how many years of her life can she bear children? If a female begins to ovulate at the age of 13, how old will she be when her possible child-bearing years are over? ✐

The average length of the menstrual cycle is 28 days. However, when menstruation first begins after puberty, it might not be regular. Even after the cycle becomes regular, it varies for each person. Cycles can be as short as 21 days or as long as 35 days.

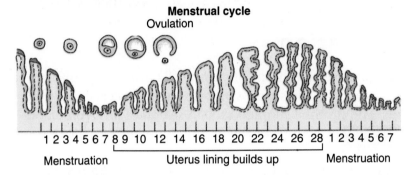

Menstrual cycle

1 2 3 4 5 6 7 8 9 10 12 14 16 18 20 22 24 26 28 1 2 3 4 5 6 7

Menstruation   Uterus lining builds up   Menstruation

Figure 20–9. The events of the average menstrual cycle occur over a period of 28 days.

If a sperm fertilizes the ovum before it reaches the uterus, menstruation does not occur. Instead, the fertilized egg attaches to the lining of the uterus, where it begins to grow. This attachment of the fertilized ovum is the beginning of pregnancy. Hormones produced by the uterus tell the reproductive system of this change. During pregnancy, the ovaries do not release new ova, and menstruation does not occur.

▼ ASK YOURSELF

Describe the menstrual cycle.

SECTION 2   REVIEW AND APPLICATION

Reading Critically
1. Why are the testes found outside the body cavity?
2. How are the functions of the male and female reproductive systems different?
3. What controls the menstrual cycle?

Thinking Critically
4. Why do you think the time of puberty is different for different people?
5. The changes that occur during puberty are called *secondary sex characteristics.* Explain why this is a good name.

Prenatal Development and Birth

Lennart Nilsson/Boehringer Ingelheim International Gmbh

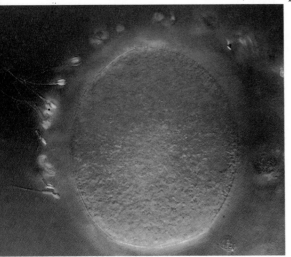

Figure 20–10.
Although many sperm surround the ovum, the ovum can be fertilized by only one sperm.

D ay 14 of the menstrual cycle: An ovum, which has been maturing in an ovary for the last 14 days, breaks away from the surface of the ovary and begins to travel down the Fallopian tube. What happens to the ovum now may be the first in a series of events that will result in the development and birth of a new human being.

Fertilization

The male releases millions of sperm cells from his penis into the female's vagina. The sperm swim through the cervix, into the uterus, and up the Fallopian tube in search of an ovum. Contractions of the uterine muscles help move the sperm along. Tiny cilia, which line the Fallopian tube, sweep the ovum toward the uterus. When a sperm cell and an ovum join, **fertilization** occurs. The new cell that results from fertilization is called a **zygote** (ZY goht).

One at a Time   Of the millions of sperm that begin the journey, only a few hundred ever reach the ovum. Of those, only one sperm can fertilize the ovum. Immediately after fertilization, a membrane forms around the zygote. This membrane keeps other sperm cells from entering the zygote.

Or Maybe Two!   Occasionally, a woman gives birth to twins. There are two kinds of twins: identical and fraternal.

Identical twins come from a single fertilized ovum. As in the development of a single baby, the zygote goes through a series of cell divisions to form a ball of cells. During the formation of identical twins, two different sections of this ball begin to act as if each were a separate entity. Each section continues to develop as a normal individual. These two sections eventually separate into two individuals. Since all the cells came from the same sperm and ovum, the two infants have identical genes. Identical twins are always the same sex.

Fraternal twins are formed differently. Occasionally, two eggs are released during ovulation. Each egg is then fertilized by a different sperm cell. The zygotes continue to develop separately. Except for their age, fraternal twins are no more similar than any other two children in a family.

ASK YOURSELF

How and where does fertilization occur?

Figure 20–11. Twins may be either identical (left) or fraternal (right).

The Developing Embryo

The process of development that takes place between fertilization and birth lasts about 266 days. After fertilization, the zygote passes down the Fallopian tube and into the uterus. During this five- to seven-day trip, cell division occurs several times. By the time it reaches the uterus, the zygote has become a ball of tightly packed cells called an **embryo**. After one week of development, the embryo is about the size of the period at the end of this sentence. The new organism is called an embryo from the second to the eighth week. From the ninth week until birth, it is called a **fetus.**

DISCOVER BY Researching

Using reference books, prepare a chart or table that describes human development between fertilization and birth. Illustrate your descriptions.

For development to continue, the embryo must attach itself to the wall of the uterus. Enzymes, given off by the embryo, break down a tiny spot in the thick wall of the uterus. The embryo attaches itself into this tiny spot. In a few days, the wall of the uterus covers the embryo.

Figure 20–12. Shown here are four stages of human embryonic development—fertilized ovum, two cells, four cells, and eight cells.

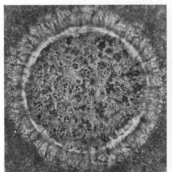
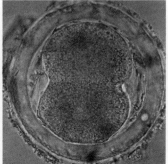
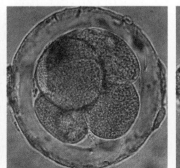
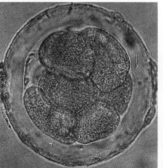

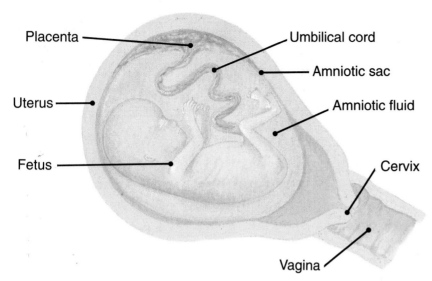

Placenta

Umbilical cord

Amniotic sac

Uterus

Amniotic fluid

Fetus

Cervix

Vagina

Figure 20–13. While in the uterus, the fetus is attached to the mother by the umbilical cord. However, the blood supplies of the fetus and the mother do not mix.

The Placenta   Cells from the embryo combine with cells from the uterus to form the *placenta* (pluh SEHN tuh), which nourishes the developing embryo. An exchange of nutrients and wastes between the mother and the embryo takes place through the placenta. The developing circulatory system of the embryo forms a network of capillaries in the placenta. The blood-rich, spongy tissue of the uterus surrounds these capillaries, but the blood of the embryo and the blood of the mother do not mix.

The placenta nourishes the embryo throughout its development. Food and oxygen molecules pass through the placenta from the mother to the embryo by diffusion. Large blood vessels collect the blood and carry it to the embryo's tissues. These blood vessels are located in the *umbilical cord*, which connects the embryo with the placenta. Wastes move through another blood vessel in the umbilical cord from the embryo back to the placenta.

Organ Development   Cell division occurs rapidly during embryonic development. At the same time, groups of cells gradually begin to differ from each other. Each cell group eventually develops into a different organ or body part.

The embryo's heart develops early. By the end of the fourth week, it begins to pump blood. Blood circulates through tiny arteries carrying nutrients and oxygen to the embryo's tissues. Carbon dioxide and other waste molecules diffuse from the body tissues into the blood. This blood moves through veins back to the placenta for removal of wastes.

The nervous system also begins to develop during this period. Since the brain requires more time to develop than other structures, at first the brain grows faster than the rest of the embryo. This uneven growth makes the head look large compared to the body. By the end of the second month, all the body's systems have begun development. The embryo is now called a fetus. By the end of the third month, the eyes, ears, and nose begin to form, and the arms and legs become visible as tiny bumps. The fetus is about 7 cm long with its head making up about half of its length.

During the next three months, the body catches up with the growth of the head. Development of the body's systems continues. During this time, the fetus begins its first movements in the uterus, and it responds to stimuli. The mother usually feels movement for the first time during the fifth month of pregnancy. By the end of the sixth month, the fetus is usually 30 to 36 cm long and weighs about as much as a head of lettuce (0.5 kg). The fetus can hear sounds and may even respond to music. The fetus floats in a sac of clear fluid. This fluid protects and cushions the fetus from sudden shocks.

By the seventh month, all the body's parts are formed. A baby born at this stage can survive with help. The fetus is active and has already begun sucking its thumb as practice for nursing after birth. The last two months are largely a period of growth. At birth, the newborn weighs about 36 000 times as much as an embryo at two months of development.

Figure 20–14. Early development is rapid. Shown here are an embryo at five weeks (left), a fetus at nine weeks (center), and a fetus at four months (right).

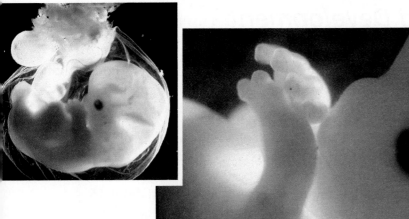

Care of Mother and Fetus
During pregnancy, a woman must be especially careful about what she takes into her body. Some chemical substances can easily damage the developing organs of the fetus. Drugs can pass through the placenta and harm the developing fetus, possibly resulting in mental and physical defects. Even commonly used medicines can be harmful. Aspirin, for example, can cause bleeding from embryonic tissues and can lead to severe birth defects.

The use of alcohol and tobacco can also harm the developing fetus. Alcohol crosses the placenta and collects in the fetal brain and liver where it can severely damage these organs. The use of tobacco reduces the flow of blood to the placenta.

Reduced blood flow means less oxygen and fewer nutrients reach the fetus. The use of either alcohol or tobacco can also result in premature birth.

Diet is also important during pregnancy. Both the mother and the fetus need a balanced diet, rich in nutrients such as calcium and iron. Calcium is needed for bone growth, and iron for blood cells. The fetus produces large amounts of protein during this period. Muscle and brain cells require large amounts of amino acids to make the proteins required for growth. If the mother's diet has been poor, nutrients are removed from her tissues to supply the fetus. This can weaken the mother and make her less able to carry the fetus. To protect her health and the health of the fetus, a pregnant woman should start prenatal care as soon as she suspects pregnancy.

 ASK YOURSELF

What is the difference between an embryo and a fetus?

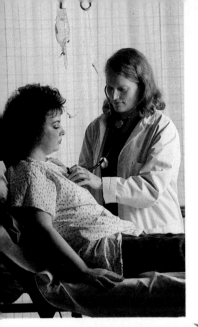

Figure 20–15. It is important for a woman to receive medical advice as soon as she knows she is pregnant.

Birth and Development

Birth usually takes place between 256 and 276 days after fertilization. During birth, the baby is forced from the mother's body by strong contractions of the uterine muscles. Although birth is a continuous process, it is often described in three stages.

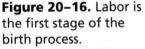

Figure 20–16. Labor is the first stage of the birth process.

The Birth Process   The first stage of birth, called *labor*, begins as the muscles of the uterus begin to contract and relax. This process causes the opening of the cervix to enlarge. At first, these contractions last for about 30 seconds and occur about 15 minutes apart. As labor continues, the contractions become stronger. This part of the birth process may require from 2 to 20 hours. Once the cervix is open enough, the contractions of the uterus can push the baby's head through the opening and into the vagina.

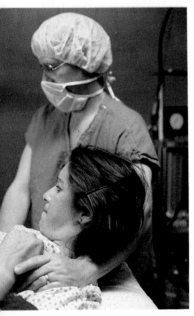

Delivery, the second stage of birth, may take from a few minutes to several hours. This stage ends when the contractions have pushed the baby completely out of the mother's body. The third stage of the birth process begins after delivery. The umbilical cord, which still connects the baby to the placenta, is tied and cut. Ten to fifteen minutes later, another series of contractions begins in the uterus. At this time, the *afterbirth,* which consists of the placenta and other membranes, is pushed out of the uterus. After the birth is complete, the smooth muscles of the uterus continue to contract, returning the uterus to its normal size.

Development After Birth   The baby's development does not end at birth. In comparison with many other newborn animals, newborn humans are helpless. Much care is necessary in order for the baby to survive. During pregnancy, the mother's body prepares itself to feed the baby during the early months of life. The mammary glands are prepared by certain hormones to produce milk to feed the newborn child.

During the nine months of development in the uterus, a human being grows from a single cell weighing less than a gram to an individual weighing an average of 3.18 kg. At no time during the rest of his or her life will growth be as rapid. However, the changes that occur during the first two years of life are very dramatic. The baby soon develops into a toddler. No longer satisfied to just sleep, eat, and be held, the toddler explores his or her world.

The toddler quickly develops into a young child. As development continues, physical and mental abilities increase. Within a space of a few years, the child becomes a teenager and then an adult. In a period of 14 or 15 years, an individual develops from a baby into a physically mature person. Within just a few more years, the individual becomes emotionally and socially mature enough for the life cycle to begin again.

Figure 20–17. Humans continue to develop as they become older, from (left to right) infant, to toddler, to child, to adolescent. What stages come after adolescence?

 ASK YOURSELF

Describe the three stages of birth.

SECTION 3   *REVIEW AND APPLICATION*

Reading Critically

1. Where does fertilization of the egg usually take place?
2. When does the heart of an embryo begin to function?
3. How does a pregnant woman know that labor has begun?

Thinking Critically

4. Why are such large numbers of sperm cells released, if only one sperm is needed for fertilization?
5. When astronauts walk in space, they are sometimes attached to the spacecraft by a line called an *umbilical.* Why do you think this line has been given this name?

SKILL

Graphing Human Fetal Growth

▶ **MATERIALS**
● graph paper ● colored pencils

▼ **PROCEDURE**

1. Make two graphs like those shown. Use intervals of 25 mm on the length graph. Extend the graph to 500 mm. Use intervals of 100 g on the mass graph. Extend this graph to 3300 g. Use intervals of 2 weeks for time on both graphs. Both graphs should extend to 40 weeks. ·

2. Study the table of data. Plot the data in the table onto your graphs. Use a colored pencil to draw the curved line that joins the points.

TABLE I: INCREASE OF MASS AND LENGTH OF AVERAGE HUMAN FETUS		
Time (weeks)	Mass (g)	Length (mm)
2	0.1	1.5
3	0.3	2.3
4	0.5	5.0
5	0.6	10.0
6	0.8	15.0
8	1.0	30.0
13	15.0	90.0
17	115.0	140.0
21	300.0	250.0
26	950.0	320.0
30	1500.0	400.0
35	2300.0	450.0
40	3300.0	500.0

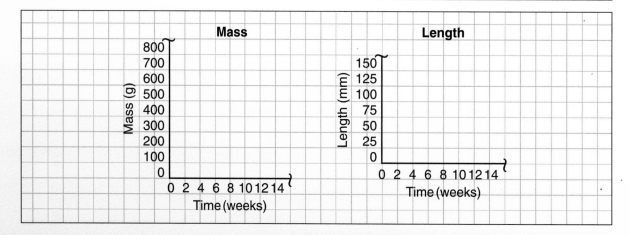

▶ **APPLICATION**
1. Describe the change in mass of a developing fetus. How can you explain this change?
2. Describe the change in length of a developing fetus. How can you explain this pattern?

✳ *Using What You Have Learned*
The average infant with a mass less than 2500 g at birth is 40 times more likely to die than an infant who has a greater mass. Look at your graph and determine in which week a fetus would have a mass of 2500 g.

The Big Idea

All organisms are born, grow, reproduce, and die. Because the process of reproduction ensures the continuation of all species, including humans, it is one of the most important life process-es. All the systems of the human body work together during the production of a new individual, but the endocrine and reproductive systems perform the most important roles.

Hormones produced by glands in the endocrine system control the functions of male and female reproductive systems. Each reproductive system has its own separate but related functions. The structures of the male reproductive system produce and deliver sperm. The structures of the female reproductive system produce ova, receive sperm, and shelter the developing offspring from fertilization to birth.

For Your Journal

Review the answers you wrote in your journal before you read the chapter. How accurate was your explanation of the process that causes identical twins? Summarize your understanding of the chapter by creating a diagram that shows how the endocrine and reproductive systems are related.

Connecting Ideas

Copy this diagram of the human life cycle into your journal. Complete the diagram by writing the terms from the box in the correct order.

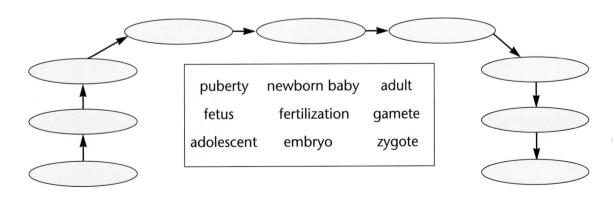

puberty   newborn baby   adult

fetus   fertilization   gamete

adolescent   embryo   zygote

Understanding Vocabulary

Explain the relationship between each pair of terms.

1. hormones (545), endocrine system (544)
2. hormones (545), puberty (550)
3. puberty (550), menstruation (552)
4. fertilization (554), gametes (551)
5. embryo (555), fetus (555)
6. zygote (554), embryo (555)

Understanding Concepts

MULTIPLE CHOICE

7. The endocrine system is an example of a feedback control system because
   a) hormones are needed for the digestive system.
   b) hormones are produced only when needed.
   c) hormones help keep emotions under control.
   d) hormones are released during puberty.

8. Hormones travel through the human bloodstream so that they
   a) create the changes that occur during puberty.
   b) enable the reproductive system to produce gametes.
   c) can affect many organs at the same time.
   d) can easily be eliminated as wastes when they are no longer needed.

9. One ovum is released in the female reproductive system during
   a) ovulation. b) menstruation.
   c) labor.   d) puberty.

10. Before the zygote attaches itself to the wall of the uterus,
    a) menstruation must occur.
    b) it goes through several cell divisions.
    c) the placenta develops.
    d) it develops into a fetus.

11. Functions controlled by the endocrine glands include
    a) the amount of sugar in the blood.
    b) growth rate.
    c) development of reproductive organs.
    d) all of these.

12. The function of the placenta is to
    a) prepare the uterus to receive a fertilized ovum.
    b) release the ova.
    c) prevent more than one sperm from fertilizing the ovum.
    d) nourish the developing embryo.

SHORT ANSWER

13. How does a fetus get nourishment and remove wastes?

14. Describe the effects of adrenalin on the body.

15. Describe the menstrual cycle.

Interpreting Graphics

16. Explain what is happening in each of these illustrations.

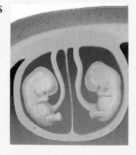 

17. *Systems and Structures*
How are the endocrine and nervous systems similar? How are they different?

18. *Systems and Structures*
Compare the structures of the male and female reproductive systems.

Thinking Critically

19. Explain why menstruation is an example of a feedback control system.

20. How might a high fever affect the male and female reproductive systems?

21. During pregnancy, the placenta produces hormones that keep the ovaries from ovulating. Why is this necessary?

22. Smoking of tobacco can cause blood vessels to narrow. How might this affect a developing fetus?

23. A person with type A blood cannot safely receive a transfusion of type B blood. Why is it safe for a woman with type A blood to bear a fetus that has type B blood?

24. What important role might adrenalin have played in the survival of human beings over thousands of years?

25. In some areas of the country, new laws require signs, such as the one in the photograph, to be displayed in some businesses. Why do you think laws of this type have been passed? Do you think the signs are necessary? Do you think they are a good idea?

Discovery Through Reading

Avraham, Regina. *The Reproductive System.* Chelsea House Publishers, 1991. Discusses aspects of reproduction, including heredity and pregnancy, and explains the stages of growth from fertilization to birth.

DAILY DECISIONS ABOUT YOUR BODY

Young people have to make dozens of choices every day about their bodies. They must decide what to eat, how to stay fit, and how to feel about their bodies. Nutrition science and exercise science provide information for good choices for your body.

STARTING OUT RIGHT

Take one of your first decisions every day: eating breakfast. Overnight your body's metabolism has fallen, and your body has used up its temporary store of fluid. Nutrition experts advise everyone to drink something soon after waking. A glass of fruit juice or water restores your digestive and excretory systems.

Even more important, scientists say, eat something for breakfast. Even a small bowl of cereal, some toast, or fruit

Drinking something soon after waking is advised.

**Regular exercise
is beneficial.**

stimulates the body's functions. Eating breakfast can raise metabolism by four to five percent for the entire day. A good decision about breakfast helps your body burn food more efficiently all day.

LUNCH CHOICE

And lunch? A young person trying to lose weight might decide to skip lunch—not a good decision, according to nutrition science. Skipping meals can trigger a "starvation response" in your body. When deprived of food, the body lowers its metabolism and burns calories more slowly. Eating regular meals will do more to control weight than skipping meals. The scientific choice is a lunch high in complex carbohydrates and low in fat. An apple and a whole-wheat roll, or a salad with low-fat dressing will keep the body's metabolic fires burning.

When school's out, what are you ready to do? Scientists who study the mind and the body recommend that you make time for moderate exercise. Scientific research has documented the benefits of physical fitness. Physical activity raises metabolism and stimulates the body's circulatory and respiratory systems. Regular exercise increases your body's immunity to illness.

Fitness also increases your mental capacity. A Canadian research study found a link between physical fitness and ability to learn math and writing skills. Perhaps best of all, studies show that young people who exercise regularly feel better about themselves. Exercise enhances self-esteem at a time in life when young people are deciding how to feel about themselves.

NEW ROUTINES

Exercise science advises that to get benefits from exercise, you don't have to sprint around a track. Benefits will come from any daily activity that raises your heart rate for 20 to 40 minutes. You might decide to bicycle to the library or store instead of going by car or bus. You can have that talk with your best friend during a brisk walk. Instead of sunning by the pool, walk through the water at chest-high depth. Or take up roller-skating, an exercise that tones hips and thighs and reduces body fat.

It helps to find easy and pleasant ways to get physical exercise. You don't need to subscribe to the philosophy of "No pain, no gain!" Exercise physiologists advise that pain during exercise may be a warning of impending injury. Besides, if exercise is dull or painful, you won't do it regularly. And regular exercise is a key to a healthy body and a positive attitude.

Sooner or later, most young people come face-to-face with the menu at a fast-food restaurant. Does anyone think of science at a time like that? Nutritionists would give you simple advice. Whenever possible, stay away from foods high in fat and choose low-fat, high-fiber foods.

To begin with, notice how a fast-food item is cooked. A flame-grilled hamburger is lower in fat than a burger fried on a griddle. A roast turkey sandwich may contain only 9 grams of fat and 300 calories. A fried fish sandwich may pack a whopping 30 grams of fat and 540 calories.

Another fast-food tip is to balance high-fat foods against low-fat foods. A broiled chicken sandwich and low-fat milk help balance that fat-ladened order of French fries. Finally, be sparing with toppings like mayonnaise and ketchup. You don't gain much by ordering a salad and then drenching it in high-fat dressing.

DIET KNOW-NO'S

Finally, a scientific word about dieting. In their early teenage years, many people start to worry about body size and shape. They often find appeal in specialized diet programs and sometimes suffer from eating disorders. The teenage years can mark the beginning of "yo-yo" dieting, a pattern of rapid weight loss followed by regained weight.

Scientific research has good advice for choices about dieting. First, low-calorie "starvation" diets may simply lower the body's metabolism. In a month, metabolism can fall by 25 percent. When calorie con-

Starvation diets do more harm than good.

sumption returns to normal, the body prepares for the next "famine" by storing extra fat. So starvation diets may actually increase body fat, not reduce it. Also, repeated weight loss and gain increase the chances of heart disease in both males and females.

Few diet programs achieve consistent, long-term success. Diets that encourage eating large amounts of a single food may actually harm your health. There are scientific alternatives. Nutritionists suggest you eat the same amount of food daily but eat less fat. Have a snack of raw vegetables instead of chips or cheese and crackers. Eat more mashed potatoes at dinner and less fried chicken. Also, try dividing your normal diet into five meals instead of three. Start an exercise routine of three to four moderate workouts per week.

The decisions that young people make about fitness and diet can last a lifetime. Today's science recommends a few basic principles for deciding how to care for your body. ◆

Is this a balanced meal?

I SLEPT FOR SCIENCE:

A Study of Sleep Problems

"How are you doing in there, Andy? Think you're about done sleeping?" The voice from the ceiling drifted down softly, gently, like the soothing sounds of a cool summer rain.

It's a good thing too. I was in no mood for screaming. I had just spent the night with 13 wires attached to various parts of my body, a metal chain wrapped around my rib cage, a clamp stuck on my fingertip, and a plastic sensor clipped to my nostril.

"Are you sure I'll be able to sleep with all these wires?" I asked Joe Brown, sleep technologist, as he "wired" my body the night before.

"Absolutely," Joe replied. "You'd be surprised at how well most people sleep when they're here."

I was more than surprised; I was shocked. Falling asleep under such strange conditions proved to be as easy for me as falling asleep in my own bed. Not everyone is so lucky. Some people—including many kids—have trouble sleeping. To find out more about kids with sleeping problems and what can be done to treat them, I decided to "sleep for science."

So I went to the New Haven (Conn.) Sleep Disorders Center. There, Dr. Robert Watson, sleep expert and director of the center, tested me for sleep disorders.

I learned two important facts. First, I learned that I don't like wearing clips in my nose. Second, I learned that about 50 million people in the U.S. suffer from one or more sleep disorders, many of which can be easily treated. If you have a sleeping problem—perhaps you snore or feel unusually tired during the day—read on. This article is for you.

I Get "Hooked Up"

When I arrived at the sleep center, Joe "hooked" me to a *polysomnograph*, a machine that records brain, heart, lung, and muscle activity during sleep. Joe first rubbed certain spots on my skin with a special solution. This solution made the wires Joe glued to my skin more sensitive to electrical signals coming from my body.

All body tissues "give off" weak electrical signals during activity. For instance, whenever a muscle moves, muscle cells produce electrical energy. The wires, or electrodes, that must be worn sense this energy and send it to the polysomnograph.

Joe also attached a small clip to my nostril to measure air flow, and a larger clip to my index finger to measure oxygen in the blood. Then he wrapped a metal chain and spring around my chest. Each time I breathed, the chain would stretch and send a signal to the polysomnograph.

Electrical signals from all the wires and electrodes produce a line of ink on a continuous length of lined paper. This is the *polysomnogram*, or sleep recording. Doctors use polysomnograms to interpret sleep patterns.

Before letting me fall asleep, Dr. Watson explained that there are two stages of sleep: Rapid-eye-movement, or REM; and Non-rapid-eye-movement, or Non-REM.

Non-REM sleep begins with stage one, that strange period between being awake and being asleep. If you've ever told your parents that you were "just about to fall asleep" when they woke you, you may have been floating through stage one sleep.

I vaguely recall going through this stage when I slept at the center. My thoughts sort of drifted away from me, and I became less and less aware of the wires attached to my skin. Stage one usually lasts from 30 seconds to about seven minutes in persons who don't have a sleep disorder. Dr. Watson says that my stage one lasted about two minutes.

Stage two is the first real sleep stage. It lasts about 15 to 45 minutes in adolescents and young adults. In stage two, brain activity begins to slow down. Thoughts come in short bursts and tend not to make much sense.

Next comes deep, or *delta*, sleep. Experts believe this stage affords the brain an opportunity to "rest." The body, on the other hand, often moves about during this stage. In fact, despite common belief to the contrary, we shift positions about every 15 minutes dur-

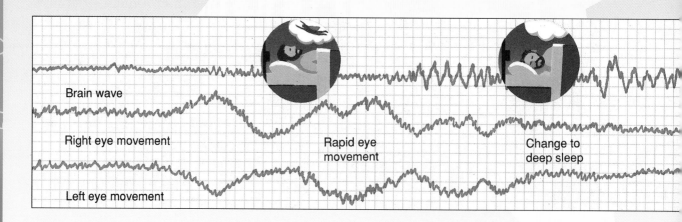

Brain wave

Right eye movement

Left eye movement

Rapid eye movement

Change to deep sleep

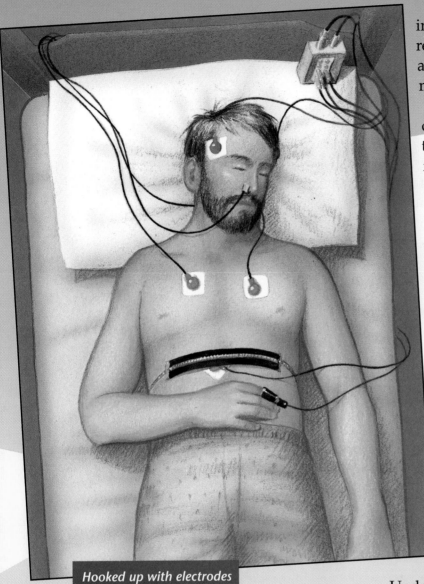

Hooked up with electrodes

ing sleep. Although I don't remember shifting position at all, Joe convinced me that I had moved several times.

After about 30 minutes of delta sleep, I returned to stage two sleep. Then I entered the first of several periods of REM sleep.

Dreaming in REM Sleep

REM sleep is a period of intense activity within the brain that experts say relates to the time we spend dreaming. Sleep experts know exactly when a person is dreaming because the person's eyes move in a typically rolling pattern, called rapid eye movement.

Unlike Non-REM, only the eyes move during REM. Most of the rest of our muscles become paralyzed, or unable to move, during REM sleep.

Under normal conditions, we drift in and out of REM sleep several times a

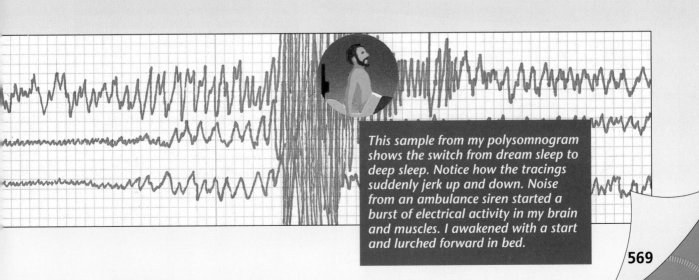

This sample from my polysomnogram shows the switch from dream sleep to deep sleep. Notice how the tracings suddenly jerk up and down. Noise from an ambulance siren started a burst of electrical activity in my brain and muscles. I awakened with a start and lurched forward in bed.

night. The first REM cycle lasts about five minutes. Experts say dreams in this first cycle tend to be rather boring.

As the night progresses, however, REM cycles lengthen and dreams become more and more vivid. The final REM cycle may last as long as one hour. Dreams at this time are the most active of the night (or day, depending on when you sleep).

After I woke up, Joe showed me the REM activity on my polysomnogram. He told me if he had awakened me during my last REM cycle, I would

Joe Brown monitors a polysomnogram. Each foot (30 centimeters) of polysomnograph paper equals one minute of time. My polysomnogram was 773 feet (236 meters) long and weighed 7 pounds (3 kilograms)!

have rambled on and on about the dream I was having. "You probably wouldn't have made much sense," Joe said, "but you would have remembered the dream. It was a whopper."

Whopper dream or not, I learned that my sleep patterns were normal. "You're lucky," Dr. Watson told me later. "Almost everyone who comes here has some kind of sleep disorder. Many have more than one." ◆

Charles Drew (1904–1950)

Millions of people are alive today thanks to the blood plasma work of Charles Drew. His blood preservation discoveries led to the establishment of today's blood banks.

Charles Richard Drew was born in 1904 in Washington, D.C. Drew was a star athlete in both high school and college. After graduating from Amherst College, he coached, and taught biology and chemistry at Morgan State College.

In 1933, Drew received his medical degree from McGill Medical College and began his blood research at Canada's Montreal General Hospital. Drew discovered that blood plasma could be given to any person and could be stored for long periods. This knowledge helped save millions of lives in World War II.

Drew became the first director of the blood plasma collection program of the American Red Cross. He then was professor of surgery and director of Freedman's Hospital at Howard University Medical School. For his achievements, Drew re-ceived the National Associa-tion for the Advancement of Colored People's (NAACP) Spingarn Medal. ◆

Irene Duhart Long (1951—)

Dr. Irene Duhart Long plans to be there when the National Aeronautics and Space Administration (NASA) space station goes into operation. As an aerospace physician, Long's goal is to be the space station's medical officer.

Irene Duhart Long was born in Cleveland, Ohio, in 1951. As a child, she was fascinated with air travel. At the age of nine, Long decided that she wanted to become a NASA physician.

Long attended Northwestern University and in 1977 received her medical degree from the St. Louis University School of Medicine. Long then earned her Master of Science degree in aerospace medicine from the Wright University School of Medicine. Long began work with NASA in 1982 and is now Chief of the Medical and Environmental Health Office in the Biomedical Operations and Research Office at the John F. Kennedy Space Center.

Long studies the effects of gravity on the health of astronauts and also provides emergency medical care. She is doing experiments on cardiovascular and endocrine reactions in space.

Long's discoveries might make it possible for you to live in space one day. ◆

Mamie Lou, Pharmacist

Many people think that all a pharmacist does is prepare the medicine prescribed by a doctor for people who are ill. However, a pharmacist's job is more that just counting out pills and filling medicine bottles. Today's pharmacist is an important part of a health care team.

Mamie Lou is a pharmacist at a medical teaching hospital in Houston, Texas. As a pharmacist, she performs many jobs. "I counsel patients about the medicines prescribed by their doctors as well as the medicines that they buy without a prescription," says Lou.

in helping pharmacists keep accurate records on patients and their medications.

Lou instructs nurses and technicians on the skills and measuring tools needed for the preparation of medicines. "I use a balance to accurately

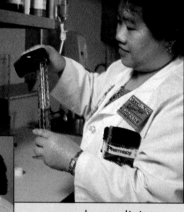

Communicating with doctors and nurses about a patient's medical treatment is also a very important part of a pharmacist's job. Lou reviews patients' charts and computer profiles to verify that they receive the correct medications that the doctor ordered. Computers are very important

measure dry medicines and a graduated cylinder to measure liquids whenever I make medicines," says Lou. Pharmacists also use a mortar and pestle to grind medicines to make sure the ingredients are evenly mixed.

A pharmacist is a licensed health care professional. To become a pharmacist, a person must complete college courses in biology, chemistry, math, and physics. He or she must also pass a state examination before receiving a license. "I have always wanted to help people and work in

a medical field," says Lou, "which is why I became a pharmacist."

Pharmacists can work in other areas besides hospitals. They can work in nursing homes, sales, government, research, education, and home health care. Pharmacists can also choose specialty fields such as oncology, emergency medicine, nutrition, and pain management. Being a pharmacist is a great profession for students who are interested in people and science. ◆

DISCOVER MORE

For more information about a career as a pharmacist, write to the

American Association of Colleges of Pharmacy
1426 Prince Street
Alexandria, VA 22314

Aging

A person born in 1900 could expect to live an average of about 45 years. Infants born in 1986, however, could expect to live an average of about 74 years. Few people can expect to live beyond 100 years.

The process of aging begins at birth.

Modern medicine and technology now enable people to live longer lives.

The Study of Aging

Scientists called *gerontologists* study aging. They are trying to answer the question, "What causes people to age?" Gerontologists are not sure exactly how aging takes place, but most agree that genes are somehow involved. They also agree that only a few people reach a very old age.

Studies show that proteins produced by old cells are like those produced by young cells. Research also shows that DNA replicates more slowly in old cells than it does in young cells. This may be due to a molecule blocking reproduction of cells at one of the phases of mitosis. In addition, repair of DNA takes longer as we get older. The combination of slower replication and slower repair of DNA could result in aging. The studies imply that there is no difference in the DNA itself, only in the amount and use of it in young and old cells.

Victims of Nature

Systems become more disorganized as time goes by. This law of nature is called *entropy*. Some scientist think that cells are victims of entropy. The disorder in cells appears in the form of "mistakes" by enzymes that are not doing their jobs. These mistakes might cripple cells and could bring on changes we see as people age.

Scientists today are searching for the causes of these mistakes. Their research includes the study of certain molecules that are byproducts of cell processes. These molecules, called *free radicals*, react with nearly every other molecule in a cell, including DNA, and they can change a harmless form of cholesterol into the form that clogs arteries. But free radicals are not the only threat to cells. Studies also show that the cell system that breaks down harmful proteins works more slowly in old cells than in young cells.

Scientists do not yet fully understand what causes aging, but progress is being made as research and discoveries continue. In the future, people can expect to live longer and healthier lives. ◆

REFERENCE SECTION

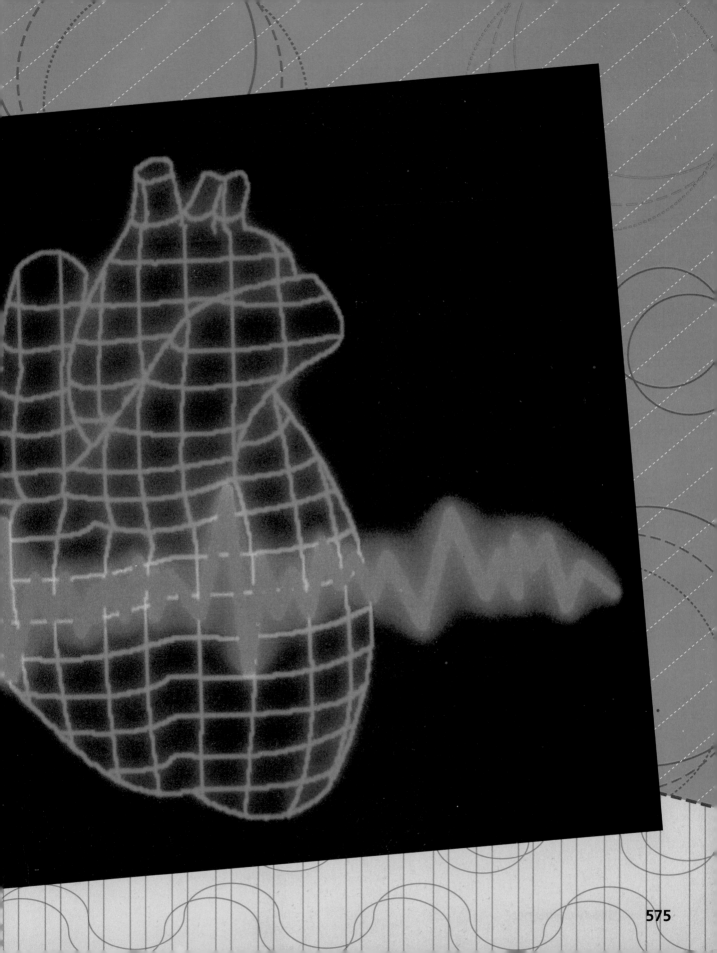

SAFETY GUIDELINES

Participating in laboratory investigations should be an enjoyable learning experience. You can ensure both learning and enjoyment from the experience by making the laboratory a safe place in which to work. Carelessness, lack of attention, and showing off are the major causes of laboratory accidents. It is, therefore, important that you follow safety guidelines at all times. If an accident should occur, you should know exactly where to locate emergency equipment. Practicing good safety procedures means being responsible for your classmates' safety as well as your own.

You will be expected to practice the following safety guidelines whenever you are in the laboratory.

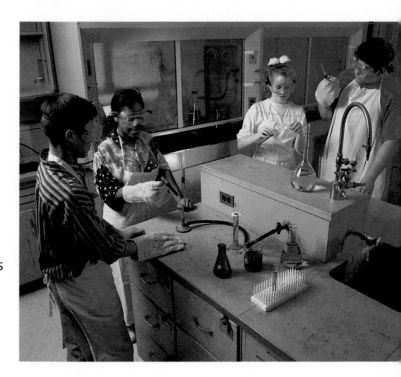

1. **Preparation**   Study your laboratory assignment in advance. Before beginning your investigation, ask your teacher to explain any procedures you do not understand.

2. **Neatness**   Keep work areas clean. Tie back long, loose hair and button or roll up long sleeves when working with chemicals or near an open flame.

3. **Eye Safety**   Wear goggles when handling liquid chemicals, using an open flame, or performing any activity that could harm the eyes. If a solution is splashed into the eyes, wash the eyes with plenty of water and notify your teacher at once. Never use reflected sunlight to illuminate a microscope. This practice is dangerous to the eyes.

4. **Chemicals and Other Dangerous Substances**   Some chemicals can be dangerous if they are handled carelessly. If any solution is spilled on a work surface, wash the solution off at once with plenty of water.

   • Never taste chemicals or place them near your eyes. Never eat in the laboratory. Counters and glassware may contain substances that can contaminate food. Handle toxic substances in a well-ventilated area or under a ventilation hood.

   • Never pour water into a strong acid or base. The mixture produces heat. Sometimes the heat causes splattering. To keep the mixture cool, pour the acid or base slowly into the water.

   • When noting the odor of chemical substances, wave the fumes toward your nose with your hand rather than putting your nose close to the source of the odor.

• Do not use flammable substances near a flame.

5. Safety Equipment   Know the location of all safety equipment, including fire extinguishers, fire blankets, first-aid kits, eyewash fountains, and emergency showers. Report all accidents and emergencies to your teacher immediately.

6. Heat   Whenever possible, use an electric hot plate instead of an open flame. If you must use an open flame, shield the flame with a wire screen that has a ceramic center. When heating chemicals in a test tube, do not point the test tube toward anyone.

7. Electricity   Be cautious around electrical wiring. Do not let cords hang loose over a table edge in a way that permits equipment to fall if the cord is tugged. Do not use equipment with frayed cords.

8. Knives   Use knives, razor blades, and other sharp instruments with extreme care. Do not use double-edged razor blades in the laboratory.

9. Glassware   Examine all glassware before heating. Glass containers for heating should be made of borosilicate glass or some other heat-resistant material. Never use cracked or chipped glassware.

• Never force glass tubing into rubber stoppers.

• Broken glassware should be swept up immediately, never picked up with the fingers. Broken glassware should be discarded in a special container, never into a sink.

10. Unauthorized Experiments   Do not perform any experiment that has not been assigned or approved by your teacher. Never work alone in the laboratory.

11. Cleanup   Wash your hands immediately after any laboratory activity. Before leaving the laboratory, clean up all work areas. Put away all equipment and supplies. Make sure water, gas, burners, and electric hot plates are turned off.

Remember at all times that a laboratory is a safe place only if you regard laboratory work as serious work.

The instructions for your laboratory investigations will include cautionary statements when necessary. In addition, you will find that the following safety symbols appear whenever a procedure requires extra caution:

 Wear safety goggles

 Biohazard/disease-causing organisms

 Electrical hazard

 Wear laboratory apron

 Flame/heat

 Rubber gloves

 Sharp/pointed object

 Dangerous chemical/poison

 Radioactive material

LABORATORY PROCEDURES

READING A METRIC RULER

1. Examine your metric ruler. The numbers on it represent lengths in centimeters. The usual metric ruler is about 30 cm long. There are 10 marked spaces within each centimeter, which represent tenths of centimeters (0.1 cm).

2. To measure the width of a piece of paper, place the ruler on the paper. The zero end of the ruler must line up exactly with one edge of the paper. Look at the other edge of the paper to see which of the marks on the ruler is closest to that edge. In Figure A, for example, the edge of the paper is nearest to the second line beyond the 7. Therefore, the width of the paper is 7.2 cm.

3. The edge of the paper might fall exactly on one of the centimeter marks. In Figure B, the edge is just on the 5-cm mark. The width of this paper is 5.0 cm. You must write in the .0 to indicate that the measurement is accurate to the nearest tenth of a centimeter; that is, it is more than 4.9 cm and less than 5.1 cm.

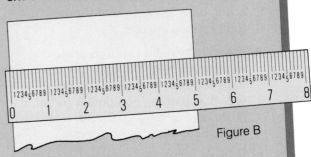

Figure B

4. Sometimes you may want to make a reading with more accuracy. It is possible to estimate readings to the nearest hundredth of a centimeter, but you must be very careful. Look at Figure A again. You can guess the number of tenths in the distance between the marks. The edge of the paper is about 3 tenths of the space between 7.2 and 7.3. The best estimate, then, is that the width of the paper is 7.23 cm.

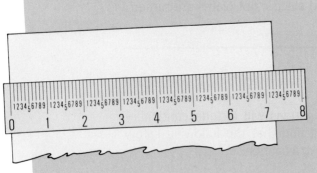

5. In Figure C, the edge of the paper falls exactly on the 8.6 mark. If you are taking careful readings, accurate to the nearest hundredth of a centimeter, you must record the width as 8.60 cm.

6. Note the general rule: You can estimate scale readings to the nearest tenth of a scale division. If the scale is marked in tenths, you can estimate the hundredths place but never more than that.

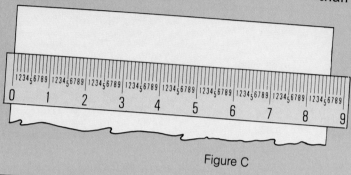

Figure C

CONVERTING SI UNITS

In SI, it is easy to convert from unit to unit. To convert from a larger unit to a smaller unit, move the decimal to the right. To convert from a smaller unit to a larger unit, move the decimal to the left. Figure D shows you how to move the decimals to convert in SI.

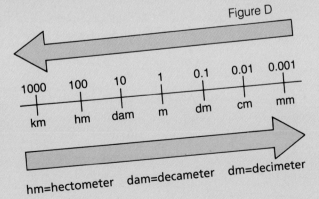

Figure D

hm=hectometer   dam=decameter   dm=decimeter

LABORATORY PROCEDURES

SI Conversion Table

SI Units		Converting SI to Customary		Converting Customary to SI	
Length		1 km	= 0.62 mile	1 mile	= 1.609 km
kilometer (km)	= 1000 m	1 m	= 1.09 yards	1 yard	= 0.914 m
meter (m)	= 100 cm		= 3.28 feet	1 foot	= 0.305 m
		1 cm	= 0.394 inch		= 30.5 cm
centimeter (cm)	= 0.01 m	1 mm	= 0.039 inch	1 inch	= 2.54 cm
millimeter (mm)	= 0.001 m				
micrometer (μm)	= 0.000 001 m				
nanometer (nm)	= 0.000 000 001 m				
Area		1 km²	= 0.3861 square mile	1 square mile	= 2.590 km²
		1 ha	= 2.471 acres	1 acre	= 0.4047 ha
square kilometer (km²)	= 100 hectares	1 m²	= 1.1960 square yards	1 square yard	= 0.8361 m²
hectare (ha)	= 10 000 m²			1 square foot	= 0.0929 m²
square meter (m²)	= 10 000 cm²	1 cm²	= 0.155 square inch	1 square inch	= 6.4516 cm²
square centimeter (cm²)	= 100 mm²				
Mass		1 kg	= 2.205 pounds	1 pound	= 0.4536 kg
		1 g	= 0.0353 ounce	1 ounce	= 28.35 g
kilogram (kg)	= 1000 g				
gram (g)	= 1000 mg				
milligram (mg)	= 0.001 g				
microgram (μg)	= 0.000 001 g				
Volume of Solids		1 m³	= 1.3080 cubic yards	1 cubic yard	= 0.7646 m³
			= 35.315 cubic feet	1 cubic foot	= 0.0283 m³
1 cubic meter (m³)	= 1 000 000 cm³	1 cm³	= 0.0610 cubic inch	1 cubic inch	= 16.387 cm³
1 cubic centimeter (cm³)	= 1000 mm³				
Volume of Liquids		1 kL	= 264.17 gallons	1 gallon	= 3.785 L
		1 L	= 1.06 quarts	1 quart	= 0.94 L
kiloliter (kL)	= 1000 L	1 mL	= 0.034 fluid ounce	1 pint	= 0.47 L
liter (L)	= 1000 mL			1 fluid ounce	= 29.57 mL
milliliter (mL)	= 0.001 L				
microliter (μL)	= 0.000 001 L				

READING A GRADUATE

1. Examine the graduate and note how the scale is marked. The units are milliliters (mL). A milliliter is a thousandth of a liter and is equal to a cubic centimeter. Note carefully how many milliliters are represented by each scale division on the graduate.

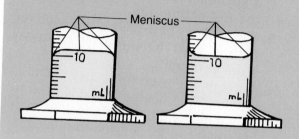

2. Pour some liquid into the cylinder and set the cylinder on a level surface. Notice that the upper surface of the liquid is flat in the center and curved at the edges. This curve is called the *meniscus* and may be either upward or downward. In reading the volume, you must ignore the curvature and read the scale at the flat part of the surface.

3. Bring your eye to the level of the surface and read the scale at the level of the flat surface of the liquid.

USING A LABORATORY BALANCE

1. Make sure the balance is on a level surface. Use the leveling screws at the bottom of the balance to make any necessary adjustments.

2. Place all the countermasses at zero. The pointer should be at zero. If it is not, adjust the balancing knob until the pointer rests at zero.

3. Place the object you wish to mass on the pan. **CAUTION: Do not place hot objects or chemicals directly on the balance pan, because they can damage its surface.**

4. Move the largest countermass along the beam to the right until it is at the last notch that does not tip the balance. Follow the same procedure with the next largest countermass. Then move the smallest countermass until the pointer rests at zero.

5. Determine the readings on all beams and add them together to determine the mass of the object.

6. When massing crystals or powders, use a piece of filter paper. First, mass the paper; then add the crystals or powders and remass. The actual mass is the total minus the mass of the paper. When massing liquids, first mass the empty container, then mass the liquid and container. Finally, subtract the mass of the container from the mass of the liquid and the container to get the mass of the liquid.

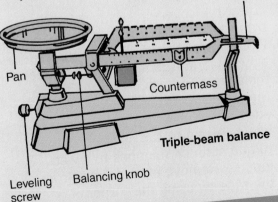

Triple-beam balance

LABORATORY PROCEDURES

USING DISSECTING TOOLS

1. Dissecting tools are used to examine the internal and external features of an organism.
2. **CAUTION: Some dissecting tools are sharp and can cause injuries when not used properly.**
3. The most commonly used dissecting tools are the dissecting pan, scissors, scalpel, forceps, needle probe, and blunt probe.

4. The dissecting pan holds the specimen in place during examination. **CAUTION: Never attempt to dissect a specimen that is not secured to the dissecting pan.**
5. The scissors and scalpel are used for cutting. **CAUTION: Always cut away from yourself.**
6. Forceps are used for grasping and holding. The needle probe is used to move delicate parts of the specimen. The blunt probe is used to move larger, less delicate parts of the specimen.
7. Always clean and dry each of your dissecting tools after using them.

MAKING A WET MOUNT

1. Use lens paper to clean a glass slide and a coverslip.
2. Place the specimen you wish to observe in the center of the slide.
3. Using a medicine dropper, place one drop of water on the specimen.
4. Hold the coverslip at the edge of the water and at a 45° angle to the slide. Position the coverslip so that it is at the edge of the drop of water. Make sure that the water runs along the edge of the coverslip.
5. Lower the coverslip slowly to avoid trapping air bubbles.
6. Water might evaporate from the slide as you work. Add more water to keep the specimen fresh. Place the tip of the medicine dropper next to the edge of the coverslip. Add a drop of water. (You also can use this method to add stain or solutions to a wet mount.) Remove excess water from the slide by using the corner of a paper towel as a blotter. Do not lift the coverslip to add or remove water.

USING A COMPOUND LIGHT MICROSCOPE

Parts of the Compound Light Microscope

- The *eyepiece* magnifies the image 10X.
- The *low-power objective* magnifies the image 10X.
- The *high-power objective* magnifies the image either 40X or 43X.
- The *revolving nosepiece* holds the objectives and can be turned to change from one magnification to the other.
- The *body tube* maintains the correct distance between eyepiece and objectives.
- The *coarse adjustment* moves the body tube up and down to allow focusing of the image.
- The *fine adjustment* moves the body tube slightly to bring the image into sharper focus.
- The *stage* supports a slide.

Compound light microscope

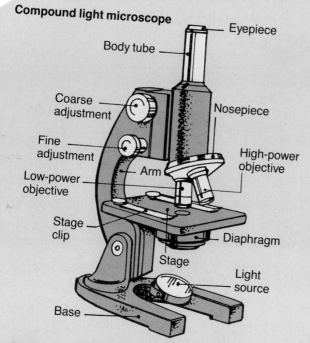

- Eyepiece
- Body tube
- Coarse adjustment
- Fine adjustment
- Arm
- Low-power objective
- Nosepiece
- High-power objective
- Stage clip
- Diaphragm
- Stage
- Light source
- Base

- *Stage clips* hold the slide in place for viewing.
- The *diaphragm* controls the amount of light coming through the stage.
- The *light source* provides light for viewing the slide.
- The *arm* supports the body tube.
- The *base* supports the microscope.

Proper Use of the Compound Light Microscope

1. Carry the microscope to your lab table, using both hands. Place one hand beneath the base and use the other hand to hold the arm of the microscope. Hold the microscope close to your body while moving it to your lab table.
2. Place the microscope on the lab table, at least 5 cm from the edge of the table.
3. Check to see what type of light source is used by your microscope. If the microscope has a lamp, plug it in, making sure that the cord is out of the way. If the microscope has a mirror, adjust it to reflect light through the hole in the stage. **CAUTION: If your microscope has a mirror, do not use direct sunlight as a light source. Direct sunlight can damage your eyes.**
4. Always begin work with the low-power objective in line with the body tube. Adjust the revolving nosepiece.
5. Place a prepared slide over the hole in the stage. Secure the slide with the stage clips.
6. Look through the eyepiece. Move the diaphragm to adjust the amount of light coming through the stage.
7. Now, look at the stage from eye level. Slowly turn the coarse adjustment to lower the objective until it almost touches the slide. Do not allow the objective to touch the slide.
8. Look through the eyepiece. Turn the coarse adjustment to raise the low-power objective until the image is in focus. Always focus by raising the objective away from the slide. *Never focus the objective downward.* Use the fine adjustment to sharpen the focus. Keep both eyes open while viewing a slide.
9. Make sure that the image is exactly in the center of your field of vision. Then switch to the high-power objective. Focus the image, using only the fine adjustment. *Never use the coarse adjustment at high power.*
10. When you are finished using the microscope, remove the slide. Clean the eyepiece and objectives with lens paper. Return the microscope to its storage area. Remember, you should use both hands to carry the microscope correctly.

FIVE-KINGDOM CLASSIFICATION OF ORGANISMS*

Kingdom Monera

Organisms in this kingdom have cells that lack a true nucleus and membrane-bound organelles; mostly unicellular.

Phylum Schizophyta: Bacteria; about 2500 species, including eubacteria (true bacteria), rickettsias, mycoplasmas, and spirochetes

Phylum Cyanophyta: Blue-green algae, or cyanobacteria; about 200 species

Kingdom Protista

Organisms in this kingdom include a diverse group of unicellular and simple multicellular organisms whose cells have a true nucleus and membrane-bound organelles.

Phylum Euglenophyta: Euglenoids; about 800 species

Phylum Zoomastigina: Flagellates, about 2500 species

Phylum Sarcodina: Sarcodines; about 11 500 species; includes amoebas

Phylum Ciliophora: Ciliates; about 7200 species; includes paramecia

Phylum Sporozoa: Sporozoans; about 6000 species; includes *Plasmodia,* the cause of malaria

Phylum Chrysophyta: Golden algae; about 12 000 species

Phylum Pyrrophyta: Fire algae; 1100 species; major component of marine phytoplankton

Phylum Myxomycota: Slime molds; about 600 species

Phylum Chlorophyta: Green algae; about 7000 species; probable ancestor of modern land plants

Phylum Phaeophyta: Brown algae; about 1500 species; includes kelps

Phylum Rhodophyta: Red algae; about 4000 species; includes multicellular seaweeds

Kingdom Fungi

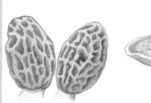

Organisms in this kingdom obtain food by absorption; most are multicellular, composed of intertwined filaments.

Division Basidiomycota: Mushrooms, bracket fungi, rusts, smuts; about 13 000 species

Division Deuteromycota: Fungi Imperfect; about 16 000 species; includes some *Penicillium,* athlete's foot fungus

* This is not a complete list of all known phyla or divisions. For each kingdom, only representative examples are given.

Kingdom Plantae

Organisms in this kingdom are multicellular and carry out photosynthesis in chloroplasts; mostly land dwellers; cell walls contain cellulose; body has distinct tissues; life cycle of alternating sporophyte and gametophyte generations.

Division Bryophyta: Bryophytes; about 15 600 species that lack vascular tissues and true roots, stems, and leaves; includes liverworts and mosses

Division Psilophyta: Whisk ferns; a few species of seedless plants lacking roots and leaves

Division Sphenophyta: Horsetails; about 15 species of seedless plants with hollow stems

Division Lycophyta: Club mosses; about 1000 diverse species of seedless plants with leafy sporophytes

Division Pterophyta: Ferns; about 12 000 diverse species of seedless plants

Division Cycadophyta: Cycads; about 100 species of palmlike plants; gymnosperms

Division Ginkgophyta: Ginkgo; 1 species only; fan-shaped leaves

Division Gnetophyta: Seed plants similar to angiosperms; about 70 species

Division Coniferophyta: Conifers; about 550 species of gymnosperms; most species are evergreens

Division Anthophyta: Angiosperms; about 235 000 species of plants that produce enclosed seeds; reproductive structures are flowers; mature seeds are enclosed in fruits

Kingdom Animalia

Organisms in this kingdom are multicellular and obtain food by ingestion; most are motile; reproduction is predominantly sexual.

Phylum Porifera: Sponges; about 5000 aquatic, mostly marine

Phylum Cnidaria (also called Coelenterata): Coelenterates; about 9000 aquatic species; tentacles armed with stinging cells; includes jellyfish and coral

Phylum Ctenophora: Sea walnut and comb jellies; about 90 species; gelatinous marine animals

Phylum Platyhelminthes: Flatworms; about 13 000 species; includes tapeworms

Phylum Nematoda: Roundworms; about 12 000 parasitic species

Phylum Acanthocephala: Spiny-headed worms; about 500 species

Phylum Rotifera: Rotifers or "wheel" animals; wormlike or spherical

Phylum Bryozoa: "Moss" animals

Phylum Brachiopoda: Lamp shells; about 250 species, 30 000 extinct

Phylum Mollusca: Mollusks; about 47 000 species of soft-bodied animals; includes snails and clams

Phylum Annelida: Segmented worms; about 9000 species; includes earthworms

Phylum Arthropoda: Arthropods; at least 1 million species; includes insects, spiders, crustaceans

Phylum Echinodermata: Echinoderms; about 6000 marine species; includes starfish, sand dollars, and sea urchins

Phylum Hemichordata: Acorn worms; about 80 species

Phylum Chordata: Chordates; about 43 000 species that at some stage have gill slits and tail; includes fish, amphibians, reptiles, birds, and mammals

GLOSSARY

A

absorption (uhb ZAWRP shuhn)   the movement of nutrient molecules into blood vessels **(463)**

active transport   the movement of molecules across a membrane from an area of low concentration to an area of high concentration **(122)**

algae   plantlike protists that contain chlorophyll and carry out photosynthesis **(236)**

alveoli (al VEE uh ly)   air sacs in the lungs **(492)**

antibiotic   chemical substance used to kill or slow the growth of bacteria **(221)**

arteries   blood vessels that carry blood away from the heart to other parts of the body **(480)**

asymmetry   absence of symmetry **(330)**

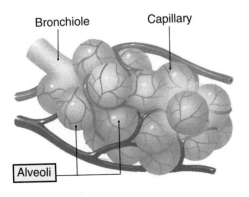

Bronchiole   Capillary

Alveoli

B

bilateral symmetry the arrangement of body parts the same way on both sides of an animal's body; two sides that are mirror images **(328)**

binomial nomenclature system of naming organisms using two Latin names **(177)**

biome large-scale land ecosystem that has a similar climate and vegetation throughout **(62)**

biosphere narrow layer near Earth's surface in which life can exist **(60)**

brain stem the place where the brain and the spinal cord meet **(518)**

bronchi (BRAHN kee) two small tubes that branch off from the trachea and carry air into the lungs; lined with mucus to trap dirt, dust, and any unwanted material brought in with the air **(492)**

bronchioles (BRAHN kee ohlz) small tubes in the lungs that branch from bronchi **(492)**

Bilateral symmetry

C

camouflage (KAM uh flahj) any marking or coloring that helps an animal hide from other animals **(400)**

capillaries small blood vessels with walls that are only one cell thick and that carry blood to all body cells **(480)**

cartilage (KAHRT uh lihj) the tough, flexible tissue from which most bones are formed **(438)**

cell the smallest unit of life; the basic unit of structure and function of all living things **(106)**

cell membrane the covering that surrounds the cell **(108)**

cell theory theory that states that all organisms are made of cells, that cells are basic units of structure and function for all living things, and that all cells come from other cells **(107)**

cellular respiration process in which energy is released when oxygen combines with sugar molecules, forming carbon dioxide and water as waste products **(134)**

cerebellum (sehr uh BEHL uhm) part of the brain that controls muscle coordination, balance, and muscle tone **(518)**

Camouflage

Climate

cerebrum (seh REE bruhm)   the largest part of the mammal brain; involved with intelligence and the organizing of information **(518)**

chlorophyll (KLAWR uh fihl)   the green material in plants that traps energy from sunlight and uses it to break down water molecules into atoms of hydrogen and oxygen **(270)**

chromosomes (KROH muh sohmz)   threadlike structures made of DNA that are in the nuclei of all cells **(127)**

chromosome theory   theory that states that genes are located on chromosomes, that traits are passed to offspring by the chromosomes, and that each gamete contains chromosomes in the nucleus **(165)**

cilia   short, hairlike structures used for locomotion in certain protozoans **(232)**

climate   the general weather pattern that occurs in an area **(62)**

coldblooded   having a body temperature that changes with the temperature of the environment **(372)**

community   different populations living together in an area **(34)**

complete metamorphosis   the changes in insect body form through four stages of development—egg, larva, pupa, adult **(360)**

compound light microscope   microscope in which two lenses are used to magnify an image **(102)**

conservation   the careful use of the earth's resources **(19)**

consumers   organisms that eat other organisms for food **(44)**

cornea   the clear part of the eye; lets light pass into the eye **(521)**

cotyledons   the first leaves, or seed leaves, to develop on a seedling **(289)**

D

dermis (DUR mihs)   the thick inner layer of skin **(502)**

diaphragm (DY uh fram)   a thick muscle found at the bottom of the chest cavity that helps in breathing **(493)**

diffusion (dih FYOO zhuhn) the movement of molecules from an area of high concentration to an area of low concentration **(121)**

digestion (dih JEHS chuhn) the process that changes food into a form that the body can use **(458)**

DNA hereditary material found within the nucleus **(127)**

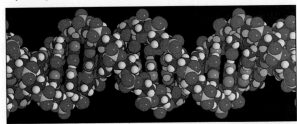

dominant (DAHM uh nuhnt) a strong factor in determining a trait; prevents the recessive trait from showing up in offspring **(162)**

drug any substance other than food, air, or water that can affect the way the body functions **(530)**

drug abuse the incorrect and unsafe use of a drug **(531)**

E

ecology the study of the relationships between organisms and their environment **(35)**

ecosystem the combination of a community and its nonliving environment **(35)**

embryo (EHM bree oh) in animals, a ball of tightly packed cells that develops from a zygote **(555)**

endocrine (EHN duh krihn) **system** group of glands that produce hormones **(544)**

epidemic the rapid spread of a disease through a large area **(213)**

epidermis (ehp uh DUR mihs) in animals, the outer layer of skin **(502)**

epiglottis (ehp uh GLAH tihs) a flap of tissue that moves over the opening to the trachea during swallowing **(491)**

erosion the carrying away of topsoil by water, wind, or glaciers **(20)**

evolution slow changes in living organisms **(150)**

excretion (ihks KREE shuhn) the process by which wastes are removed from the body **(497)**

exoskeleton (ehk soh SKEHL uh tuhn) an outer skeleton covering an arthropod's body **(358)**

Embryo

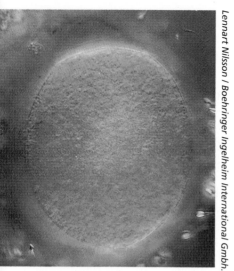

Fertilization

F

fermentation (fur muhn TAY shuhn) process that gives off energy without using oxygen **(135)**

fertilization the joining of a sperm cell and an ovum **(554)**

fetus in humans, the embryo after two months of development **(555)**

flagella (fluh JEHL uh) long, whiplike structures used for locomotion in certain protozoans **(233)**

fossils traces of once-living organisms **(152)**

fungi (FUHN jy) one of the five kingdoms of organisms; organisms that cannot move from place to place, have no chlorophyll, and absorb food from their surroundings **(240)**

G

gamete a reproductive cell; an egg or a sperm cell **(165)**

genes factors that determine hereditary characteristics **(163)**

gestation the period during which a young mammal is developing within its mother's body **(393)**

gland a group of cells that make special chemicals for the body **(544)**

gram the basic unit of mass in SI **(100)**

H

habitat (HAB uh tat) the place in which a population lives **(25)**

hormones (HOHR mohnz) chemicals produced by endocrine glands and certain body functions **(545)**

host an organism invaded by a virus or by another organism **(212)**

hypothesis (hy PAHTH uh sihs) a possible answer to a question **(5)**

I

immune system   system by which the body fights infection **(214)**

incomplete metamorphosis   changes in insect body form through three stages of development—egg, nymph, adult **(360)**

incubation   keeping eggs warm until they hatch **(388)**

invertebrate   any animal without a backbone **(332)**

involuntary muscles   muscles that work without signals from the brain **(448)**

Invertebrate

J

joint   place where two or more bones come together **(442)**

K

kidney   main organ of the excretory system **(498)**

kingdom   the largest category in the classification system of living things **(185)**

L

lens   the part of the eye that focuses light entering the eye **(522)**

lichen   organism that is part fungus and part alga **(244)**

ligaments (LIHG uh muhnts)   tough strips of connective tissue that act like strong rubber bands to hold together the bones in movable joints **(442)**

liter   the basic unit of volume in SI **(100)**

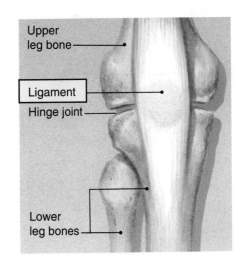

Upper leg bone

Ligament

Hinge joint

Lower leg bones

M

marrow (MAR oh)   the soft, living part of bone **(435)**

meiosis (my OH sihs)   the type of cell division in which gametes are formed **(165)**

menstruation (mehn STRAY shuhn)   the breakdown of the thickened lining of the uterus and the movement of that material out of the body through the vagina **(552)**

meter   the basic unit of length or distance in SI **(99)**

mimicry (MIHM ihk ree)   an animal's ability to copy the appearance or behavior of another unrelated organism **(403)**

mitosis (my TOH sihs)   the type of cell division by which two identical daughter cells are formed **(127)**

monerans   one of the five kingdoms of organisms; organisms that do not have nuclei in their cells **(218)**

muscular system   muscles of the body working together to coordinate the movements of the body **(445)**

N

natural selection   the process by which those organisms best suited to their environment survive and reproduce **(155)**

nerves   groups of neurons that carry messages between the different parts of the body and the central nervous system **(513)**

neuron   individual nerve cell **(514)**

niche (NIHCH)   the function an organism plays in its habitat **(43)**

nicotine (NIHK uh teen)   a psychoactive drug found in tobacco **(534)**

nucleus   that part of a cell that controls the cell's activities **(109)**

nutrient (NOO tree uhnt)   a chemical substance that the body needs to build new cells and to keep those cells alive **(458)**

O

organ   group of tissues that work together **(113)**

organelle (AWR guh nehl)   structure within a cell that has a certain job to do in the cell **(108)**

osmosis (ahs MOH sihs)   a special type of diffusion that occurs when water moves across a selectively permeable membrane **(123)**

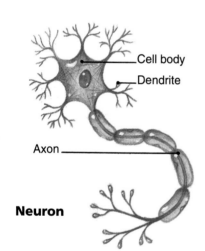

Cell body
Dendrite
Axon
Neuron

P

passive transport   process that requires no energy for the movement of molecules across a membrane from an area of high concentration to an area of low concentration **(122)**

pheromones (FEHR uh mohnz)   chemicals serving as signals that are picked up by animals of the same species **(412)**

phloem (FLOH ehm)   in vascular plants, tissue that transports nutrients that are made in the leaves to all parts of the plants **(260)**

photosynthesis (foht uh SIHN thuh sihs)   the process by which green plants use chemicals from the environment and energy from the sun to make their own food **(263)**

pigment   chemical that gives color to the tissue of living organisms **(223)**

pistil   the female part of a flower **(278)**

plasma (PLAZ muh)   the liquid part of blood **(473)**

pollution   making the environment unclean with waste products **(10)**

population   organisms of the same species living together in a particular place and at a particular time **(34)**

producers   green plants that produce their own food **(44)**

protozoan (proht uh ZOH uhn)   microscopic organism that is a member of the protist kingdom **(230)**

pseudopods (SOO duh pahdz)   false feet; projections made of cytoplasm used for locomotion and food getting in certain protozoans **(231)**

puberty   the stage of life during which a human matures sexually **(550)**

R

radial symmetry   the arrangement of body parts around a center point **(327)**

receptors   specialized nerve cells that receive information from their surroundings **(513)**

recessive (rih SEHS ihv)   a weak factor in determining a trait; prevented from showing up in offspring by the dominant trait **(162)**

reflex   an automatic response **(517)**

Pollution

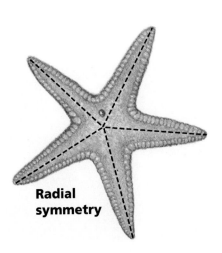

Radial symmetry

regeneration in plants, the ability to grow or replace missing parts **(294)**; in animals, the growth of a part of an organism into a complete organism **(344)**

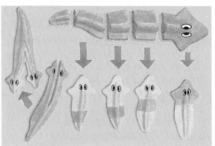

replication (rehp luh KAY shuhn) the process by which DNA makes copies of itself **(132)**

retina (REHT uhn uh) the inside layer of tissue on the back of the eyeball **(522)**

S

skeletal system the framework of the body **(434)**

slime mold organism that looks and reproduces like a fungus, moves and eats like an amoeba **(244)**

species a group of organisms that naturally mate with one another and produce fertile offspring **(184)**

stamen the male part of a flower **(278)**

succession the slow changes that take place when one community replaces another **(48)**

symmetry (SIHM uh tree) balanced arrangement of body parts around a center point or line **(327)**

system group of organs that work together **(113)**

T

taxonomy the science of classification **(176)**

tendon (TEHN duhn) type of connective tissue that holds muscles to bones and keeps different muscles together **(447)**

theory (THIR ee) a hypothesis that is supported by the work of many scientists **(8)**

tissue a group of similar cells working together to perform a specific job **(112)**

torpor an inactive state in which an animal's body functions slow down **(413)**

Stamen

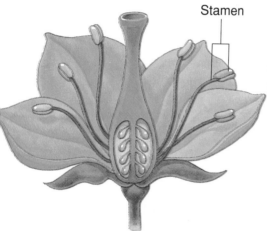

Torpor

trachea (TRAY kee uh) in vertebrates, tube that transports air to the lungs **(491)**

tropism (TROH pihz uhm) the response of a plant to a stimulus **(291)**

V

vaccine a substance made of dead or weakened viruses used to prevent a specific disease **(215)**

vascular cambium in a plant, the growth tissue that produces the xylem and the phloem **(263)**

vegetative propagation asexual reproduction in which cuttings of roots, stems, or leaves grow into new plants **(292)**

veins blood vessels that carry blood to the heart **(481)**

vertebrate (VUHR tuh briht) any animal that has a backbone **(332)**

virus a very small particle that can reproduce only inside a living cell **(211)**

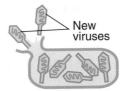

New viruses

voluntary muscles muscles that can be controlled by the brain; skeletal muscles **(447)**

W

warmblooded having a nearly constant body temperature, regardless of the temperature of the environment **(384)**

X

xylem (ZY luhm) in vascular plants, tissue that transports water and dissolved minerals from the roots to the leaves **(260)**

Z

zygote (ZY goht) the new cell resulting from fertilization **(554)**

Vaccine

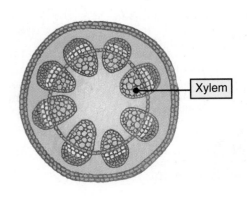

Xylem

INDEX

E

Ear, anatomy of, 524, **524**
Ear canal, 524
Eardrum, 524
Earphones, **524**
Earth, **210**
Earthworms, 351–352, **351, 352**
   anatomy of, **352**
Echinoderms, 356–357
Echolocation, **322**
Ecology, 35
Ecosystems, 35, 42
   changing, 48–53
   freshwater, 66–69
   land, 62–63
   marine, 63–69
Egg-eating snake, **382**
Eggs
   of amphibians, 378
   of birds, 387–389, **387, 388**
   of chickens, **388,** 389
   fertilization of, 554
   of fishes, 376
   of flamingos, 389
   of frogs and toads, 378, 379,
      379
   of green sea turtle, **39**
   incubating, 388–389
   of insects, 360
   of larks, 389
   of mallard ducks, 389
   of reptiles, 381
   of royal albatross, 389
   of salamanders, 380
Ejaculation, 550
Electrical impulses
   and nervous system, 544
Electron microscope, 103, **103,
   181**
Elephants, **25,** 391
   behavior of, 405
Embryo, 393
   care of mother and, 557–558
   described, 555
   development of, 555–558, 555
   placenta of, 556
   plant development, 288–289,
      288
Emigration, 37
Endangered species, 24
Endocrine glands, 544–547, **545,**
   table 546
   of female reproductive system,
      551
   of male reproductive system,
      549
Endocrine system, 544–547
Endoplasmic reticulum (ER),
   110, **110**
Energy
   from carbohydrates, 464–466

and cellular work, 134
and fermentation, 135
flow of, 43–46
from lipids, 466
from respiration, 134
from simple sugars, 464–465
Energy cycle, 137
Energy pyramid, **46**
Energy transfer
   of plants, 274
Environment
   community dependence on,
      35
   pollution of, 10–17
   and population size, 38–39
   protecting the, 18–27
Enzymes, 468
   of saliva, 459
   of stomach, 461
Epidemic, 213
Epidermal tissue, 112
Epidermis, 264, 502, **502,** 503,
   504
Epiglottis, 491, **491**
Erosion, 20
   defined, 20
Esophagus
   of birds, 386
   of earthworms, 352
   of fish, 375
   human, 460, **460,** 491, **491**
Estivation, 413
Estuary, 69, **69**
Ethyl alcohol, 536
Euglena, **187,** 237, **237**
Eustachian tube, 524
Evening primrose, **178**
Evergreens, 75, 264
Evolution, 150
   defined, 150
   process of, 150
   theories of, 155, 158
Excretion, 497–500
   defined, 497
Excretory system, 498–500, **498**
Exercise, 136
   and muscle tone, 451
Exoskeleton
   of arthropods, 358
Experimental group, 6
Experiments
   conducting, 6
Extinction, 23
Eyes
   anatomy of humans, 521, **521**
   of crayfish, **364**
Eyespots, 348

F

Fallopian tubes, 551, **551**
   function of, 554
Family, 185

Farsightedness, 523
Fat-soluble vitamins, 468
Fats, 466
   and digestion, 463
Feathers, 384–385, **385**
Feces, 463
Feedback control, 546–547, **547**
Felis concolor. See Mountain lion
Female reproductive system, 551
Fermentation, 135, 242
   comparison with respiration,
      table 136
Ferns, 266–267, **266**
   life cycle of, **266**
Fertilization, 279, 554–555
Fertilizer, 21, 124, 223
Fetus, 555, 556–557, **556**
Fever blister, 213
Fiber, 466
Fibrous roots, 260, **261**
Filament, 278
Finches, 149, **149**
Fingerprints, **503**
Fins
   of sharks, **373**
Fish, 372–376
   anatomy of, **375**
   bony, 375–376
   cartilaginous, 373–374
   jawless, 373
Five-kingdom classification,
   582–583
Fixed joints, 442, **443**
Flagella, 233, 237, 343
Flagellates, 233, **233**
Flamingo, **155**
Flat bones, 436, **436**
Flatworms, 333, 348–349
Fleming, Alexander, 243
Floating ribs, 441
Flounder, 375
Flowers
   parts of, 278, **278**
Flukes, 348
   life cycle of, **349**
Fly, **359**
Food chain, 45, **45**
Foods
   and health, 470
   high fiber, 466
   protein-rich, **467**
   pyramid, **470**
Food supply, 38–39
Food vacuole, 231, **232**
Food web, 45–46, **45**
Forest ecosystem, **35**
Forests, 71–75
Forsyth, Adrian, 400, 402, 403,
   407–408, 410, 414, 415
Fossil fuels
   greenhouse effect and, 17
Fossils, 152, **152**
Fragmentation, 344

CREDITS

PHOTOS

Abbreviations used: (t) top, (c) center, (b) bottom, (l) left, (r) right, (bkgrd) background.

Page: ii, Laurence Parent; vi (t), NASA/Science Source/Photo Researchers; vi (bl), Ed Reschke/Peter Arnold; vi (br), Roger Wilmshurst/Bruce Coleman, Inc.; vii (tl), David Scharf/Peter Arnold; vii (tr), HRW photo by James Newberry; vii (b), Dwight Kuhn; viii (tl), Art Wolfe/Allstock; viii (tr), Montagnier/ Instit Pasteur/SPL/Science Source/Photo Researchers; viii (b), David Doubilet; ix (t), Alfred Pasieka/Bruce Coleman, Ltd.; ix (bl), Kerry T. Givens/Tom Stack & Associates; ix (br), Runk/Schoenberger/Grant Heilman; x (tl), Michael P. Gadomski/Bruce Coleman, Inc.; x (tr), Art Wolfe/Allstock; x (b), Hans Reinhard/Okapia/Photo Researchers; xi, Breck Kent/Animals Animals; xii (tl), Frans Lanting/Allstock; xii (tr), Shelby Thorner/David Madison; xii (b), Mary Gow; xiii, Park Street; xiv (tl), HRW photo by Richard Haynes; xiv (tc), Don Fawcett/Science Source/Photo Researchers; xiv (tr), M. Wurtz/ Biozentrum, University of Basel/Photo Researchers; xiv (b), John Walsh/Photo Researchers; xv (t), Eric Grave/Photo Researchers; xv (b), HRW photo by Richard Haynes; xvi (tl), M. I. Walker/ Photo Researchers; xvi (tr), Terry Domico/Earth Images; xvi (b), Marty Snyderman; xviii (t), HRW photo by James Newberry; xviii (b), HRW photo by Henry Friedman; xix, HRW photo by Dennis Carlyle Darling; xxii, David Parker/Science Photo Library/Photo Researchers; 1 (inset), Will & Deni McIntyre/Allstock; 2, Photo Library International/Nawrocki Stock Photo; 3 (t), Susan McCartney/Photo Researchers; 3 (b), Simon Fraser/ Science Photo Library/Photo Researchers; 4(l), Frederica Georgia/ Photo Researchers; 4 (r), Dave Brown/ Nawrocki Stock Photo; 5 (t), E.R. Degginger; 5 (b), Joe Branney/Tom Stack & Associates; 6, both HRW photos by Richard Haynes; 8, HRW photo by Griff Smith; 10 (l), Gerhard Gscheidle/Peter Arnold, Inc.; 10 (c), Craig Aurness/Woodfin Camp & Associates; 10 (r), Chuck O'Rear/ Woodfin Camp & Associates; 11, Paul Dix/Tony Stone Worldwide; 12 (l), Tompix/Peter Arnold; 12 (r), Ray Pfortner/Peter Arnold; 13 (l), Ray Pfortner/Peter Arnold, Inc.; 13(r), Runk/Schoenberger/Grant Heilman

Photography; 15, NASA/Science Source/Photo Researchers; 17, Dr. Nigel Smith/Earth Scenes; 18 (l), Grant Heilman/Grant Heilman Photography; 18 (r), Laurence Parent; 19, Peter Miller/Photo Researchers; 20, Ray Ellis/ Photo Researchers; 21, Grant Heilman/ Grant Heilman Photography; 23 (t), Randall Hyman/Stock Boston; 23 (b), Culver Pictures; 24 (t), Rick Sullivan/ Bruce Coleman, Inc.; 24 (b), Tom & Pat Leeson/Photo Researchers; 25 (l), Bruce Davidson/Animals Animals; 25 (r), Steve Turner/Oxford Scientific Films/ Animals Animals; 26 (t), Tom McHugh/ Photo Researchers; 26 (b), 27, Ron Garrison/Zoological Society of San Diego; 29, Photo Library International/ Nawrocki Stock Photo; 31, Park Street; 32, William Townsend/Photo Researchers; 33, Tim Childs/Leo de Wys; 34, Johnny Johnson/DRK Photo; 35 (l), David Cavagnaro/DRK Photo; 35 (r), John Trott/Animals Animals; 36, D. & M. Zimmerman/VIREO; 37 (l), John Walsh/Photo Researchers; 37 (r), David Cavagnaro/DRK Photo; 38, Kevin Schafer/Peter Arnold; 39, J.A.L. Cooke/Oxford Scientific Films/Animals Animals; 40 (tl), Arthur Panzer/Photo Researchers; 40 (tr), Norman Myers/ Bruce Coleman, Inc.; 40 (bl), Peter Ward/Bruce Coleman, Inc.; 40 (br), Adrian Davies/Bruce Coleman, Inc.; 41, Bill Wood/Robert Harding Picture Library; 43 (tl), E. R. Degginger; 43 (tr), Jeff Lepore/Photo Researchers; 43 (bl), Hans & Judy Beste/Animals Animals; 43 (br), Keith Gillett/Animals Animals; 44, Stephen J. Krasemann/DRK Photo; 46, E. R. Degginger; 48 (l), Jed Wilcox/Tony Stone Worldwide; 48 (r), William E. Ferguson; 49, Larry Ulrich/DRK Photo; 50 (l), Tom Bean; 50 (c), (r), Tom & Susan Bean/DRK Photo; 51, Tom Bean; 52, William E. Ferguson; 53, R. Carr/Bruce Coleman Ltd.; 55, Wm. Townsend/Photo Researchers; 57 (l), Tom & Susan Bean/DRK Photo; 57 (r), John Gerlach/Tom Stack & Associates; 58, both photos by NASA; 60, Park Street; 61, NASA; 62 (l), Laurence Parent; 62 (tr), Steve Vidler/Nawrocki Stock Photo; 62 (br), David C. Fritts/ Earth Scenes; 64, Frans Lanting/Photo Researchers; 65, both photos by D.P. Wilson/Science Source/Photo Researchers; 66, Peter David/Photo Researchers; 67 (l), Jeff Apoian/ Nawrocki Stock Photo; 67 (r), Ronald Toms/Oxford Scientific Films/Earth Scenes; 69, Doug Wechsler/Earth Scenes; 71 (l), Laurence Parent; 71 (r),

Brian Lovell/Nawrocki Stock Photo; 72, Peter Veit/DRK Photo; 73, David Hiser/Photographers Aspen; 74, Roger Tully/Tony Stone Worldwide; 75, Jim Brompton/Valan Photos; 76, Roger Wilmshurst/Bruce Coleman, Inc.; 77, Linda Dufurrena/Grant Heilman; 79 (l), John Cancalosi/Peter Arnold; 79 (r), Ed Reschke/Peter Arnold; 81, NASA; 82, Charlie Ott/Photo Researchers; 83 (l) Bill Bachman/Earth Images; 83 (r), Grant Heilman/Grant Heilman Photography; 84, Gregory G. Dimijian, M.D./Photo Researchers; 85, David Barnes/Allstock; 85 (inset), Joy Spurr/Bruce Coleman, Inc.; 86 (l), Gregory G. Dimijian, M.D./Photo Researchers; 86 (tr), Michael Fogden/Animals Animals; 86 (cr), E.R. Degginger, Bruce Coleman, Inc.; 86 (br), Tom McHugh/Photo Researchers; 91 (t), Erich Hartman/Magnum; 91 (b), Marine Science Research Center; 92, both HRW photos by James Newberry; 93, NASA; 94, 95 (inset), Figaro Magazine/Gamma–Liaison; 96, Dr. Jeremy Burgess/Science Photo Library/Photo Researchers; 98, HRW photo by Eric Beggs; 99, 102, HRW photos by Richard Haynes; 103, Science Photo Library/Photo Researchers; 104 (t), (b), Jim Zuckerman/Westlight; 105, HRW photo by Eric Beggs; 106 (l), Bettmann Archives; 106 (r), Runk/Schoenberger/ Grant Heilman Photography; 107, Arthur M. Siegelman; 109 (tl), Don Fawcett/Photo Researchers; 109 (tr), Bill Longcore/Science Source/Photo Researchers; 109 (b), Don Fawcett/ Photo Researchers; 110, Don Fawcett/ Photo Researchers; 112 (t), Michael Abbey/Photo Researchers; 112 (c), Kevin Morris/Allstock; 112 (bl), (bc), G.W. Willis, M.D./Biological Photo Service; 112 (br), Jim Solliday/Biological Photo Service; 114 (l), Alfred Owczarzak/Biological Photo Service; 114 (r), Runk/Schoenberger/Grant Heilman; 115, Dr. Jeremy Burgess/ Science Library/Photo Researchers; 116, P. Dayanandan/Photo Researchers; 117, David Scharf/Peter Arnold, Inc.; 120, all HRW photos by Richard Haynes; 121, HRW photo by Eric Beggs; 122, both HRW photos by Richard Haynes; 124, both HRW photos by Eric Beggs; 126, Dwight Kuhn; 127, Dr. Lloyd M. Beidler/Science Photo Library/Photo Researchers; 128, 129, all photos by Arthur M. Siegelman; 130, Barr-Brown/Camera Press; 131, Dan McCoy/Robert

Langridge/Rainbow; 133, HRW photo by Richard Haynes; 134, Leonard Harris/Leo de Wys, Inc.; 135 (l), Uselmann/H. Armstrong Roberts, Inc.; 135 (r), HRW photo by Richard Haynes; 136, Paul J. Sutton/ Duomo; 141, Tom Stack & Associates; 147, both photos by Tom Brakefield; 150 (l), Breck P. Kent; 150 (r), Eric Hosking/Bruce Coleman, Inc.; 152 (t), Runk-Schoenberger/Grant Heilman Photography; 152 (bl), Stephen J. Krasemann/DRK Photo; 152 (br), J. Koivula/Photo Researchers; 153 (l), Ed Cooper; 153 (r), William E. Ferguson; 154, UPI/Bettmann Newsphotos; 155 (t), Stephen J. Krasemann/Photo Researchers, Inc.; 155 (b), M. P. Kaul/VIREO; 156, Zig Leszczynski/Animals Animals; 158, Culver Pictures; 160, Ken Brate/Photo Researchers; 161, Culver Pictures; 163, HRW photo by Mary Gow; 165, Manfred Kage/Peter Arnold; 167, Culver Pictures; 169 (l), Tom McHugh/Photo Researchers; 169 (cl), Art Wolfe/Allstock; 169 (cr), (r), Park Street; 171, Tom McHugh/Photo Researchers; 172, HRW photo by James Newberry; 174, HRW photo by Mary Gow; 175 (tl), Ralph A. Reinhold/Animals Animals; 175 (tr), M. Wendler/Okapia/Photo Researchers; 175 (c), Alinari/Art Resource; 175 (b), Culver Pictures; 176 (l), Larry Lefever/Grant Heilman Photography; 176 (l, inset), Leonard Lee Rue III/Animals Animals; 176 (c), Park Street; 176 (c, inset), Kristin Finnegan/Allstock; 176 (r), Vince Streano/Allstock; 176 (r, inset), Charles Krebs/Allstock; 177 (t), The Mansell Collection; 177 (b), Robert Winslow/Tom Stack & Associates; 178, E. R. Degginger; 179, all HRW photos by Eric Beggs; 181 (t), A. B. Dowsett/Science Photo Library/Photo Researchers; 181 (b), HRW photo by Eric Beggs; 182 (tl), Rod Planck/Photo Researchers, Inc.; 182 (tr), Renee Lynn/Photo Researchers, Inc.; 182 (bl), Jeff Foott/Bruce Coleman, Inc.; 182 (br), Laura Riley/Bruce Coleman, Inc.; 183 (tl), Laurence Parent; 183 (tr), George E. Jones III/Photo Researchers; 183 (b), Art Wolfe/Allstock; 184 (l), L. L. T. Rhodes/Animals Animals; 184 (tr), Rod Allin/Tom Stack & Associates; 184 (c), Porterfield/Chickering/Photo Researchers; 184 (b), Grant Heilman Photography, Inc.; 185, HRW photo by Eric Beggs; 187 (t), Dr. Tony Brain/Photo Researchers; 187 (bl), Eric V. Grave/Photo Researchers; 187 (bc), (br), E.R. Degginger/Bruce Coleman, Inc.; 188 (t), E.R. Degginger; 188 (b), Laurence Parent; 189 (l), (c), E.R. Degginger; 189 (tr), Jeff Rotman; 189 (cr), John Shaw/Tom Stack & Associates; 189 (br), Breck P. Kent/Animals Animals; 191, HRW photo by

James Newberry; 194, Figaro Magazine/Gamma–Liaison; 194 (bkgr), Ron Slenzak/Westlight; 195, Gamma-Liaison; 196 (tl), (tc), (tr), David Brill; 196 (b), Yoichi R. Okamoto/Photo Researchers; 197, all photos by David Brill; 198, (l), (r), David Doubilet; 198 (bkgrd), Ralph A. Clevenger/Westlight; 200, David Doubilet; 201 (t), The Mansell Collection; 201 (b), Erich Hartmann/Magnum; 202, HRW photo by Henry Friedman; 203 (l), Runk/Schoenberger/Grant Heilman Photography; 203 (r), Louis Bencze Photo/Allstock; 204 (t), Grant Heilman/Grant Heilman Photography; 204 (b), Thomas C. Boyden; 205 (t), Chip Porter/Allstock; 205 (b), Ken Graham/Allstock; 206, CDC/Science Source/Allstock; 208 (tl), CNRI/SPL/Photo Researchers; 208 (tr), Dwight Kuhn; 208 (b), Jean–Loup Charmet/Science Photo Library/Photo Researchers; 209, Jane Burton/Bruce Coleman, Inc.; 210 (t), Alfred Pasieka/Bruce Coleman, Ltd.; 210 (bl), (bc), HRW photos by Eric Beggs; 210 (br), NASA; 211, Dr. James E. Duffus/U.S. Department of Agriculture; 211 (inset), U.S. Department of Agriculture; 212 (l), CDC/RG/Peter Arnold; 212 (c), M. Wurtz/Biozentrum, University of Basel/Photo Researchers; 212 (r), Omikron/Photo Researchers; 213 (t), Susan Gibler/Tom Stack & Associates; 213 (b), Bettmann Archives; 214, Montagnier/Instit Pasteur/SPL/Science Source/Photo Researchers; 215 (t), Nathan Benn/Woodfin Camp & Associates; 215 (b), Zeva Oelbaum/Peter Arnold; 216, Bettmann Archives; 218, David M. Phillips/Visuals Unlimited; 219 (l), J. Robert Stottlemyer/Biological Photo Service; 219 (r), R. Knauft/Biology Media/Photo Researchers; 221 (tl), CNRI/Science Photo Library/Photo Researchers; 221 (tr), Laurence Parent; 221 (bl), Centers for Disease Control; 221 (bcl), (bcr), Arthur M. Siegelman; 221 (br), Martin Rotker/Phototake; 222, Hank Morgan/Rainbow; 223 (l), T. E. Adams/Visuals Unlimited; 223 (r), Sinclair Stammers/Science Photo Library/Photo Researchers; 225, CNRI/SPL/Photo Researchers; 226 (l), J. Robert Stottlemeyer/Biological Photo Service; 226 (r), Park Street; 228 (l), C. James Webb/Bruce Coleman, Inc.; 228 (r), Stephen Dalton/Photo Researchers; 229, Brown Brothers; 230, Laurence Parent; 231, Manfred Kage/Peter Arnold, Inc.; 232 (tl), (tr), (cl), (cr), Michael Abbey/Science Source/Photo Researchers; 232 (b), Manfred Kage/Peter Arnold, Inc.; 233, Arthur M. Siegelman; 235 (l), M. I. Walker/Photo Researchers; 235 (c), E. R. Degginger/Bruce Coleman, Inc.; 235 (r),

Biophoto Associates/Photo Researchers; 236, Ed Degginger/Bruce Coleman, Inc.; 237, Eric Grave/Photo Researchers; 238 (t), Manfred Kage/Peter Arnold, Inc.; 238 (bl), Brian Parker/Tom Stack & Associates; 238 (bc), Tom Stack/Tom Stack & Associates; 238 (br), E.R. Degginger/Bruce Coleman, Inc.; 240, Runk/Schoenberger/Grant Heilman Photography; 241, Barry L. Runk/Grant Heilman Photography; 242 (tl), (tr), (cl), E.R. Degginger; 242 (cr), Grant Heilman/Grant Heilman Photography; 242 (bl), Manfred Kage/Peter Arnold, Inc.; 242 (br), Kerry T. Givens/Tom Stack & Associates; 243 (l), Arthur M. Siegelman; 243 (r), A. & F. Michler/Peter Arnold, Inc.; 244 (t), Grant Heilman/Grant Heilman Photography; 244 (b), Kerry T. Givens/Tom Stack & Associates; 245, Stephen Dalton/Photo Researchers; 246, Grant Heilman/Grant Heilman Photography; 247, D.C. Lowe/Allstock; 248 (t), Will & Deni McIntyre/Allstock; 248 (bl), Will & Deni McIntyre/Photo Researchers; 248 (br), NIBSC/Science Photo Library/Photo Researchers; 249 (l), Wil Phinney; 249 (c), Reuters/Bettmann; 249 (r), Will & Deni McIntyre/Photo Researchers; 250, Science Photo Library/Photo Researchers; 250 (bkgrd), Ed Reschke/Peter Arnold; 251 (t), (c), Biophoto Associates/Photo Researchers; 251 (b), Mary Evans Picture Library/Photo Researchers; 251 (bkgrd), Manfred Kage/Peter Arnold; 252 (t), Dan McCoy/Rainbow; 252 (b), Bettmann Archives; 253, all HRW photos by James Newberry; 254 (t), William E. Ferguson; 254 (c), (b), David J. Sams/Texas Imprint; 255 (t), (b) , D.P. Wilson/Science Source/Photo Researchers; 255 (c), M. I. Walker/Photo Researchers; 256, M. Thonig/Allstock; 258, Terry Madison/The Image Bank; 259, HRW photo by Eric Beggs; 260, Laurence Parent; 261 (l), Runk/Schoenberger/Grant Heilman; 261 (r), Walter Chandoha; 265, David Ball/Allstock; 266, Phil Degginger; 266 (inset), Wayne Lankinen/Bruce Coleman, Inc.; 267 (l), Bob & Clara Calhoun/Bruce Coleman, Inc.; 267 (r), Professor R. C. Simpson/Valan Photos; 268 (l), Larry West/Bruce Coleman, Inc.; 268 (c), E. R. Degginger/Bruce Coleman, Inc.; 268 (r), E. R. Degginger; 270, Will McIntyre/Photo Researchers; 271 (l), Marty Snyderman; 271 (r), E.R. Degginger; 272 (t), Runk/Schoenberger/Grant Heilman Photography; 272 (bl), Ed Cooper; 272 (bc), Norman Owen Tomalin/Bruce Coleman, Inc.; 272 (br), Michael P. Gadomski/Bruce Coleman, Inc.; 274 (l), Dan McCoy/Rainbow; 274 (r), Paul Conklin/TexaStock; 275 (l), Hans Reinhard/Okapia/Photo Researchers;

Coleman, Inc.; 465, 466, Barry L. Runk/Grant Heilman Photography; 467 (t), HRW photo by James Newberry; 467 (b), Will McIntyre/Photo Researchers; 468, HRW photo by Park Street; 469, James Newberry; 471, Laurence Parent; 472, Manfred Kage/Peter Arnold, Inc.; 473, Steve Dunwell/The Image Bank; 474 (l), Sklar & Peiper/Photo Researchers; 474 (tr), James White/University of Minnesota; 474 (br), Don Fawcett/Science Source/Photo Researchers; 475, Phillip A. Harrington/Peter Arnold; 476, Park Street; 478, Jean Claude Revy/Phototake; 483 (t), CNRI/Science Photo Library/Photo Researchers; 483 (b), Biophoto Associates/Photo Researchers; 486, Manfred Kage/Peter Arnold, Inc.; 488, David Lissy/Nawrocki Stock Photo; 489 (t), Cindy Lewis; 489 (b), David Madison; 490, Guido Alberto Rossi/The Image Bank; 491, Clark Overton/Phototake; 492, Carol Rosegg/Martha Swope Associates; 494 (t), Mary Gow; 494 (bl), (br), Herbert Wagner/Phototake; 496, H. Armstrong Roberts; 497, HRW photo by Richard Haynes; 500 (l), (r), Dianora Niccolini/Medichrome; 503 (l), Runk/Schoenberger/Grant Heilman Photography; 503 (r), Martin Dohrn/Science Photo Library/Photo Researchers; 506, Peter Menzel/Stock Boston; 507, David Lissy/Nawrocki Stock Photo; 508, both photos by Ira Wyman/Sygma; 509, Astrid & Hanns-Frieder Michler/Science Photo Library/Photo Researchers; 512, Shelby Thorner/David Madison; 514, Everett C. Johnson/Leo de Wys Inc.; 515, Biophoto Association/Photo Researchers; 518, Alexander Tsiaras/Science Source/Photo Researchers; 520, P.R. Dunn; 524, Mike Maple/Woodfin Camp & Associates; 525, Paolo Koch/Photo Researchers; 528, Gabe Palmer/Tony Stone Worldwide; 530, 531, all HRW photos by Richard Haynes; 532, 533, James Newberry; 534 (t), Matt Meadows; 534 (b), James Stevenson/Science Photo Library/Photo Researchers; 536, L. O'Shaughnessy/Allstock; 537 (l), Frederick C. Skvara/Peter Arnold; 537 (r), Alfred Pasieka/Bruce Coleman, Ltd.; 541, NHPA; 542 (both), 543, John Ficara/Woodfin Camp & Associates; 544, Renee Lynn/David Madison; 548, Mary Gow; 550 (t), HRW photo by Richard Haynes; 550 (b), John Walsh/Photo Researchers; 551 (t), John Giannicchi/Science Source/Photo Researchers; 551 (b), HRW photo by Richard Haynes; 555 (tl), Tony Freeman/PhotoEdit; 555 (tr), Mary Kate Denny/PhotoEdit; 555 (bl), (bcl), (bcr), (br), 557 (all), Petit Format/Nestle/Science Source; 558 (t), Kindra Clineff/Allstock; 558 (b), Jeff Reed/Medichrome; 559 (l), Tony Freeman/PhotoEdit; 559 (cr), Myrleen Ferguson/PhotoEdit; 559 (cr), Mary Kate Denny/PhotoEdit; 559 (r), Gabe Palmer/Tony Stone Worldwide; 561, John Ficara/Woodfin Camp & Associates; 563, 564, James Newberry; 565, 566 (t), (b), Mary Gow; 571 (t), NIH Visitors Information Center; 571 (b), NASA; 572, all HRW photos by P.R. Dunn; 573 (l), HRW photo by James Newberry; 573 (r), Judy Allen-Newberry; 574, David Wagner/Phototake; 576, HRW by James Newberry; 585, Cosmos Blank/Photo Researchers; 586 (t), David C. Fritts/Earth Scenes; 586 (b), Steve Vidler/Nawrocki Stock Photo; 587 (t), Tom Stack & Associates; 587 (b), Dr. Lloyd M. Beidler/Science Photo Library/Photo Researchers; 588, Mitch Reardon/Photo Researchers; 589, NHPA; 590, all photos by Arthur Siegelman; 591, Craig Aurness/Woodfin Camp & Associates; 592, Wolfgang Bayer/Bruce Coleman, Inc.

ILLUSTRATIONS

Bego, Dolores 333, 335(t)
Benner, Beverly 241, 242, 343, 349(b), 386, 387, 391
Bordelon, Melinda 118, 119
Botzis, Ka 349(t), 351, 352, 355(b), 373, 441, 452
Warren Budd & Associates 154, 295, 377, 383
Byers, Scott 434
Collins, Don 42, 74, 76, 77, 78, 82, 113, 149, 162, 163, 166, 170, 180, 233, 266, 273, 276, 278, 279, 284, 346, 348, 350, 353, 355, 366, 396, 409, 416, 470
Colrus, Bill 385
Cooper, Holly 7, 142, 143, 144, 145, 146, 148, 149, 150, 316
David, Susan 22, 68
Erickson, Barry 14
Evans, Tom 456, 457

Fischer, David 47, 87, 88, 89, 90, 108, 157, 193, 432, 433, 436, 437, 438, 440, 442, 443, 445, 446, 447, 448, 454, 455, 462, 482, 491, 493, 495, 498, 515, 516, 517, 518, 521, 524, 540, 545, 549, 551, 568, 569
Frank, Robert 359, 363, 364, 380, 381
Gardner, Sharon Carter 101, 168, 173, 308, 309, 310, 312, 313, 314, 318, 319, 363, 390, 398, 399, 437, 451, 469, 481, 504, 512, 513, 514, 521, 527, 530, 535
JAK Graphics 16, 30, 49, 63, 64, 67, 81, 101, 165, 107, 110, 111, 128, 186, 187, 212, 264, 282, 291, 326, 404, 406, 458, 459, 460, 461, 463, 479, 484, 495, 501, 502, 525, 527, 538, 547
Katz, Joel 132, 231, 233, 237
Kelvin, George 20, 52, 121, 126, 134, 212, 239

Lebo, Narda 513, 522, 526(b)
LeMonnier, Joe 100, 102, 103, 224
Longacre, Jimmy 510, 511
Merrilees, Rebecca 261, 262, 263, 265, 269
Morgan-Cain & Associates 9, 28, 38, 54, 70, 78, 123, 125, 159, 164, 190, 217, 220, 243, 298, 302, 331, 336, 411, 416, 450, 465, 477, 504, 519, 529, 538
Network Graphics 36, 137, 444, 553
Phillips, Harriet 480, 481, 491(t), 526(t)
Reid, Fiona 45, 46, 329
Skorpil, Judy 375, 378, 379, 499, 550, 556
Smith-Griswold, Wendy 160, 161